INSTRUCTOR'S MANUAL WITH TEST BANK

to accompany

APPLIED CALCULUS

SECOND EDITION

Deborah Hughes-Hallett
University of Arizona

Andrew M. Gleason
Harvard University

Patti Frazer Lock
St. Lawrence University

Daniel E. Flath
University of South Alabama

et al.

Prepared by

Elliot J. Marks

Bradley E. Garner

Carrie J. Garner

Guadalupe I. Lozano

Kelly Molkenthin Fuller
Nazareth College

Natasha Speer
Michigan State University

JOHN WILEY & SONS, INC.

COVER PHOTO © Lester Lefkowitz/Corbis Stock Market.

To order books or for customer service call 1-800-CALL-WILEY (225-5945).

ISBN 0-471-21352-7

Printed in the United States of America

10 9 8 7 6 5 4 3 2

Printed and bound by Victor Graphics, Inc.

Table of Contents

PART I

THE BOOK

A New Curriculum With a Variety of Options

The second edition of *Applied Calculus* has the same vision as the first edition and provides instructors with additional choices through the *Focus On* sections and additional chapters containing new material. Instructors can select a focus for their course which reflects their interests and the needs of their students. In particular:

- All *Focus On* sections are optional.
- For an earlier start on calculus, Chapter 2 may be covered after Section 1.6.
- In Chapters 4 and 6, instructors may select the sections relevant to their students.
- Chapters 9 (multivariable calculus) and 10 (differential equations) may be covered in either order.

Because different users often choose very different topics to cover in a one-semester applied calculus course, we have designed this book for either a one-semester course (with a lot of flexibility in choosing topics) or a two-semester course. Sample syllabi outlining various options are provided in the third part of this manual.

The Rule of Four

The Rule of Four is that, where appropriate, concepts should be studied graphically, numerically, analytically, and verbally. This means giving equal time to all four ways of understanding concepts. It means expecting graphical explanations; it means accepting numerical calculations properly done; it means requiring students to communicate their understanding in practical terms, such as colleagues or employers without calculus training might understand. We still expect students to carry out the correct algebraic manipulations that might be necessary, but we accept graphical and numerical reasoning as well.

The Rule of Four requires changes in approach:

- numerical methods such as approximating derivatives, estimating integrals, and fitting curves to data become essential topics
- calculators and computers make possible a shift in emphasis from constructing to interpreting graphs
- student assessment should include verbal explanations

The Problems

The problems are the heart of the text. Assign fewer problems than you have been accustomed to, and give the students more time. Many require careful thought and the ability to write clearly. There are some essay questions, questions that require graphical and numerical work rather than algebra, and questions that require a calculator or computer.

Verbal explanations are an essential part of the course. Composing written answers helps students clarify their own understanding and reading these explanations will provide you with valuable insight into the students' thinking processes.

Many of the problems in the text do not have unique correct answers. If you use student graders or lab assistants, urge them to be open to a variety of interpretations on these problems. Be flexible. Any reasonable answer, with reasonable justification, should be acceptable.

Although most of the problems are not difficult, students are not accustomed to reading and interpreting math problems nor to modeling real-world situations. They expect to see problems that mimic the examples in the text and have unique answers in the back of the book. The initial shock at finding these conditions changed will have to be dealt with. You may need to reassure students frequently that this new approach is difficult at first but that their honest efforts will pay off handsomely in the end. Comments on some of the problems are included in the second part of this guide.

Technology

This text assumes that students have access to computer programs or graphing calculators that can be used to

- estimate a zero of a function

- draw the graphs of a function in an arbitrary viewing window

- estimate a definite integral

It is helpful (but not essential) to have access to a computer or calculator that will

- estimate the derivative of a function at a specified point

- draw a contour diagram for a function of two variables

- compute a regression line or exponential regression function for a set of data

Programs for several models of calculators are given in Chapter Five of this manual.

The advantage of using technology is that it can enable students to experiment, make conjectures, and check special cases of general results. Inappropriate uses of the technology are those that make it a crutch. It should not be used when the desired result is more quickly obtained mentally. There are no icons telling students to use a calculator or computer, as we believe that students should learn to decide when to use technology and when not to use it.

A Book to be Read

This text is neither a reference book nor an encyclopedia, but a presentation of the basic ideas of calculus in an applied context. The writing is informal because we want students to read it. It was not written to facilitate what students usually do, that is, look first at homework problems, then look for a worked example that fits the template, and only as a last resort read the text between the examples. It is well to warn students that this text will be different.

Overview of Book: What's in Each Chapter

Chapter 1: Functions and Change introduces functions as representations for how one quantity depends on another and investigates rate of change. Included is a brief study of key properties of most basic functions used in the text: linear, power, and exponential functions, as well as natural logarithms, polynomials and some periodic functions. Though the functions themselves should be familiar to students, their graphical and numerical properties and the practical uses introduced here are usually not. Functions represented by tables of data receive particular attention, including an introduction to curve fitting at the end of the chapter.

Chapter 2: Rate of Change: The Derivative discusses the key concept of derivative and derivative function according to the Rule of Four. Differentiation formulas are delayed until Chapter 3 in order to better develop the mathematical ideas and their practical uses. Limits and the symbolic definition of the derivative are in the Focus on Theory section at the end of the chapter.

Chapter 3: Short-Cuts to Differentiation presents the symbolic approach to differentiation.

Chapter 5: Accumulated Change: The Definite Integral discusses the key concept of the definite integral in the same spirit as Chapter 2.

Chapter 6: Using the Definite Integral presents applications of definite integrals in the same spirit as Chapter 4. Techniques of integration are omitted in favor of applications; technology is used to estimate the definite integrals that result.

Chapter 7: Antiderivatives discusses analytical, graphical and numerical approaches to antidifferentiation. Included is the Fundamental Theorem of Calculus and Integration by Substitution.

Chapter 8: Probability introduces density functions and cumulative distribution functions. This chapter discusses some probability and includes a section on mean and median.

Chapter 9: Functions of Several Variables introduces functions of two or more variables using contour diagrams, tables and formulas, followed by differentiation of functions of two variables, with applications to optimization.

Chapter 10: Differential Equations introduces differential equations, their solutions and applications to mathematics modeling. Systems of differential equations are introduced and treated graphically.

Chapter 11: Geometric Series focuses on geometric series and their applications in business and the life sciences.

Key Concepts: Chapters 1, 2, 5

These chapters illustrate the spirit of the text. It is important not to go too rapidly through these chapters. We know from experience how important it is for students to have a thorough grasp of the material. Although much of the material in Chapter 1 may be familiar to the students, it is presented in a sufficiently different manner that most will find it quite new. It is crucial that students become comfortable with graphical and numerical work early on. The idea of Chapter 1 is to make students familiar with functions from every point of view. For example, a student should be able to identify the family to which a function belongs from its graph alone, to identify a linear or exponential function by looking at a table of values, and to describe in graphical terms the relative growth rates of functions.

Chapters 2 and 5 introduce the key concepts of derivative and definite integral and the relation between them. You should cover these chapters slowly enough that students have time to think about what the key concepts really mean.

"Focus On" Sections and Projects

These sections are optional. They are designed for supplementary reading or for incorporating into the regular class sessions as the instructor needs.

What Students Find Difficult

Students are often reasonably proficient at using rules to manipulate formulas. They are much less proficient at understanding and interpreting mathematics critically, as well as applying it to practical situations. This means that many students initially find the material in this book difficult. In particular, if you have students who have done well in the past in courses emphasizing manipulation, both you and they may be surprised at the difficulty they have. Dealing with their apprehension will probably require your repeated reassurance, particularly early on.

Students have more difficulty with tables and graphs than they do with formulas. They confuse functions with their formulas, the rules that generate them. Tables are usually foreign to them; they need time and practice to get used to them. Graphs can also be difficult; while most students will have seen graphs before, many students will have difficulty interpreting them. For example, some students will have difficulty interpreting a graph of distance versus time, often confusing it with a trajectory. A graph of velocity versus time can also cause them trouble. This difficulty with interpretation is one of the reasons for the emphasis on interpreting the derivative and the definite integral. However, students in your class whose algebraic skills are weak may welcome the opportunity to learn calculus with the assistance of graphs and tables.

Students often have considerable difficulty thinking geometrically. Many, for example, cannot read the slope of a line from its graph. Try to wean them away from depending on formulas to estimate such quantities. They may also have difficulty interpreting geometric objects; many confuse the secant line itself with the average rate of change (which is the slope of the secant line) or the tangent line with the derivative (which is the slope of the tangent line).

Many students will have difficulty with basic material, such as exponential functions and logarithms, or even percents and fractions. We suggest that you not spend too much time at the beginning of the course going over manipulative rules but instead review as you go along whenever it is needed.

Students may also misinterpret results on their graphing calculators or computers. This is often due to the computer's tendency to deal badly with the very large (say, near an asymptote) and the very small (roundoff error). Finding an appropriate viewing region for a graph can be difficult for students as well. They often give up too easily when graphing to find points of intersection, especially when large values are involved (for example, $y = e^x$ and $y = 2 + 100x$). Students sometimes also assume that if the cursor is on top of the point of intersection then the (x, y) values are exact, whereas these values can actually be off by large amounts, depending on the window settings.

Teaching Style

In the spirit of the book, try to make your class interactive; make it a class rather than a lecture. Since many of the problems can be done by several different methods, a useful class discussion can result from having students explain to the rest of the class the method they used and why. As you develop new ideas, ask questions to the class (and give them time to answer) to get students into the habit of thinking ahead and asking themselves "what if." Encourage interaction among the students, even outside of class; suggest that their classmates can help them figure out if they have done the problems satisfactorily.

The philosophy of this book is that mathematics is a way of making sense out of the world we live in. In this sense it richly deserves the name "natural philosophy" that it had at the time of Newton.

PART II

THE CHAPTERS

This part of the manual contains more detailed information about each chapter and section of the book. If you have trouble knowing what to do with a particular section or deciding which points to emphasize, look at the information on that section. There you will find information on timing (50-minute periods are assumed), suggestions on how to introduce the topic, extra examples, tips on what students find hard, ideas for computer or calculator use, and a discussion on the problems in each section. Note that you may not be able to cover every section of the book carefully in one semester, so be prepared to make some choices.

Suggested problems are selected to reflect the Rule of Four, wherever possible. We do *not* recommend that you assign all the suggested problems; you may need to adjust the number upward or downward depending on the needs of your class and the speed at which you intend to cover the material.

CHAPTER ONE

Ten to twelve classes.

Overview

This chapter sets the background and the tone for the whole book. It is a book about mathematical modeling of real world phenomena. From the first page there is a focus on things that change and their rate of change. The topics in the first chapter are all precalculus topics, but the graphical, numerical, and modeling approach is probably new to most of the students. Keep the flow lively and try not to get bogged down in any particular topic, but don't skip anything, either.

The functions used in this book are introduced in this chapter, the linear, exponential, and power functions as well as natural logarithms, polynomials and some basic periodic functions. Our purpose is to acquaint the student with each function's individuality (its graphical shape, identifying properties, comparative growth rates, and general uses), to train students to read graphs and think graphically, to read tables and think numerically, and to apply these skills to the mathematical modeling of the real world. Students are being asked to think about mathematics in new ways. Be assured, and assure them, that time and effort spent in this chapter will pay off well in subsequent chapters.

1.1 WHAT IS A FUNCTION?

One class.

Key Points

- The Rule of Four: algebraic, graphical, numerical, and verbal descriptions of functions.
- Function notation and intercept interpretation.

Ideas for the Class

Many students at this stage, depending on their backgrounds, regard functions as formulas. A function may not have a formula such as daily high temperatures of a city. Emphasize the much more general concept of a function as a rule that uniquely assigns one number to another number, and lead off with numerical and graphical examples of defining a function, in accordance with the Rule of Four.

For example, you might start with the following table:

Table 1.1.1

t	-3	-2	-1	0	1	2	3
$f(t)$	16	96	144	160	144	96	16

Then represent the data graphically, and ask students to come up with a formula for the function. Then indicate that $f(t)$ represents the height above ground, in feet, of an object thrown into the air so that its highest point is reached at time $t = 0$, in seconds. Have students discuss salient features of the table, the graph, and the formula $f(t) = 160 - 16t^2$, noticing how those features are reflected in the other representations of the function. For example, the maximum value in the table is 160; see how that fact is supported in the graphical, algebraic, and verbal descriptions of the function. Tell them that we want them to learn how to read the same properties of a function from any of its representations. Functions that arise naturally from real situations have more appeal as the Way of Archimedes claims, if you can find them.

It may take some students a bit of time to get comfortable with the use of function notation, so it does not hurt to begin with extremely simple examples. Suggest the example of a child's height vs. time (can refer back to it later). Discuss vertical intercept (no horizontal intercept here, why?)

Example   (This works well in groups, done in the middle or at the end of the class.) The value of a car goes down as the car gets older, so we can think of the value of a car, V, in thousands of dollars, as a function of the age of a car, a, in years. We have $V = f(a)$.

(a) Draw a possible graph of V against a. You don't need scales on the axes, but label each axis as V or a.

(b) What does the statement $f(5) = 6$ tell you about the value of the car? Be sure to use units for 5 and for 6. Label this as a point on your graph, and mark the 5 and the 6 on the appropriate axes.

(c) Put a vertical intercept of 15 on your graph of the function. Explain the meaning of this vertical intercept in terms of the value of the car.

(d) Put a horizontal intercept of 10 on your graph of the function. Explain the meaning of this horizontal intercept in terms of the value of the car.

Problems

Problems 2, 3, 6–9, 11, 12, 18–23: deal with graphical representations of a function, which are essential for students to develop intuitive reasoning. Assign as many as you can, be flexible on the answers.

Problems 1, 4, 5, 10, 13–17, 24, 25: deal with function evaluating, algebraic formulas, vertical and horizontal intercepts, etc. Designed for students to integrate what they just learned to their knowledge of functions and changes.

Suggested Problems

1–3, 5, 7, 9, 13, 14, 15, 16, 17, 18, 19, 21, 23.

1.2 LINEAR FUNCTIONS

One class.

Key Points

- Linear functions imply equal changes in independent variable correspond to equal changes in dependent variable. Equivalently, a table of values with regularly spaced entries will show constant differences.
- Slopes and difference quotients.
- Introduce Δ notation.
- General forms and families of linear functions, parameters.
  New vocabulary: increasing, decreasing, family of functions, parameter.

Ideas for the Class

Follow the Rule of Four. Start with a table of values and ask the students to give the slope or the formula. It is best to use real data such as the pole vault example in this section. Otherwise, students think you are just torturing them by creating a table and then hiding the formula that created it. Have the class focus on how the formula comes from the table, not the other way around. Students may have trouble at first realizing that they can read the slope as a change in the y-value divided by the corresponding change in the x-value. You might want to include an example in which the x-values are not evenly spaced.

Also, give students a verbal description of a linear function and ask them to come up with a formula. For example: For every mile over 36,000 I go on my leased car, the dealership charges me 0.15 (or 15¢). If I have t miles on my car when my lease is up, assuming that $t > 36,000$, write a formula for the amout I will owe the dealership when I return my car. Again emphasize that the characteristic property of a linear function was embodied in this verbal description in the constant rate of decrease in price.

Explain to students that slope is just another way of describing the theme topic of the course, which is rate of change. Given a linear function described either numerically or graphically, ask for the slope of the function over an interval. Explain how we will interpret this as "rate of change," which will be discussed in more detail later in this chapter.

The Δ-notation is one of the simplest in common use by mathematicians. Tell the students that Δ stands for "difference" and make it a point to use the notation in many of your examples. Total change in a quantity is a *difference* Δy and rate of change is a *difference quotient* $\Delta y / \Delta t$.

Keep in mind that in real applications the two axes of a graph are usually scaled quite differently; even the units of measurement are often different. Try to reflect this in your choice of examples, say on occasion marking off the x-axis in units of 20 and the y-axis in units of 10,000. Students should be able to write the formula for any linear function given its graph. If you want to calculate some slopes from points, have different students use different pairs of points and point out that the slope is independent of the pair of points chosen.

The terms increasing and decreasing are introduced in this section. Emphasize that these terms apply as the independent variable increases, moving from left to right. Some students want to say that the function $y = x^2$ is increasing for negative x, and the graph does rise as you move away from the origin in a negative direction, but point out that this is not what the definition means.

This section includes the first use of the terms *family of functions* and *parameter*, both of which are fundamental concepts in modeling situations. Encourage students from the first to think in terms of families of functions rather than of individual functions.

Be sure to emphasize how to recognize negative, positive and zero slope, and how to tell a slope of 2 from a slope of 1/2 (when the scales are equal). You may want to sketch some quick lines on the board and have the students play a "match" game similar to Problem 9. Developing a good understanding of slope now will pay off in Chapter 2.

Example 1 on solid waste is a good one to do in class, because they will see the connection with rate of change. Be sure to discuss the units of slope.

Example   (Good example for them to do at the end of class, because it ties together the ideas using the rule of four.) Tickets for a concert go on sale, and 100 tickets sell immediately. The tickets then sell at a rate of 20 tickets every day. Let N represent the total number of tickets sold t days after the tickets went on sale.

(a) Make a table of values for N and t, with $t = 0, 1, 2, 3, 4, 5$.
(b) Plot the points from part (a) and sketch a graph of N against t.
(c) Find a formula for N as a function of t.
(d) What is the vertical intercept in your formula? Why does this make sense? What is the slope in your formula? Why does this make sense?

Problems

Problems 1– 8: involve finding or using the equation of a line

Problems 9– 11: involve matching graphs with formulas. Students should feel comfortable with these. Be sure to assign at least one or two.

Problems 12– 14, 24– 30: require more thought and are good for group work. Encourage students to answer similar problems in which they are interested.

Problems 15– 18, 21– 23: involve getting a linear function from a table of data. Students would benefit from doing a few of these.

Problem 19– 20: involve the graphical interpretation

Suggested Problems

1– 10, 13– 16, 19, 21, 23, 30.

1.3 RATES OF CHANGE

One class.

Key Points

- Average rate of change
- Increasing and decreasing functions
- Secant lines
- Concavity
- Distance and velocity

Ideas for the Class

Students will want to know from the first day how calculus will be relevant to their own goals, which for students of this course are not fundamentally mathematical goals. Remind them that calculus is the mathematical basis for the careful study of things that change over time. Have them name in class examples of changing phenomena that interest them.

Students understand immediately that the first mathematical step in an analysis of many of their examples is to quantify the change. They may be surprised to find that there are two ways to do it. In this section you want to bring the students to an appreciation of the distinction between total change and (average) rate of change, using real world examples that they already have some familiarity with. A good lead-off example could be the growth of a child. It shows the distinction very clearly. Point out that change and rate of change are measured in different units, in the growth example in 'inches' and in 'inches per year' respectively. Students may notice the analogy with the familiar concepts of distance and velocity.

It is also important to be able to interpret a statement as an average rate of change or as a total change. For example: "The concentration of salt in the water changed during the first five minutes by 10 mg/ml." Is this rate of change or just change? Once they have identified this as just change, ask them to rephrase this as rate of change.

Example At the start of class, tell the class that the world's population was 2.555 billion people in 1950 and was 5.295 billion people in 1990. Ask them to calculate the change in the population over this 40 year period. (What are the units?) Then ask them to calculate the average annual rate at which the population is increasing, in billion people per year. This will get them thinking about the difference between the two ideas. When they realize that the population has been going up by 0.0685 billion people per year, translate this to 68,500,000 people per year or a net increase of about 130 people every minute.

The terms increasing and decreasing are introduced in this section. Emphasize that these terms apply as the independent variable increases, moving from left to right. Some students want to say that the function $y = x^2$ is increasing for negative x, and the graph does rise as you move away from the origin in a negative direction, but point out that this is not what the definition means.

Concavity is also introduced in this section. Be sure to give examples of increasing and decreasing functions with all types of concavity (including none). An in-class excercise that both amuses and instructs the students is to draw a circle on the board and ask them to find sections of it that exhibit each of the four possible combinations of increasing/decreasing and concave up/concave down.

Problems

Problems 1– 7, 29, 30: deal with increasing, decreasing, and concavity.

Problems 8, 10, 12, 14, 17, 19– 21: data are given in words or tables. Students should become familiar with such presentations. They would benefit greatly by doing a number of these questions. In particular, Problem 21 asks students, by knowing the average rate of change, to predict the change in the future.

Problem 9, 15, 18, 27, 28: data are given in a graph. A good point to assure students of the validity of graphical approximations and solutions.

Problem 11: a good problem for generating a discussion. Students could be asked to add to the list the questions of their own interest.

Problems 13, 16: deal with function evaluation and average rate of change.

Problem 23: gives a better understanding of the average rate of change.

Problems 22, 24– 26: involve graphical representation of average rate of change. Problems 22 and 25 are good for discussion: *slope = average rate of change.*

Suggested Problems

5– 9, 11, 12, 14, 15, 17, 21– 23, 27, 30.

1.4   APPLICATIONS OF FUNCTIONS TO ECONOMICS

One to one-and-a-half classes.

Key Points

Introduction of the terms cost, revenue, profit. Depreciation, budget constraints, and supply and demand curves, mostly linear, and equilibrium price, with effects of taxation. Vocabulary: fixed cost, variable cost, break-even point.

Ideas for the Class

Many of the ideas in this section may be familiar to the students, and even if not they are intuitively grasped. This section illustrates the use of linear functions in economics.

Supply and demand curves provide a natural opportunity to discuss functions graphically. They will make many further appearances in the text, so it is a good idea to treat them explicitly in class now.

Work out some simple, common-sense examples of cost/revenue/profit, straight-line depreciation, budget constraints, and equilibrium points in supply/demand. Students may have difficulty thinking through the effects of taxes on supply/demand equilibrium, so a couple of examples are in order. In particular, if the tax is paid by the producer, the effect is to reduce the supply and drive up the equilibrium price, but if the tax is paid by the consumer, the effect is to reduce the demand and drive down the equilibrium price. You might get students to discuss these facts from a purely graphical point of view, noting how the tax changes the supply or demand curve. In each case, the equilibrium supply is lowered, and both producer and consumer end up sharing the tax burden.

It is mentioned that even though many applications require cost and revenue functions to have discrete domains (whole-number units q), we are treating them as though they were continuous and smooth and defined for all $q > 0$. If you have a class of economics students, they may be familiar with the convention of considering the cost function $C(x)$ as the cost in dollars per week of producing x units per week rather than as the cost in dollars of producing x units, a convention in which the independent variable is naturally continuous rather than discrete.

Problems

Problems 1–13: involve algebraic, numeric and graphical representations of cost, revenue and profit functions for developing skills of modeling such variables with linear functions. Using the Rule of Four, students should have a good grasp of fixed cost, variable (marginal) cost and break-even point.

Problems 14, 15: linear depreciation.

Problems 16, 17, 25: deal with budget constraint and its variants. Problem 25 is good for discussion or group work.

Problems 18–24, 29, 30: deal with supply and demand curves and equilibrium point.

Problems 23, 29, 30: good for discussion or group work.

Problems 26–28: involve tax effect on the equilibrium price. This is an interesting but difficult concept. A good idea would be to draw graphs to explain what happens to equilibrium prices.

Suggested Problems

#2, 4, 5, 9, 12, 13, 15, 17, 21, 23, 25, 27, 29, 30.

1.5 EXPONENTIAL FUNCTIONS

One class.

Key Points

After linear functions, exponential functions are the most important for applications. The values of these functions change by constant ratios over equal intervals.

- Contrast with the way linear functions change

- Growth and decay rate

- Percent growth rate

Ideas for the Class

Start with the following example. This is a good group work problem and you can walk around and help them understand why the multiplying factor in part (b) is 1.05. Part (a) will reinforce the work done in linear functions and will highlight the distinction between these two important classes of functions.

Example In each situation below, make a table of values of the population of the town at time $t = 0, 1, 2, 3,$ and 4 years, and then find a formula for the population, P, of the town in year t.

(a) The town has a population of 1000 in year 0 and grows by 50 people a year.

(b) The town has a population of 1000 in year 0 and grows by 5% a year.

Emphasize that if the town is growing by 5%, then we multiply by 1.05. Emphasize also that in part (a), we are adding a constant amount each time (a linear function), but in part (b) we are multiplying by a constant amount each time (an exponential function). When they have finished with this example, have them graph the two functions ($P = 1000 + 50t$ and $P = 1000(1.05)^t$)

together on larger and larger windows in the first quadrant. They will quickly see that exponential growth outstrips linear growth.

Next start with a table of values for a simple exponential function, and ask the students how they know it is not the table of a linear function. Then let them notice that the ratios of consecutive values are constant, and ask them to find the formula for the exponential function.

Then take a look at the population of Mexico as in the text. Let the students take ratios to find the growth factor 1.026, and then ask them to tell you the percent rate of growth (2.6%). Students sometimes wonder how experts make population predictions, and so are usually pleased to be let in on one of the simplest of methods, which is based on the fact that many populations grow exponentially. Perhaps you could ask them to find some data for other countries or even states or cities and report to the class on whether growth was exponential, and if so what the growth factor and what they predict for twenty years into the future. Spend a little time with population growth now, especially if you plan to discuss Section 4.7 on fancier growth models later in your class. It is a good idea at this point to mention the dangers of extrapolating too far into the future.

Then discuss decreasing exponential functions, and characterize growth and decay by the size of the base. Note the distinction between growth factors and growth rates; if $P = P_0(1+r)^t$, then r is the growth rate over one time period, and $1 + r$ is the growth factor, and if $P = P_0(1-r)^t$, then r is the decay rate over one time period, and $1 - r$ is the growth (or decay) factor. In Section 1.7, we will contrast r with *continuous* growth and decay rates, which are instantaneous rates.

Take some time to discuss exponential functions as another *family* of functions in the students' toolbox, adding to the families of linear and power functions.

It is very natural for students to want to discuss percentage increase in a population. Ask them if the percent increase is a change in the population or a rate of change in the population. In fact it is neither - since it is not computed by either a difference or a difference quotient. It is a percent growth rate, a new concept introduced in this section.

Problems

Problems 1, 2, 20, 21: to recognize exponential growth or decay, and find percent growth or decay and initial values.

Problems 3, 4, 16: emphasize differences in lenear and exponential models.

Problems 5, 6, 8, 17, 19, 22, 24, 25, 27, 28: involve doubling times and half-lives, and predictions.

Problem 7: reinforces the concepts of relative and absolute growth rates.

Problems 10– 15, 23, 26: involve matching graphs and data from tables and formulas to exponential growth or decay. By now, students should feel comfortable doing these.

Problems 9, 18: deal with annual percent increase/decrease

Suggested Problems

1, 3, 7– 9, 12, 14, 15, 17, 23– 25.

1.6 THE NATURAL LOGARITHM

One class.

Key Points

- Natural logarithms and their use in solving exponential equations
- Relationship between a^t and e^{kt}

Ideas for the Class

It is easy to state the point of this section. In problems involving exponential growth and decay, time appears as an exponent. If the problem asks for the time when something happens, the easiest way to solve for it is to use logarithms.

In this text only the natural logarithm is used. Thus $y = \ln x$ if and only if $e^y = x$. Focus on the idea that the logarithm is an exponent. Your students may need some practice with the basic rules for computing with natural logarithms, and you might even like to give some indication of why they hold. But spend as much time as you can with the applications examples and problems. Students do not find them easy, but they find them interesting and will enjoy a classroom discussion after they have been given a chance to struggle with them overnight on their own.

Do one or two examples in class similar to Example 4, showing the students how to convert between e^{kt} and a^t.

Problems

Problems 1–20: it is essential for students to develop a "feel" for logarithms, so assign at least ten of these.

Problems 36, 38, 42: good "How long?" "How many?" and "When?" questions with exponential growth or decay.

Suggested Problems

2–20 (evens), 34, 36, 38, 42

1.7   EXPONENTIAL GROWTH AND DECAY

One class.

Key Points

- Doubling time and half-life
- Compound interest
- Present and future value
- "Rule of 70"

Ideas for Class

It is essential for students to develop a "feel" for exponential growth and decay. Explain why every exponentially growing function has a fixed doubling time, why every exponentially decaying function has a fixed half-life, and that they do not depend on the initial value of the quantity.

Students of different backgrounds react very differently to the materials of present and future values. It is good to start with the point that $100 now is not the same as $100 a year from now.

Problems

Problems 1–4: deal with exponential growth or decay quantities involving doubling time or half-life. After learning natural logarithms from last section, students should feel comfortable dealing with these problems.

Problems 6–9, 15, 19, 22, 23, 25, 28: are business applications.

Problems 11, 13, 14, 16, 17, 20, 21, 24, 26, 27: are life science applications.

Problems 35–39: deal with present and future values. Encourage students to apply common sense.

Suggested Problems

3– 5, 8, 14, 24, 25

1.8   NEW FUNCTIONS FROM OLD

One-half to one class.

Key Points

- Shifting, stretching, and reflecting graphs of functions
- Sums of functions
- Composition of functions

Ideas for the Class

The theme for the day can be families of functions; every function $f(x)$ fits into several families. Your students should become comfortable with: $f(x) + k$, $f(x - h)$, and $Af(x)$, the families of vertical shifts, horizontal shifts, and vertical stretches/compressions (with flip if $A < 0$) of f. Students can demonstrate the meaning of the parameters h, k, and A by sketching lots of graphs. The function $f(x) = x^2$ is a good starting place for the shifts, and $f(x) = x^3 - x$ shows up the stretches very nicely. The effects of the shift and stretch transformations are also easily demonstrated numerically by creating tables of values. Tables are especially helpful in explaining the direction of a horizontal shift, which students often find confusing. Algebraic work can consist of starting with a function given by a formula and a graph, then shifting and stretching to produce a new graph, and finally getting the formula for the new graph. This also provides a good review of function notation.

Have students do a few examples of composition of functions on their own. Although this will be a review for most of your students, the extra practice will help them later on.

Problems

Problems 2– 8, 19, 21– 23, 25– 33: deal with composite functions. Problems 19 and 21– 23 should be assigned.

Problems 9– 18, 20: deal with shifts and stretches of functions.

Suggested Problems

1– 3, 9– 12, 19, 21– 23

1.9   PROPORTIONALITY, POWER FUNCTIONS, AND POLYNOMIALS

One to one and one-half classes

Key Points

- Proportionality
- Power functions
- Polynomials and graphs of polynomials

Ideas for the Class

Start the class with Problem 20 on the circulation time of a mammal. Emphasize the fact that the one word "proportional" in that problem gives us a great deal of information. Define proportional and inversely proportional and give some quick examples as in the book. Then finish the circulation time problem by writing the function $T = kB^{1/4}$ and finding the constant of proportionality.

This leads nicely to power functions. Spend some time helping students recognize power functions and write power functions in the form kx^p. Have them do some problems at their seats similar to those in Example 4. Students have a surprisingly hard time with this, and time spent here will pay off in Chapter 3.

If there is time, have students use their graphing calculators to look at the graphs of power functions. They will see that the odd integer powers all have the same basic shape and the even integer powers all have the same basic shape. Then have them look only in the first quadrant and and compare the graphs of x, x^2, x^3, and x^4, and discuss how the graphs compare.

This section is short, and you should not get bogged down in the details. You want students to come away with some understanding of the graphs of polynomials. They should understand that the number of possible "turning points" is one less than the degree of the polynomial. Sketch Figure 1.89 on the board to illustrate this. They should also understand that the global shape of the polynomial is the same as that of the leading term. Have them do Example 7 in class, or use a similar example. Ask them to explain what they expect the graph of $y = x^5 - 5x^4 + 2x^3 - 12$ to look like. They should say that (since the global picture is the same as x^5) the ends of the function will be in the first and third quadrants, and that (since the degree is 5) the curve connecting these two ends can turn around at most 4 times. Sketch a picture on the board illustrating this.

Finally, an example similar to Example 6 is a wonderful example to do in class because it reinforces the material on linear functions, on demand equations, and on the revenue function while showing the students an application of polynomials.

Problems

Problems 1–12: involve recognizing and working with power functions. Assign most or all of these. Students will need this skill in Chapter 3, but they have a surprisingly hard time with it.

Problems 13–17: simple exercises, assig most or all of these

Problems 18–20, 29, 30, 33, 34: a good collection of real-world problems of proportionality, assign at least two or three of these.

Problems 21–28, 35: deal with the significance of degrees and coeffects of the leading term. Problem 35 should be assigned

Problems 31, 32, 39, 40: involve graph sketching, use technology as aid

Problems 36–38: involve application of polynomial to economics.

Suggested Problems

1–16, 18, 19, 21–30, 35–39.

1.10 PERIODIC FUNCTIONS

One class.

Key Points

- Periodic phenomena
- Amplitude and period
- Sine and cosine functions

Ideas for the Class

Note that this section is *not* titled "Trigonometry." Sine and cosine are used here only for modeling periodic phenomena. There are no angles, triangles, or circles mentioned in the text, and your students will not need to know how they relate to sine and cosine. Work in radians; make sure the students' calculators are set in radians.

Start by showing examples of periodic phenomena—cardiogram, annual climate data, oscillating time series, etc., similar to the examples in the book. Then show sine and cosine, the 'pure' oscillations. It will be clear from the examples that for mathematical modeling you will need to have available pure oscillations of all amplitudes and periods. In short, you will need a whole family of pure oscillations. Move to sketching graphs of $y = A \sin Bt$, or $y = A \cos Bt$ if you prefer. If students can do it for themselves on their calculators or on a computer, so much the better. Then introduce a vertical shift, a baseline different from 0, by considering $y = C + A \sin Bt$. Then show graphs of some of these, and have the students discover the amplitude and period and hence recover formulas for the functions.

Then come back to modeling. Example 7 makes for a great class discussion. If you plan to take a little time with this topic, Exercises 12 and 31 are well worth the effort.

Students may ask about modeling the more complicated phenomena that are not pure sines. You might tell them that more complex periodic functions can be approximated by combining several simpler ones. Have them try graphing, for example, $y = 4 \sin 3x - 3 \cos 2x$ or $y = 4 \sin(x/2) \cos(5x)$. Encourage them to experiment, or announce a prize for the most interesting periodic graph on a calculator.

Problems

Problems 1, 13: to recognize periodic functions and their periods and amplitudes from graphs.
Problems 2–4, 12, 15, 28–32: a good collection of modeling with periodic functions.
Problems 5, 14: involve tables of periodic functions. Assign one of these.
Problems 6–11, 16–27: deal with shifts and stretches of sine and cosine functions. Assign as many as you can of these questions.

Suggested Problems

2, 5, 7–9, 12, 13, 17–19, 21, 25, 30, 31.

FOCUS ON MODELING: FITTING FORMULAS TO DATA

Optional; if covered, one class.

Key Points

- Fitting lines and curves to data
- Using technology to handle data

Ideas for the Class

In this text there are many many graphs of real data that are taken as the starting point for a mathematical interpretation or analysis. However, sooner or later students will very wisely ask where the graphs themselves come from. You can tell them that it is a long and interesting story at the intersection of the subjects of statistics and mathematical modeling, and you can suggest with this section one of the simplest techniques that is used, namely regression.

If you or your students do not have access to technology to carry out regression, you may want to go very lightly on this section. On the other hand, with technology available, this section can reinforce the students' understanding of linear and exponential functions. The emphasis is on interpreting the results rather than on the calculations: let the technology carry out the details.

Start by putting a data set on the board: the data on hourly earnings from Problem 23, page 106 of the book works fine. Ask the students if the data is linear. Then plot the points with hourly earnings as a function of year since 1965, and have students come to the board and try to draw the "best line" on the board. (They can vote on the "best"—they will like this!) Tell them that there are mathematical formulas for finding exactly the best line, and show them how to input the data and find the line on their calculators. Then have them interpret the slope of the line, and plot it to see how it fits the data.

Do the same thing with a data set that looks exponential: the data on health care expenditures in Table 1.52 on page 80 in the book works well. Plot the points and ask if a line is the best way to model the data. They should suggest an exponential function. Show them that the calculator will also do exponential regression—but that the human must decide which type of regression to do. Have the calculator find the best exponential fit, and then interpret the results in terms of annual growth rate.

Problems

Problems 4, 7, 13, and 18 do not ask the student to find regression curves, but to reason about them. The other problems are a mixture of linear, exponential, and logarithmic regressions. Once the technology has been mastered, the emphasis should be on the interpretation, rather than on the calculations.

Suggested Problems

1, 4, 5, 7, 10, 13, 15, 17, 18.

FOCUS ON MODELING; COMPOUND INTEREST AND THE NUMBER e

Optional; if covered, one-half class.

Key Points

- Compounding
- Effective annual yield
- The definition of e

Ideas for the Class

Section 1.7 discusses annual compounding and continuous compounding. If you want to also discuss compounding such as quarterly or monthly, then include this section. Start with an example of a 6% annual interest rate compounded monthly. Ask them how much interest is paid *per month* to lead them to the formula in the box on page 84. This formula, with Example 6, will lead nicely to the definition of the number e.

The effective annual yield is an important concept. Give them several options to compare (such as a 7% return compounded annually, a 6.9% return compounded monthly, and a 6.8% return compounded continuously) and ask them to find and compare the effective annual yield in each case.

Problems

Most of the problems are straightforward, but students may not read the problems carefully enough to avoid mistakes. You might have them set up some problems in class, working in groups, and then discuss the types of errors that arose commonly. Problems 10, 11, 12, and 13 are good examples.

Suggested Problems

1, 3, 4, 8, 10, 11, 12, 13

CHAPTER TWO

Five to Six classes.

Overview

This chapter presents the concept of the derivative as a limit of difference quotients, focusing on the interpretations of the derivative as rate of change and as slope of a curve. The emphasis is on understanding where the derivative comes from, what it means, and how it can be applied. A concerted effort is made throughout the chapter to give graphical, numerical and verbal considerations equal billing. Since derivatives will be computed approximately by difference quotients or by estimating slopes, students can work with functions presented verbally, by tables of values, and by graphs as well as those defined by formulas. Because so much emphasis has traditionally been given to derivatives as formulas, it may seem that algebraic considerations are being given short shrift; wait until Chapter 3 where they are given full play.

2.1 INSTANTANEOUS RATE OF CHANGE

One class.

Key Points

- Instantaneous rate of change of a function at a point as the limit of average rates of change over shorter and shorter intervals.
- The slope of the tangent line.
- Estimating rate of change with difference quotients.
- Definition of derivative at a point as an instantaneous rate of change.
- Graphical and numerical computations.
- Notation for the derivative.

Ideas for the Class

Students learned how to find and interpret the average rate of change over an interval in Section 1.3. In this section, we extend the concept of rate of change to instantaneous rate of change at a single point. A thorough understanding of average rate of change will help enormously—spend time at the beginning of class reviewing average rate of change. Remind the students that the average rate of change is a difference quotient: if $y = f(x)$,

$$\text{Average rate of change of } y \text{ between } x = a \text{ and } x = b = \frac{\Delta y}{\Delta x} = \frac{f(b) - f(a)}{b - a}.$$

Students should recognize this formula also as the formula for the slope of a line. Remind them of the units and interpretation of average rate of change.

Have students compute several average rates of change using the table in the book of the grapefruit thrown straight up in the air. The students should recognize the answers as average velocities, and the units as height units (ft) over time units (seconds). This is a good example to start with because velocity at a single point makes sense to the students. Have students reflect on the difference between each of the average velocities they compute and the velocity at a single point in time. You might characterize the average velocity of a car, for example, as the distance traveled divided by the time spent and the instantaneous velocity as the speedometer reading. Then ask students to speculate on what the speedometer reading might be in a car that traveled 30 miles in one hour (no way to tell), one mile in two minutes (not much better able to tell, but likely close to 30 mph some of the time), or 44 feet in one second (very close to 30 mph).

The key to instantaneous velocity is computing average velocities (with difference quotients) over smaller and smaller time intervals. On the one hand, students must see the need for going to average velocities over an interval to estimate the velocity at an instant. This can be explained visually, if you wish, using some pictures of animals in motion taken by Edward Muybridge [*Animals in Motion*, Dover, 1957] over a century ago. As you point to a particular frame and ask what the average velocity is there, it quickly becomes apparent that you need two frames to answer the question, and it is natural to select adjacent frames. On the other hand, students must appreciate the need to go to shorter and shorter intervals to improve accuracy.

The word limit is introduced but not formally defined. This is usually not a problem for students in this course. You might tell them that when you say that the instantaneous rate of change is the limit of average rates of change, this means that the instantaneous rate can be approximated by an average rate. It is vital though that students bear in mind that difference quotient approximations are not exact. It is also good for them to understand that in practical situations often an approximation is as good as you can do because the data they have to work with is itself not exact and may be incomplete as well - for instance table values are rounded off and skip many values. Fortunately a close approximation is often adequate for purposes at hand.

As you move to rates of change other than velocity, it is important to have students computing average rates, interpreting signs and other outcomes in practical terms.

Be sure to include some graphical examples. The slope of a curve at a point is well illustrated by having the student zoom in on a smooth curve until it appears straight. Compute some slopes of curves this way, and make sure the student makes the connection between average rates of change and slopes of secant lines on the one hand, and instantaneous rate of change and slopes of tangent lines on the other.

The derivative of a function at a point is defined as the rate of change of the function at that point. In the beginning of this section, there are many examples of real world functions, the students are asked to work in this part with some 'pure' functions defined by formulas, graphs, or tables, but not arising from an applied context. The methods available to them for computing derivatives at this point in the course, though, are just the same. The only things new to the student in this section are the use of the word *derivative* and the notation $f'(x)$.

If a function is defined by a formula or a table of values, students must approximate with a difference quotient over a short interval (or in the case of a table as short an interval as they have data for). If the function is defined by a graph they can compute the slope of a tangent line either by drawing a tangent line on the graph with straightedge or by zooming in on a calculator till the portion of the graph they are seeing appears straight.

Once again, emphasize the distinction between the derivative itself and its approximations by difference quotients.

Problems

Problems 1, 2, 4, 5: relatively simple problems to reinforce the numerical and graphical interpretations of the rate of change at a point. It is a good idea to work through a few of these in class before assigning others.

Problems 3, 6, 13, 22: involve recognizing slopes from a graph. Assign at least one of these.
Problems 7, 14, 15, 17–19, 21, 24: involve graphical interpretation.
Problems 8–10, 16, 20, 23: involve numerical estimations.
Problems 11, 12: are more substantial and it would be wise, not only to assign most of these,

but to discuss them as well. The concept of instantaneous rate of change is so fundamental that every effort should be made to ensure that students comprehend this fully.

Students might be more comfortable with the graphical interpretations than with the numerical estimations. Problems such as 8, 9, 10, 14, and 15 should be assigned. Time spent on these will pay off later. Beware the difference in scales on the two axes in Problem 17. Problems 18, 19, and 23 are good for discussion in groups.

Suggested Problems

1– 5, 8– 10, 14– 17, 19, 23.

2.2   THE DERIVATIVE FUNCTION

One class.

Key Points

- Understanding the derivative as a function.
- Finding the derivative function graphically and numerically.
- The derivative and increasing and decreasing functions.

Ideas for the Class

Point out that the derivative of a function at each point defines a new function, the derivative function. This is best illutrated graphically, and you may want to spend most of your class on graphical differentiation. Start with a smooth curve with some ups and downs on the board, and go through the details of sketching the derivative. Students will need lots of practice; Problems 7– 12 are good ones to have students work through in groups. In case the graph is given on a grid, as in Problems 7, 8, and 10, students can estimate slopes by imagining tangent lines and then plotting points representing the slopes. Some easy points to plot first are the points where the derivative is zero and thus touches the x-axis. Point out, or have students notice, that the derivative is positive where the function is increasing and negative where it is decreasing, and thus is zero where the function changes direction. Once the students understand where the derivative is positive, negative, or zero, they will have to draw a graph with these characteristics. Point out explicitly that points where the derivative is zero are the x-intercepts of the derivative graph, that intervals where the derivative is positive lie above the x-axis on the derivative graph, and that intervals where the derivative is negative lie below the x-axis on the derivative graph. For some students, this is the hardest part. Emphasize the global properties and don't get bogged down in details. You might sketch the graph of a function and its derivative (such as in Figures 2.20 and 2.21) on the board, and ask students which is the function and which is the derivative. Make liberal use of "number lines" such as on the bottom of these two figures.

An activity that produces a lot of mathematical discussion is to have each student sketch the graph of a function on a piece of paper, put his or her name on it, and pass the paper to a neighbor. The neighbor draws the graph of the derivative of the function on another piece of paper and copies the name of the first person on it. The original is passed back to its author and the derivative is passed to a third student. The third student sketches the graph of the original function from the derivative and then compares it with the original. This activity can be used for review later; even at the end of the term, many students may find it difficult.

Invent similar activites with tables of values. You want students to look for global properties, such as intervals where the function is increasing or decreasing.

Since it takes some time for students to come to terms with the graph of the derivative function, resist the temptation to make too many connections for them at first. For example, unless someone in class points out the relationship between zeros of the derivative and extrema of the original function, or between extrema of the derivative and inflection points of the original, don't do it for them. If the issues do arise, just note them as perceptive observations that merit future consideration and move on.

Note that finding formulas for the derivatives of functions given by formulas is reserved until Chapter 3. There is one example in the text, $f(x) = x^2$, that can be used to suggest the existence of formulas if you wish, or it can be omitted. For the sake of students who already know the algebraic shortcuts, point out that the way you show the shortcuts work is to use the definition of the derivative, starting with difference quotients. Some students have had the shortcuts hammered into them so thoroughly that they believe such shortcuts constitute all of calculus.

Problems

Problems 1– 4: involve matching the graph of a function with its derivative.

Problems 5, 6, 15, 30: allow reasoning about the derivative from numerical data.

Problems 7– 12 and 18– 25: give practice in sketching the derivative function from a graph. Motivate the students to look for qualitative, global features rather than detail.

Problems 13, 14, 16, 17, 28, 29, 31: give practice in reading properties of the derivative from a graph or in constructing a graph, given properties of the derivative.

Problems 26, 27: refer to algebraic derivation.

Suggested Problems

1– 4, 8, 11, 14, 15, 16, 27, 29, 31.

2.3 INTERPRETATIONS OF THE DERIVATIVE

One class.

Key Points

- Using the difference quotient and units to discover the meaning of the derivative of real world functions.
- Using the derivative to estimate values of a function.
- Local linear approximation.
- The dy/dx notation.

Ideas for the Class

This is one of the most important sections of the text! What is the meaning of a derivative in a real world situation?

The mathematical content of this section is not increased from that of the previous section. But the emphasis here is different, on using the difference quotient and units to give meaning to the derivative, especially as applied to rates. The following example is a good one to include:

Example   The population of the world, P, in billions of people, is a function of the year t. We have $P = f(t)$. Explain in terms of the population of the world the meaning of the statements $f(1990) = 5.295$ and $f'(1990) = 0.086$.

Remind the students that the function f gives the total population of the world at a given time, while the derivative f' gives the *rate of change* of the population. The statement $f(1990) = 5.295$ tells us that the world's population is 5,295,000,000 people in 1990. The statement $f'(1990) =$

0.086 tells us that the population, in 1990, is increasing at a rate of 0.086 billion people per year, or 86,000,000 people every year (an increase of about 164 people every second!) Ask the students what a negative derivative would mean for world population. Emphasize the notation dP/dt for the derivative, and encourage them to use this Leibnitz notation to understand the units: P-units over t-units, or billion people per year, in this case. Have the students use the information given to estimate the world population in 1991. It is possible to get a great deal of mileage out of this example.

Make sure the students understand and use the Leibnitz notation, and the results in the box on page 108. Many students find it difficult to come up with their own explanations on the homework, and this will help. Plan to spend some time going over the homework after the students have had a chance to work on it.

Students may find some of the history of the derivative notation interesting. Newton's *dot* notation (which uses a dot over the x to represent $x'(t)$), introduced in the 1660s, is not now widely used. Leibniz's *double-d* notation [dx/dt for $x'(t)$], introduced in the 1670s, is widely used. The *prime* notation $x'(t)$ was introduced by Lagrange over a century later. Emphasize that in dx/dt, the dx is to remind us of a small change in x, and the dt is to remind us of a small change in t. Thus dx/dt is to remind us of the difference quotient which approximates the derivative and from which the derivative comes as a limit. To think of dx/dt as a fraction with products is wrong; canceling the d's would be akin to canceling the 2's in the fraction 27/28.

Problems

It is essential for students to verbalize as well as write down answers to problems in this section. Be picky as to how they answer the questions. When students can articulate their answers to this type of questions, they are well on their way to understanding the concept of the derivative as a rate of change.

A variety of interesting problems are presented in this section. It might be a good idea to do one or two together so that they realize how important the correct interpretation, correctly articulated, is.

Suggested Problems

1, 2, 5, 6, 8, 14, 15, 21, 23

2.4   THE SECOND DERIVATIVE

One half to one class.

Key Points

- Interpretation of the second derivative in terms of concavity.
- The second derivative as a rate of change.

Ideas for the Class

The most important thing for students to keep in mind about second derivatives is that they are derivatives in their own right and hence always measure a rate of change, namely the rate of change of the first derivative. The second derivative thus tells how the first derivative is changing, and for this to make any sense the meaning of the first derivative must be very clear. For instance, in a graphical situation the first derivative is a slope and so the second derivative tells how the slope is changing. In a distance-time situation, the first derivative is a velocity and so the second derivative tells how the velocity is changing. You might ask students in class to name some quantities that change over time, and in each case have them state what the first derivative means (as in the preceding section of the text), and then ask them what it would mean if the second derivative were positive or what it would mean if the second derivative were negative.

Go through the graphical interpretation of the second derivative in terms of concavity very carefully. The goal is for students to see the geometry so clearly in their minds that they are not even tempted to try to rely on memory to relate signs of first and second derivatives to shapes of graphs. Treat all four cases separately:

- increasing and concave up, like e^x, is increasing at an increasing rate;
- increasing and concave down, like $\sqrt{x}$, is increasing at a decreasing rate;
- decreasing and concave up, like e^{-x}, is decreasing at a decreasing rate;
- decreasing and concave down, like $-e^x$, is decreasing at an increasing rate.

The second paragraph on page 114 under "Interpretation of the Second Derivative as a Rate of Change" or any of the problems based on quotes (Problems 19, 20, or 21) make excellent in-class discussion or group work problems. Have students sketch a graph of the situation in each case, and discuss the interpretations of the first and second derivatives.

Acceleration has not been explicitly mentioned in this section or in the exercises. If you feel that the concept is important for your students, it is not difficult to invent some simple examples that make the idea clear.

Problems

Problems 1–3, 7–12, 14, 15, 22, 23: involve graphical interpretation of the sign of the second derivative. These should be well represented as assigned problems.

Problems 4–6, 18, 24: give practice in sketching and using the graph of a function, given information about its derivative.

Problems 13, 16, 17, 25: emphasize describing the behavior of the second derivative from a table of data.

Probelms 19–21: are based on quotes and are good discussion or group work problems.

Suggested Problems

2, 4, 5, 7, 11, 14–19, 24, 25.

2.5 MARGINAL COST AND REVENUE

One class.

Key Points

- Derivatives and the concept of marginality.
- Marginal cost and marginal revenue.
- Maximizing profit.

Ideas for the Class

A discussion of what a cost curve means, as well as interpretations of its slope and concavity, is a good place to start. How is economy of scale reflected in the shape of the graph? Follow with a discussion of revenue curves, both the straight line (fixed price) case and the concave down (quantity discount) case. Then put cost and revenue together and discuss profit. Students sometimes forget about costs, wanting to maximize revenue rather than profit, so keep reminding them that revenue and costs must always be considered together. Note that in this section we consider only cost and revenue, not demand, supply, or other things, so that our analysis is not entirely realistic.

It is easy to memorize without understanding the fact that marginal cost of 100 items is the approximate cost of producing the 100th item. Be sure your students understand why this is true. You might ask them to explain it clearly in writing using a difference quotient approximation for the derivative of the cost function.

When discussing profit, explain why (or have students explain why) increasing production increases profit when marginal revenue exceeds marginal cost, and why decreasing production increases profit when marginal cost exceeds marginal revenue. (Assume revenue exceeds cost.) Use the tangent lines to conclude that when profit is maximized, $R'(q) = C'(q)$. This is an entirely graphical explanation, but it is both clear and convincing to the students.

Problems

Problems 1, 2, 13: require written or verbal interpretations. These are essential for developing verbalization and written thoughts
Problems 3–8, 11, 12: involve graphs and form the backbone of this set of problems.
Problems 9, 10: involve tables of data and should be assigned.

Suggested Problems

1, 4, 6, 8, 9, 10, 11, 13.

FOCUS ON THEORY: LIMITS, CONTINUITY AND DERIVATIVE DEFINITION

Optional; if covered, one class.

Key Points

Formal definition of the derivative. Limits graphically, numerically, and algebraically. Continuity.

Ideas for the Class

Remind the students that $\lim_{x \to 0} f(x)$ is the number approached by $f(x)$ as x approaches 0, and then ask the class to find $\lim_{x \to 0} (1 + x)$. After a discussion of this limit, have them consider the trickier case of $\lim_{x \to 0} \dfrac{e^x - 1}{x}$. Ask the class for suggestions on how to find this value, or have them work in groups on this problem. Since they get an undefined value when they plug in $x = 0$, hopefully it will occur to some of them to try plugging in values *close to* 0, or to look at a graph of the function. After they have had a chance to think about each of these approaches, try them both in class, with this example and others.

Derive the formal definition of the derivative here and make sure they understand each of the symbols, but don't spend lots of time on this—we will return to it in more detail in the Focus on Theory section in Chapter 3. To relate the symbols back to something they have seen, try the following: if we estimate $f'(2)$ using a small interval of width $h = 0.001$, then we have

$$\frac{f(x+h) - f(x)}{h} = \frac{f(2.001) - f(2)}{0.001} = \frac{f(2.001) - f(2)}{2.001 - 2} = \frac{f(b) - f(a)}{b - a}.$$

This may help give meaning to the symbols.

When discussing continuity, keep the emphasis on graphical examples. Most of the students will not have a hard time with this.

Problems

Problems 1, 2: develop graphical understanding of the definition of the derivative.

Problems 3–8: involve finding limits graphically and numerically.

Problems 9–23: involve continuity.

Problems 24–29: use the definition of the derivative to obtain a derivative formula.

Suggested Problems

#1, 3, 4, 6, 7, 9, 11, 12, 14, 16, 17, 19, 20, 22, 24, 28.

CHAPTER THREE

Five classes.

Overview

This chapter develops the basic differentiation formulas. Included are the standard combination rules (sum, difference, product) and the chain rule, as well as formulas for differentiating powers, polynomials, exponentials, logarithms, and the sine and cosine. Informal justifications are given, using graphical and numerical reasoning where appropriate.

The title of the chapter is intended to remind students that the formulas do not constitute the definition of the derivative; plenty of problems throughout the chapter review the concept and interpretations of the derivative.

3.1 DERIVATIVE FORMULAS FOR POWERS AND POLYNOMIALS

One class.

Key Points

- Differentiation formulas for power functions and polynomials.
- Graphical and verbal interpretations of the derivative using formulas.

Ideas for the Class

Students who have seen calculus before will be relieved to have come to the easy stuff (maybe you, too!). Don't forget that many are seeing it for the first time, however.

The differentiation formulas introduced in this section are very easy, but it may take some time and effort for the students to connect them with the graphical, difference quotient, and verbal (rate of change) concepts of the derivative that they know from Chapter 2. Making this connection should be the main work of the section. To facilitate this process the algebraic techniques for proving the formulas have been de-emphasized, and are given in the Focus on Theory at the end of this chapter.

Use graphical arguments when you can, such as to predict the derivative of a linear function. It is usually worth asking what the graph of the derivative looks like (starting with the graph of the function), before you calculate the derivative. In a good many problems, maybe over half the ones you assign and discuss, ask for something more than the formula for a derivative. It is almost always a good idea to ask for an evaluation at a point, which can be done directly by asking for a comparison with a difference quotient approximation or indirectly by asking for the equation of a tangent line. And finally be sure to include some word problems in which the value of the derivative at a point must be both computed and interpreted in practical terms.

Problems

Problems 1– 26: routine practice of which students should do as many as it requires to fully master this rule.

Problems 27– 31, 34: specific derivatives are required. A selection of these problems should follow Problems 1– 26.

Problems 29, 35, 36: higher derivatives form the theme of these problems.

Problems 32, 38, 39: involve finding equations of tangent lines.

Problems 41, 42, 44, 45, 48: refer back to previous topics such as supply and demand and cost and revenue.

Problems 33, 37, 43, 49: application type problems.

Problems 40, 46, 47: involve finding points for specific values of the derivative.

Problems 50: is difficult and should be discussed.

Suggested Problems

1, 3, 8, 17, 19, 21, 29, 34, 36, 38, 41, 43, 45, 46

3.2 EXPONENTIAL AND LOGARITHMIC FUNCTIONS

One-half to one class.

Key Points

- Differentiation formulas for e^x, a^x, and $\ln x$.
- The significance of the number e.

Ideas for the Class

It is easy to see graphically that the derivative of an exponential function resembles the graph of an exponential function—for exponential growth both graphs start near zero and rise faster and faster as you move to the right. Using a graphical plausibility argument, the text gets to the equation $\frac{da^x}{dx} = c \cdot a^x$ for some constant c. If your students have calculators that will graph the derivative of a function, have them graph 2^x with its derivative and 3^x with its derivative. They should notice that the derivatives have the same shape, but that the derivative of 2^x lies below the function while the derivative of 3^x lies above the function. What about the derivative of $(2.5)^x$? By continuing in this way, have them try to find a base a such that the derivative of a^x lies exactly on the function. They will be re-discovering the number e. The derivative formulas follow nicely after this activity.

Don't forget that the point of the derivative formulas for a^x and $\ln x$ is to use them! There are some applied examples and problems in this section that will make for good discussions.

Problems

Problems 1– 22: again, are routine practice of which students should do as many as it requires to fully master this rule.

Problems 23, 24, 31, 25– 30, 32: technical problems (23, 24, 31) that provide the necessary basis for tackling the applications (25– 30, 32). In an overview of these, let students refresh their knowledge on exponential growth and decay by pointing to what problem represents what.

Problems 33– 38: are of somewhat difficult technical natures and should be reserved for students who seek further enrichment.

Also point to specific percentage increases or decreases.

Suggested Problems

2, 12, 14, 21, 23, 24, 26, 30, 32.

3.3 THE CHAIN RULE

One class.

Key Points

- Recognizing when the chain rule is necessary and how to apply it.

Ideas for the Class

The list of functions whose derivatives the students know at this point is very short—sums (with coefficients) of powers, exponentials, and natural logarithms. Now they will increase the list dramatically. You can help them to see this for themselves after reviewing composite functions just by asking them each to list three examples of functions constructed by composition from the simpler ones.

Most of the class period should be spent with lots of examples and a good bit of drill. Help the students learn how to tell when they need to apply the chain rule, or what is the same thing, how to recognize a composite function. Give them some problems to work out at their desks to check that each one has really understood the technique. It is a good idea to have them evaluate some of the derivatives at specific points, and from time to time you might have them check such an answer numerically with a difference quotient (using their calculators). You don't want them to forget that every derivative they will ever see can be approximated by a difference quotient!

Include at least one example where the derivative must be interpreted, as in Example 6.

Problems

Problems 1–30: routine problems, a number of these should be done in class by the students and a number should be assigned. The chain rule takes a while to become second nature.

Problems 31–33: involve tangent lines

Problems 34–41: application problems mainly involve exponential functions and are, as always, of utmost importance.

Note: Differentiation of periodic functions also makes use of the chain rule but is dealt with separately in Section 3.5.

Suggested Problems

1, 5, 9, 15, 25, 30, 31, 35, 37, 38, 40.

3.4 THE PRODUCT AND QUOTIENT RULES

One half to one class.

Key Points

- Differentiation formulas for products and quotients of functions.

Ideas for the Class

Your first task may be to convince students that there is a need for a product rule, that the derivative of a product is not the product of the derivatives. A good example is $x^2 \cdot x^3 = x^5$, whose derivative equals $5x^4$ and not $2x \cdot 3x^2 = 6x^3$, which is wrong in both the coefficient and exponent.

A justification for the product rule is given in the Focus on Theory at the end of this chapter, which you can refer students to if they ask, or go over quickly in class. But you will want to spend most of your time with examples and more examples. Consider getting students up at the board, differentiating as you throw them functions. Or give them several one minute quizzes to do at their desk, with correct answers revealed to them immediately thereafter. This will help them know for themselves whether they really understand the product rule, and generate questions if they do not. Start simple, but work up gradually to products of composite functions. Your students are probably still just getting comfortable with the chain rule, and the more practice with it, the better.

For decades students have laughed at the description of the product rule as the HiHo rule, a la Hoagy Carmichael. "The derivative of HiHo equals Hi dee Ho + Ho dee Hi". They may enjoy this, and have fun identifying Hi and Ho in a couple of problems.

The quotient rule is also included and a justification is given in the Focus on Theory at the end of this chapter.

Problems

Problems 1– 31: routine problems, a number of these should be done in class by the students, and a number of them should be assigned.

Problems 32, 34, 37: involve finding tangent lines.

Problem 33: requires finding higher order derivatives.

Problems 35, 36, 39: application problems.

Problem 38: is an excellent problem which students find quite difficult. It is worth spending some time on. It is an excellent problem to assign to small groups and then bring everyone back together to talk about.

Suggested Problems

1, 3, 4, 6, 9, 20, 32, 33, 35, 36, 38

3.5   DERIVATIVES OF PERIODIC FUNCTIONS

One class.

Key Points

- Differentiation formulas for $\sin x$ and $\cos x$.

Ideas for the Class

Start class by putting a graph of $y = \sin x$ on the board. Ask students to discuss the graph of its derivative. Where will it be zero? Positive? Negative? Students quickly see that the derivative of $\sin x$ must look like $\cos x$. This graphical approach is generally the best motivation for the formulas of the derivatives of $\sin x$ and $\cos x$. Make sure the students know that these formulas are only valid when x is in radians!

Do several straightforward examples on the board (such as in Example 4) and then have the students work some at their seats. This section gives the student additional practice with the chain rule and the product rule, which they will appreciate.

Problems

Problems 1- 19: routine problems, a number of these should be done in class by the students and a number of them should be assigned.

Problems 20, 22– 25 application problems. Emphasize Problem 20, especially the interpretation of the function and derivative values.

Probelm 21: involves increasing, decreasing, and concavity

Suggested Problems

1, 2, 7, 13, 19, 20, 21, 23, 24

FOCUS ON THEORY: ESTABLISHING THE DERIVATIVE FORMULAS

Optional; one class, if covered.

Key Points

- Proving derivative formulas with difference quotients.

Ideas for the Class

It is time for the students to see how some of the derivative formulas that they have been using can be proved. This is your chance also to remind them that, in the end, everything there is to know about derivatives goes back to the definition, which is in terms of difference quotients. Keep it simple. Students in this course can usually follow simple algebraic arguments, but it is often best to work up to the algebra through some preliminary concrete computations. You have to convince them first that algebra is their friend.

For example, suppose you want to prove that the derivative of x^2 equals $2x$. You might begin with a classroom project of making a huge table of approximate values by computing tons of difference quotients. Do, or have students do, $\frac{(2+0.1)^2 - 2^2}{0.1}$, then $\frac{(3+0.1)^2 - 3^2}{0.1}$, then $\frac{(4+0.1)^2 - 4^2}{0.1}$, etc., approximations for the derivative at 2, 3, and 4. Students will begin to find this tedious and repetitive, and so they will be glad to hear that they can do an infinite number of these computations all at once — it's the power of algebra!

So go back to redo and extend your table but using the formula $\frac{(x+0.1)^2 - x^2}{0.1} = 2x + 0.1$. But the table is only approximate, and we want to improve its accuracy. Start over again, this time with $\frac{(x+0.01)^2 - x^2}{0.01} = 2x + 0.01$, then again with $\frac{(x+0.001)^2 - x^2}{0.001} = 2x + 0.001$.

Your students will be begging for the chance to do all values of x and all values of h at once with more algebra: $\frac{(x+h)^2 - x^2}{h} = 2x + h$. And that is really all there is to it. They can all now see that the smaller h is, the closer the difference quotient is to $2x$, and so they have their first proof.

Proofs of the product and quotient rules are included in this section, but you may choose to omit them.

Problems

Problems 1–7: simple applications of the definition of a derivative.

Problem 8: students will probably call for guidance.

Problems 9–12: involve proving general derivative rules using the definition of the derivative.

Suggested Problems

3, 5, 7, 8, 10, 11

CHAPTER FOUR

Five to seven classes.

Overview

In this chapter students will see how the derivative can be used in a variety of settings. The analytical tools developed in Chapter 3 are brought into play, but the emphasis throughout remains on the concepts themselves and their graphical and verbal interpretations.

The derivative is used to explore the shapes of graphs given by formulas. Optimization is studied in both theory and applications. The relationship between maximum profit and marginal cost and revenue is developed more fully than was possible in Chapter 2. Average cost is studied graphically and a connection is made with marginal cost. Applications in elasticity of demand, logistic growth of populations, and surge functions of drug concentrations are introduced.

No student should have to leave this course wondering whether there are any applications of calculus to the real world.

4.1 LOCAL MAXIMA AND MINIMA

One class.

Key Points

- The derivative: critical points and local extrema.

Ideas for the Class

This section is about using calculus to uncover the relationship between equations and graphs.

Students may think at first that with function-graphing technology they don't need these ideas. Certainly they will not see the point of laboriously using calculus to graph fifteen functions by hand that can be graphed easily on a machine. But the first example in the text illustrates one way we can use calculus to lessen our frustration with the technology, namely to help us find a suitable calculator window for an unfamiliar graph. A couple of other examples in the same vein to try are $y = -7x^3 + 23x^2 - 19x - 12$ and $y = xe^{-3x} + x^2 + 50$. (You might remind students that they can look for zeros of the derivative with their calculator, too!)

Make sure students are confidently thinking geometrically to distinguish a local minimum from a local maximum. If a derivative changes from negative to positive then the function graph must first go down then back up, so the transition point must be a local minimum. To ensure that your students really see this picture rather than trying to memorize a 'test' for local minima, you might ask them to write a paragraph with illustrations. Show them the graph of a derivative crossing the x-axis and have them sketch a tiny piece of the function at x-values near the crossing.

One of the main purposes of this section is to get students comfortable with the words "critical point" and "local max/min." Have the students do the following examples in class:

Example   Draw the graph of any function which has exactly 4 critical points: at $x = -3$, $x = 0$, $x = 2$, and $x = 5$.

Example   Draw the graph of any function which has exactly 2 local maxima, exactly 2 local minima, and exactly one additional critical point which is neither a local maximum nor a local minimum.

It is good to do at least one example involving parameters. Example 5 is particularly nice, since many of the students have seen the formula for the vertex of a parabola before.

Problems 22 and 26 are excellent problems for group work.

Problems

Problems 1– 4, 17: straightforward problems to make sure they understand the terminology.

Problem 5: revisits the second derivative test.

Problems 6, 7: ask the students to recognize a critical point from a verbal description.

Problems 8, 15, 16, 23, 24: ask students to estimate critical points given a graph or table of values of a *derivative*. Make sure they recognize the distinction between the graph of a function f (Problem 1) and the graph of a derivative f' (Problem 15).

Problems 9– 14, 25: involve graphing with technology to identify different features of the functions.

Problems 18– 21: involve functions with parameters. Students find these difficult. If you assign some of these, be sure to spend some time in class doing similar problems.

Problems 25, 27: ask the student to use critical points (found algebraically) to gain knowledge about the function.

Problems 22, 26: recommended for discussion or group work.

Suggested Problems

1, 4, 5, 6, 15, 17, 18, 23, 25, 26.

4.2 INFLECTION POINTS

One class.

Key Points

- The second derivative: concavity and inflection points.

Ideas for the Class

The verbal interpretation of the second derivative: the derivative of the first derivative.

The graphical significance of the inflection points: where the second derivative changes sign.

Another approach to this material that admits from the start that graphing technology is available is to start with *both* a formula and its graph. Ask the students first to describe what they see in the graph — intercepts, increasing/decreasing sections, concave up/down sections, critical points, inflection points, local maxima and minima, asymptoptes. Then ask where all these visual features of the graph are buried in the formula. They should be able to uncover them by analyzing the function and its first and second derivatives. If you will discuss the surge function in Section 4.8, you should definitely discuss Example 4 on critical points and inflection points of $y = xe^{-x}$.

Problems

Problems 1– 4, 12, 13: are graphical exercises of finding inflection points.

Problems 5– 11, 17: involve an algebraic approach to finding inflection points.

Problems 14, 18– 22: involve practical interpretations of inflection points.

Problems 15, 16, 23– 29: demand a better understanding of graphical interpretations of the derivative, second derivative and local extremes.

Suggested Problems

4.3 GLOBAL MAXIMA AND MINIMA

One-half to one class.

Key Points

- Global extrema.
- Life science applications

Ideas for the Class

Students already know that local maxima are found at critical points, and so they will easily understand that to find a global maximum they need to just search through the list of critical points (and endpoints of domain if any) for the one that gives the greatest value. You can either just make this point with one or two examples, or drill a bit, depending upon how important working with formulas is for your class.

The emphasis of the applications in this chapter are in the life sciences. Profic, cost, and revenue is visited in Section 4.4.

Problems

Problems 1– 14: are practice of skills. Do not underestimate the importance of these.

Problems 15– 33: involve applications in the life sciences. Do not hesitate in assigning a bunch of these. A good group of these should be assigned, as well as discussed in class. Problems 16, 21, 23, and 29, are particularly good for discussion. Problems 31– 33 are more challenging and are good for group work.

Suggested Problems

4.4 PROFIT, COSTS, AND REVENUE

One-half to one class.

Key Points

- Maximizing profit and revenue
- Marginal cost and revenue

Ideas for the Class

Of more interest to some students will be the discussion of profit maximization that is a follow-up to Section 2.5, Marginal Cost and Revenue. Example 1 is worth discussing in detail. In particular, note that in the graph of Figure 4.40 there are *two* production quantities at which marginal cost equals marginal revenue, a situation discussed often in economics classes. Students should be able to explain clearly which of the two points corresponds to maximal profit and why. You might try graphing part of a similar graph, showing only one crossing of marginal cost and revenue, and asking the students to make a recommendation of whether to increase or decrease production from the crossing point.

When discussing maximization of revenue, make sure that the students understand that it is maximization of profit that is really of interest. Examples have been chosen for the text (bus company, amusement park) where the cost is thought of as independent of the quantity of sales, and so in these cases maximum profit and maximum revenue occur at the same point.

Problems

Problems 1– 14: involve cost and revenue. Explanations should be emphasized.

Problems 6, 16, 17: require graphical interpretations and are suited for generating discussions.

Problem 8: is interesting in that the cost function is explicitly given as a cubic polynomial. Encourage students to graph this function as well as the revenue function. This problem would be good for group work.

Problem 15: is based on the Cobb-Douglas function which is discussed extensively in Chapter 9. Assign this problem only to students who would like enrichment work.

Suggested Problems

1, 2, 4, 5, 6, 7, 8

4.5 AVERAGE COST

One class.

Key Points

- Average cost and its graphical interpretation.
- Minimal average cost occurs when marginal cost equals average cost.

Ideas for the Class

This section is really an extended example. It demonstrates the power of calculus and graphs to reveal hidden relationships between two different concepts. In this case, it is a relationship between average cost and marginal cost. The link is made graphically via slope. Students will see graphically that average cost of production is the slope of a line from the origin to the cost curve, and they already know that marginal cost is the slope of a tangent line to the cost curve. You might ask them if there is a production level where the two lines are the same. If they can answer the question and explain its significance, they will have understood the point of this section of the text.

Problems

Problems 1, 2, 6, 8, 9, 13: are based on graphs. The aim here is for the students to visualize the difference between marginal cost and average cost.

Problems 3, 4, 11 and 12: require the formulas for marginal cost and average cost.

Problems 5, 7, 15, 16: investigate the relationship between marginal cost and average cost.

Suggested Problems

1, 2, 3, 8, 9, 12, 13

4.6 ELASTICITY OF DEMAND

One-half to one class.

Key Points

- Elasticity of demand.

Ideas for the Class

Start class by asking students "If the price of lightbulbs goes up, do you think there will be a significant change in the demand for lightbulbs?" and "If the price of a certain brand of candy bar goes up, do you think there will be a significant change in the demand for this brand of candy bar?" Explain that the way economists measure this is with a concept called *elasticity*. Demand for a product is called inelastic if a change in the price does not have a large effect on the demand for the product, and it is called elastic if a change in the price does have a large effect on the demand for the product. Perhaps you could have students "vote" on whether the demand for each of the following items will be elastic or inelastic: furniture, baby formula, jewelry, milk, batteries, vacation cruise packages, gasoline, new cars. Don't worry if they disagree—the important thing is that they can make valid arguments for why demand for a certain item might be elastic or inelastic.

When they understand the idea, derive the formula for elasticity, and explain that inelastic means $E < 1$ and elastic means $E > 1$. They might find it interesting to look at Table 4.6, which gives the elasticity for certain items.

Include the discussion on elasticity of demand and maximum revenue if you have economics students in your classes. The result that revenue is maximized when elasticity is 1 is elegant and not hard to understand.

Problems

Problems 1–9, 17, 18: straightforward problems involving the general idea of elasticity of demand.
Problems 10–12, 14–16: involve finding elasticity given a demand equation.
Problem 13: is similar, but the demand equation is given by a table of values.
Problems 19–23: more difficult problems and can be used for enrichment.

Suggested Problems

1, 3, 4, 5, 9, 10, 12, 13

4.7 LOGISTIC GROWTH

One class.

Key Points

- The mathematical modeling process illustrated with the US population.
- The logistic function and its applications.

Ideas for the Class

Students are already familiar with the exponential growth population model. You might ask them how realistic they think it is, especially for the long term. Can the population of the earth grow forever? They may suggest that it will eventually level off, and so see that the exponential model has some limitations. This is demonstrated quite clearly in the text with actual US population data. In the period 1790–1860 an exponential model works extremely well, but after that rate of growth slows. A different model is needed.

The family $y = \frac{L}{1+Ce^{-kt}}$ of logistic functions models growth in an environment with limited carrying capacity. Students can gain familiarity with the family by graphing with different choices for the parameters L, k, and C, each one of which has a graphical significance. (L is the carrying capacity, k controls the rate of convergence to the carrying capacity, and C causes horizontal shifts in the time scale.) The fun is in the applications, where logistic functions are used not only for population studies but to track sales of new products (market eventually saturates) and physiological responses to doses of medicinal drugs (beyond a certain point more of a drug has no additional effect).

Problems

Problems 1, 2: involve modeling the US population.

Problems 3, 4, 8, 12: model different phenomena using a logistic function.

Problems 5, 7: include tables of data. Both emphasize the fact that the inflection point occurs at a height of half the carrying capacity.

Problems 6, 18: ask for a verbal explanation, with Problem 18 investigating the effect of one of the parameters.

Problems 11, 13– 17 involve the biological application of dose-response curves.

Problem 9– 10: excellent problems to use for group work or discussion in class.

Problem 19: should be reserved for those students with excellent algebraic skills.

Suggested Problems

4, 5, 6, 8, 10, 13, 18

4.8 THE SURGE FUNCTION AND DRUG CONCENTRATION

One class.

Key Points

- The family of surge functions.
- The use of surge functions in modeling biological phenomena.

Ideas for the Class

The family $y = ate^{-bt}$ of surge functions has been chosen for the final section of this chapter because of its widespread use in modeling drug concentration in the blood over time. It will be a new family of functions for your students, and so you will want to begin by encouraging them to explore the graphical significance of the parameters a and b. But the bulk of the time should be spent with the models themselves and the data they model. The students should see that making the model amounts to finding suitable choices for the parameters. They may object that the graphs from the models do not exactly trace the curves of experimental data, but they need to know that it is in the nature of a mathematical model to make simplifying approximations. The art is to simplify enough to bring out the features of interest without simplifying so much that important information is lost. Finally, and perhaps most importantly, students should learn to read the drug concentration graphs for practical information, such as drug absorption properties and minimum effective concentrations.

Problems

These problems require very few calculations. The emphasis is on interpreting general trends. Students normally find these problems interesting and do not mind spending time on them. Students do, however, lean towards scanty explanations and interpretations—alert them to that. It might be wise to do a few of these (5, 8 and 12) in class in order to encourage well-motivated answers, assign others as group work (9 and 20) and a few of the remaining ones as assigned problems.

Suggested Problems

1, 2, 4, 7, 11

CHAPTER FIVE

Five or six classes.

Overview

The purpose of this chapter is to give the student a practical understanding of the definite integral as a limit of Riemann sums and as a total change computed from a rate of change. The emphasis is on intepreting the meaning of the definite integral. We use the same method as in Chapter 2, introducing the concept graphically and numerically without going into analytical techniques. Definite integrals can be computed approximately by Riemann sums or by estimating areas, and so students can work throughout with functions presented by tables of values and by graphs as well as those defined by formulas. For functions given by formulas, the student is encouraged to use technology to evaluate definite integrals. The student should finish the chapter with the ability to estimate a definite integral and an understanding of how to interpret it as area under a curve and as total change in a quantity given its rate of change.

5.1 ACCUMULATED CHANGE

One class.

Key Points

- Given the velocity function, approximate the distances traveled over short time intervals by using the formula

  $$\text{Distance} = \text{Rate} \times \text{Time}$$

  (This leads to the idea of a Riemann sum).
- Graphical interpretation (area under the velocity graph).
- Extension to rates and changes other than velocity and distance.

Ideas for the Class

This section is your chance to help the students see that a Riemann sum is not anything mysterious, that it is a natural generalization of the formula Distance = Velocity × Time with which they have long been comfortable. To help them discover these sums on their own, the text begins with a very simple problem: I traveled 30 miles/hour for 2 hours, then 40 miles/hour for 1/2 hour, then 20 miles/hour for 4 hours. How far did I go? Instantly the students can figure this out, and so get

$$\text{Distance} = \text{velocity} \times \text{time} + \text{velocity} \times \text{time} + \text{velocity} \times \text{time}$$

It is clear why sums are necessary—because velocity changes over time! And it is clear why the problem is easy—because the velocity is not changing all the time!

Students at this point are just one step away from a full Riemann sum, which will come out of considering the thought experiment in the text, How Far Did the Car Go?, in which velocity does change all the time. The big difference is that they will be forced to accept an approximate rather than an exact distance for the answer.

Go over the thought experiment in the text, but using different numbers. Better yet, have students tackle a similar problem before you talk about it, any of Problems 2, 10, 13, or 16 is appropriate. For a more colorful problem you might like to have students get into groups of four or so and try a problem like the following:

Example War Emblem, ridden by Jockey Victor Espinoza, was the winner of the 2002 Kentucky Derby. These thoroughbreds can reach speeds of almost 50 mph! War Emblem's speed was clocked every 30 seconds during this race, and the results are given in Table 5.1.1.

Table 5.1.1 *Can you determine the length (in miles) of the track?*

Time (sec)	0	30	60	90	120
Speed (mph)	0	40	38	35	37

Start with just the table of velocities, and have the students give upper and lower bounds for the distance traveled by the horse and jockey, using left- and right-hand sums (see if they notice that time is in seconds, but speed is in miles per hour). If they suggest averaging the two for an imporved answer, let them know their idea is a good one. Demonstrate that more data (recording velocity more often) will make the upper and lower bounds closer together. As much as possible, get the students to give estimates. Just for the record, the actual distance is 1.25 miles. Then have the students translate the table (and estimates) into graphical form.

When you make a graph of velocity vs. times, be prepared for confusion. Students are accustomed to graphing distance vs. time and are likely to want to interpret this graph in the same way. They may be confused by the fact that an area under a graph can represent a variety of quantities, they don't have much experience letting an area represent anything but area. It may help to remind them that on the graph, horizontal lengths represent time and vertical lengths represent velocity, so the units of area are (time)(distance/time) = distance.

For another approach to the ideas of this section, couched in a more general rate of change setting, you might want to use the following problem instead of a velocity problem.

Example Land management officials notice that an introduced species of tree is making serious inroads into an ecosystem. (An example is the tamarisk tree in the Great Basin of the western United States). In a certain area, the number of new trees per year is increasing every year. Some of the growth rates are given in Table 5.1.2.

Table 5.1.2 *Rate at which new trees are appearing*

Year	1950	1960	1970	1980	1990
Trees/year	337	371	408	448	493

We want to estimate the total number of new trees that have appeared between 1950 and 1990. Answering these questions will show you how such an estimate can be made.

(a) What is the minimum number of new trees that could have appeared between 1950 and 1960? The maximum number?

(b) What is the minimum number of new trees that could have appeared between 1960 and 1970? The maximum number?

(c) During these four decades, from 1950 to 1990, what is the minimum number of new trees that could have appeared? The maximum number? Did you use the assumption that the number of new trees each year is increasing? Where?

(d) If you had to guess how many new trees appeared between 1950 and 1990, what would be your guess? Suppose that some additional growth figures are obtained, and presented in Table 5.1.3.

Table 5.1.3 *Rate at which new trees are appearing*

Year	1955	1965	1975	1985
Trees/year	343	389	418	467

(e) Is the second set of information consistent with the first? In what interval would the number of new trees in 1955 have to be in order to be consistent?

(f) Recalculate the left and right hand sums in the light of this new information. Make a new guess for the total number of new trees that appeared between 1950 and 1990.

Problems

Problems 1– 3, 9, 10, 13, 15, 16: based on tables of data. Note that in Problem 13 the speed is given in *mph* and time is given in *min*. Most students will not notice this at first (hopefully all if you did the Kentucky Derby example in class).

Problems 3, 4, 10, 18: non-velocity-distance problems and are good for discussion in groups. Problem 3 should be assigned.

Problems 5, 6: involve graphing and estimating.

Problems 7, 8, 11, 12, 17: based on graphs and students may need some guidance on these.

Problem 14: contains a temptation of using formulas for students who may have had calculus before. Such students may want to use the Fundamental Theorem. Sketching and explaining are required which should steer them in the right direction.

Suggested Problems

2, 3, 7, 8, 10, 11, 13

5.2   THE DEFINITE INTEGRAL

One class.

Key Points

- Left- and Right-hand sums, and Sigma notation.
- The definite integral of a function as a limit of Riemann sums.
- Estimation of an integral using left and right hand sums.
- Evaluation with calculator.

Ideas for the Class

Begin the class by computing some *left-hand* and *right-hand sums:* one for a table of values and one for a graph. Students shouldn't have much trouble with this since the groundwork was laid in the previous section. Introduce the notation of n for the number of intervals/rectangles and Δt for the width of the interval. Ask the students to give you n and Δt for the two examples you started with.

Point out to the students that the left- and right-hand sums found earlier are sums of areas of rectangles = sums of "height times width" = sums of terms in the form $f(t_i)\Delta t$. If you wish, you can use the sigma notation, or you can immediately define the definite integral as the limit as $n \to \infty$ of such a sum. The details are less important than the general idea that the definite integral $\int_a^b f(t)\,dt$ represents a sum of height ($f(t)$) times width (dt).

Reinforce this idea that the definite integral is approximated by left- and right-hand sums by immediately doing a table of values example similar to Example 3. Explain that if we have a formula for the function, we can find left- and right-hand sums that are more and more accurate by taking n larger and larger as seen in Figure 5.9. This is tedious, however, and we let our calculators do the dirty work. Finish the class by showing the students how to compute definite integrals on their graphing calculators. (If using the TI-80 series, use the "fnint" command under the calculus menu.)

Problems

Many of these problems require a calculator or computer.

Problems 1, 2, 11–14, 26: deal with estimations using graphs.

Problems 3–6, 15–24: do not underestimate these. It is worth assigning a good selection of them.

Problems 7–10: involve the use of tables.

Problem 25: provides the opportunity to discuss functions that are not monotonic over the given intervals.

Suggested Problems

#1–3, 5, 8, 12, 19, 22, 25, 26

5.3   THE DEFINITE INTEGRAL AS AREA

One class.

Key Points

- Basic interpretation of the definite integral as (signed) area.

Ideas for the Class

The section begins by reinforcing the idea of the definite integral as area under the curve, when $f(x)$ is positive. What if $f(x)$ is not positive? Try the following class example, with graphs labeled as follows.

Example   In each case below, find $\int_0^5 f(x)\,dx$.

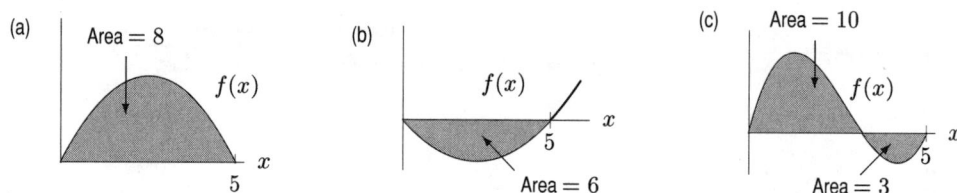

Figure 5.3.1

This example is nice, because it makes the key point without lots of extraneous information. Part (a) reinforces the idea of the definite integral as area, so $\int_0^5 f(x)\,dx = 8$. To motivate part (b), remind them that the definite integral is defined as the limit of a sum of terms in the form $f(x)\Delta x$. Since $f(x)$ is negative for every x in the interval, the sum (and hence the integral) will be negative. If $f(x)$ is negative, the integral gives the *negative* of the area and we have $\int_0^5 f(x)\,dx = -6$. What if $f(x)$ is part positive and part negative, as in part (c)? They should guess at this point that the integral counts the area above the x-axis as positive and the area below the x-axis as negative, and so we have $\int_0^5 f(x)\,dx = 10 - 3 = 7$. Point out clearly to them that the total shaded area in part (c) is $10 + 3 = 13$, but that the integral is $10 - 3 = 7$. Make sure that the student distinguishes between the value of a definite integral (a number, which can be positive, negative, or zero) and an area (a number which can never be negative).

Do a quick example similar to Example 5 in class.

It is very helpful at some point during this class to do an example similar to Example 1, where a grid is given. Have the student approximate the area by counting the number of boxes and multiplying by the area of the individual boxes. This method is much simpler for the students than finding a Riemann sum, and is used later in the course. For example, you could use the function given in Figure 5.4 on page 224. The area under the graph between $x = 0$ and $x = 6$ includes about 14.5 boxes. Since each box has area $(10)(1) = 10$, the total area under the curve (and hence the integral) is about 14.5 times 10, or 145. Figure 5.4 is available as an overhead transparency master in Part IV of this Instructor's Manual.

Problems

All the problems require the use of graphs (with and without calculators and computers).

Problems 1, 9, 17, 18, 22, 28, 29: involve estimation of area and integrals using graphs without calculators or computers.

Problems 2–4, 10–16, 20, 21, 30: require a calculator or computer to find area and integrals.

Problems 5–8, 23–28: ask students to estimate area and integrals and decide whether they are positive, negative, or near zero.

Problem 19: involves the use of a table

Suggested Problems

1, 3, 5, 7, 9, 12, 15, 19, 23, 26, 29, 30

5.4 INTERPRETATIONS OF THE DEFINITE INTEGRAL

One class.

Key Points

- Using units and the interpretation as total change to discover the meaning of the definite integral of real world functions.
- Bioavailability of drugs.

Ideas for the Class

Start the class with the emphasis on the units of the definite integral. Set up a table of values such as the following:

Table 5.4.4

t (seconds)	0	5	10
v (ft/sec)	100	60	35

Ask the students what the units are for a left-hand sum for this table. Next set up a graph for the students such as the following, and ask them what the units are for the area under the curve.

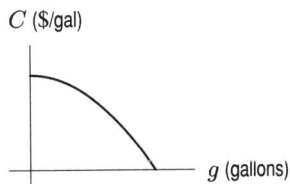

Figure 5.4.2

Now you are ready to explain that because the integral $\int_a^b f(x)\, dx$ is a sum of terms in the form "$f(x)$ times a difference in x," the units of this integral are units for $f(x)$ times units for x. Make sure that the students understand the units of the integral, and understand the meaning of the integral as stated in the box on page 237. The following class examples might be helpful.

Example   If $r(t)$ represents the rate at which the heart is pumping blood, in liters per second, and t is time in seconds, give the units and meaning of the following integral: $\int_0^{10} r(t)\, dt.$

(Make sure they understand that this integral represents the total amount of blood (in liters) pumped by the heart between time $t = 0$ and $t = 10$.)

Example   Assume $f(t) = 60\sqrt{t}$ gives the rate of change of the population of a city, in people per year, at time t years since 1990. If the population of the city is 5000 people in 1990, what is the population in 1999?

This example emphasizes that the integral is the *change* in the population, not the actual population. Students should work on this at their seats. They should use technology to find that $\int_0^9 60\sqrt{t}\, dt = 1080$, and they should be able to interpret this as "The change in the population of the town between 1990 and 1999 was 1080 people." Emphasize that since the change is positive, the town's population has *increased* by this amount. Since the population was 5000 people in 1990, it is $5000 + 1080 = 6080$ people in 1999.

Any of the following make an excellent in-class group activity: Example 3 in the text, Problem 17 or Problem 18.

If you are teaching a class with life sciences students in it, the bioavailability example at the end of the section is an excellent application to include.

Problems

Problems 1–5: concern the interpretation of the integral as a rate of change.

Problems 6, 8–10, 12, 15, 16, and 24: involve interpretation of the integral numerically or using a calculator or computer. The numerical interpretations should serve to refresh Riemann-sum calculations.

Problems 7, 11, 13, 14, 17, 18, 21–23: do the same thing graphically. They are confusing to students at first, and they might need more guidance here.

Problems 19, 20: deal with peak concentration, speed of absorption and total bioavailability of drugs.

Suggested Problems

#1, 3, 8, 10–12, 14, 15, 18, 19

5.5   THE FUNDAMENTAL THEOREM OF CALCULUS

One-half to one class.

Key Points

- The Fundamental Theorem of Calculus, explained by considering the definite integral of a rate of change as total change.
- Marginal cost and change in total cost.

Ideas for the Class

There are two new ideas in this section. The first is that we introduce the derivative notation $F'(t)$ for the rate of change. This is the only thing new in Example 1, but it helps to reinforce what the students have learned in both Chapters 2 and 5. The material on marginal cost and change in total cost is important for students in business and economics. Begin with a quick reminder of the concept of marginal cost. If you cover Example 2, you might start by asking the students to tell you the marginal cost at $q = 100$ and to interpret the meaning of the number. This will reinforce the units (marginal cost at 100 is 6 *dollars per item*) and help students tie together the ideas on cost/marginal cost in Chapters 2 and 5.

Figure 5.51 is available as an overhead transparency master in Part IV of this Instructor's Manual.

Problems

Problems 1, 3–8: involve cost and marginal cost.

Problem 2: gives another example of change from rate of change.

Problems 9–11: are good for discussion in groups.

Problems 12: is a straightforward application of the Fundamental Theorem.

Suggested Problems

1–6, 8–10, 12

FOCUS ON THEORY: THEOREMS ABOUT DEFINITE INTEGRALS

Optional: if covered, one-half to one class.

Key Points

- Second fundamental theorem of calculus.
- Sums and multiples of definite integrals.

Ideas for the Class

In the Second Fundamental Theorem of Calculus, the main point for the students is that the derivative of the integral of a function is that function. The derivative and integral "undo" each other.

When discussing sums and properties of definite integrals, use the area interpretation of the definite integral, and lots of graphs, to motivate the ideas.

Problems

Problems 1–4: involve the Second Fundamental Theorem of Calculus.

Problems 5–7: involve understanding $F(x)$ given information about $F'(x)$.

Problems 8–11: investigate sums and multiples of definite integrals.

Suggested Problems

1, 4, 6, 10, 11

CHAPTER SIX

Four classes.

Overview

Chapter 6 presents a variety of applications of the definite integral. It is not necessary to cover them all, or to cover them in the order given. You can be very flexible with this chapter, choosing topics based on the interests of your students and the time you have available.

Section 6.1 discusses average value. This topic is important for all students, and this is a good application to lead off with. Section 6.2 (Consumer and Producer Surplus) and Section 6.3 (Present and Future Value) are excellent sections for students in business and economics. Section 6.4 (Population Growth) is particularly good for students in the life sciences. Students will still use a calculator or computer to compute all definite integrals in this chapter.

6.1 AVERAGE VALUE

One half to one class.

Key Points

- Average value of a function.
- Graphical interpretation and computation.

Ideas for the Class

Average value of a function should be explained both graphically and in terms of Riemann sums. Probably students will like the graphical explanation best. Ask them to imagine that the region under the graph of a function between high walls $x = a$ and $x = b$ is made of wax. Then the average value of the function between a and b is the level the wax would reach if it were melted and allowed to flow under the influence of gravity, but always staying between the walls. Give some graphical examples in which students are asked to guess the average value from the graph before calculating it numerically. They enjoy getting picky, trying to place a horizontal line on the graph so that the area above the line but beneath the function exactly matches the area below the line but above the function.

Problems

Problem 1: has the students use the formula for average value while computing the definite integral graphically. This is a good problem to remind them of the concepts in Chapter 5.

Problems 2, 7, 8, 14, 15: involve estimating the average value graphically. Problem 2 is particularly valuable because it links the intuitive visual notion of average value and the formula for average value.

Problems 3–6, 9–13: require the students to use the formula for average value (and to use technology to evaluate the definite integrals). Problems 3–6 are simple forerunners of the more substantial Problems 9–13.

Problems 16–19: require additional thought from the students. Problems 18 and 19 are valuable problems because the notions of derivative and integral are dealt with in the same problem. Note that Problem 16 refers to the absolute value of f.

Suggested Problems

1, 2, 5, 7, 8, 10, 11, 14, 16

6.2 CONSUMER AND PRODUCER SURPLUS

One class.

Key Points

- Supply and demand curves.
- Consumer surplus and producer surplus.
- The effects of wage and price controls.

Ideas for the Class

You might open class with a discussion of the nature of trade. When two persons trade items, what are the relative values of the items? Often their "common sense" tells students that two persons trade items of equal value. But that is not correct! Students must understand that when one person buys from another, each person winds up with more than they had before and so both persons gain from the trade (which is why they are willing to trade). This may seem paradoxical to students, but a simple example can help make the point.

Suppose I plan to pay up to $20 for a compact disc that I find in a store for sale at $15. The disc is worth $20 to me but I have traded for it with only $15, so I have gained $5 from the trade. Suppose the store would have been willing to sell for anything above $12. Then the disc is only worth $12 to the store, but they traded it for $15, so the store has gained $3 from the trade. And this is how trade enriches a country.

Once you are sure the students understand how people gain from single trades, you can proceed to the goal of this section, which is finding the total gains from trade (the consumer surplus) of a group of consumers who place different values on the same good or the total gains from trade (the producer surplus) of a group of suppliers who place different values on the same good. Do your students see that the demand curve keeps track of the way consumers value the good? Try asking them what the demand curve would look like if all consumers valued a good at less than $20. (We want them to say that the demand curve must intercept the price axis below $20.) Or how would the demand curve change if an advertising campaign caused consumers to value the good more highly? (The same quantities could be sold at higher prices, so the new demand curve would be higher than the old one.)

In this section, the graphical interpretations of consumer and producer surplus are of greater importance than the formulas. The formulas should be downplayed and the interpretations given greater weight. In fact, if your students are weak, it is possible to teach this section using only the graphical interpretations. It is still possible to include the applications to wage and price controls, which are interesting and should definitely be covered.

Any one of Problems 7, 8 or 10 would be an excellent in-class group work problem.

Problems

Problems 1, 2, 3, 8– 10: require only a graphical understanding of the concepts.

Problems 4, 6, 7: require an ability to work with the formulas.

Problem 5: involves data given in tables.

Problems 11– 13: are more difficult and require the students to understand the Riemann sum derivation of the formulas.

Suggested Problems

1– 3, 5, 6, 8, 9, 10

6.3 PRESENT AND FUTURE VALUE

One class.

Key Points

- Present and future values of a continuous stream of payments.

Ideas for the Class

Begin by reviewing the ideas of present and future value of a single payment, as discussed in Section 1.7. In particular, remind the students that if the interest rate is r, then the present value, P, of a payment, B, made t years in the future is $P = Be^{-rt}$. Derive the formula for the present value of an income stream by doing an example in class, as follows. Assume that the interest rate is 6%. Suppose we have $1000 a year coming in over the course of a year. If we wait until the end of the year to deposit the money, then the present value is $P = 1000e^{-(0.06)(1)} = 941.76$. However, if we deposit the money every 6 months, then at the end of 6 months, we deposit $500 and this deposit is made 0.5 years in the future. Six months later, we deposit the other $500 and this deposit is made 1 year in the future. Therefore the present value of the two payments is $P = 500e^{-(0.06)(0.5)} + 500e^{-(0.06)(1)} = 956.10$. What if we deposit the accumulated money every three months (or quarter-year)? Have the students figure out this present value, and then talk about the concept of a continuous income stream where the money comes in throughout the year, and we deposit it as it comes in. At this point, you are ready to derive the formula for the present value of a continuous income stream, as on page 265 of the text. Finish the example above by finding the present value of the $1000, if we assume that it comes in (and is deposited) continuously. We have

$$P = \int_0^1 1000e^{-0.06t}\, dt = 970.59.$$

It is best to follow this with an easy example such as Example 1 in the text. After the students have computed the future value in this example, ask them how much of this future value was actually deposited ($1000 every year for 20 years, or $20,000) and how much represents interest earned (all the rest, or $38,668.62 − $20,000 = $18,668.62.) This is an easy question that nonetheless helps determine whether the students understand the idea of an income strem and a future value. (It also helps point out the positive effects of compound interest!)

For more advanced students, here are some more complicated examples:

Suppose you are a writer working on a best seller. Publisher X offers you an advance of $50,000, and 5% royalties on sales, which, starting two years from now, are expected to increase exponentially for five years according to the function $S(t) = 1000000 \cdot e^{0.1t}$ dollars/year, and then abruptly die away. Publisher Y offers you no advance but 6% royalties, with the same expectations for sales. Which deal is better, assuming an interest of 10% for the next 7 years? (Answer: Both deals have the same present value.)

Here is another problem: the typical high school graduate will get a job starting at $20,000 a year, and the typical college graduate will get a job starting at $30,000 a year. Both can expect 5% annual pay raises. College tuition runs at about $10,000 a year. Does it pay to go to college? Assume an annual interest rate of 10%. This problem can be made more complicated and realistic, which would be suitable if you wanted to make it into a project. You can have different rates for the pay raises, have ceilings on the salaries, and take into account the shorter working life of the college graduate.

Problems

Problem 1: simple graphical understanding
  Problems 2– 8: standard questions on present and future values.
  Problems 9– 12: deal with real-world examples.
  Problems 13- 16: develop students' modeling ability.

Suggested Problems

2, 4– 6, 8, 9, 12, 13

6.4 RELATIVE GROWTH RATES

One-half to one class.

Key Points

- Population growth in general: absolute growth rate, relative growth rate.
- Computing changes in population from the relative growth rate.

Ideas for the Class

Begin the class by asking the students to consider the following problem.

Example   A town has a population of 5000 at time $t = 0$, with t measured in years. Write a formula for the population of the town, P, as a function of the year, t, if

1. the town's population grows at 100 people a year.
2. the town's population grows at 2% a year.

Follow this with a discussion of absolute rate of change (in people/year) versus relative rate of change (in percent/year). Make sure the students understand that a linear function has constant absolute change while an exponential function has constant relative change. Remind students that the derivative gives the absolute rate of change at a point. How do we find the relative rate of change? Since it is a percent, it is the absolute rate of change divided by the population.

   This is enough to convey to students the important distinction between absolute and relative growth rates, and it is possible to cover only this first part of the section (which will take about half a class). The rest of the section discusses how to compute the change in a population given a relative growth rate. This requires facility with logarithms, and students may find this difficult. Go slowly as you work through an example such as Example 1.

Problems

Problems 1– 6: are relatively simple problems involving the concepts of absolute and relative rate of change.
  Problems 7– 15: require an understanding of finding the change in a population given a relative growth rate.
  Problem 16: compares the growth of exponential functions and power functions.

Suggested Problems

1– 4, 7, 9, 11, 15

CHAPTER SEVEN

Four classes.

Overview

Until this point, the book uses calculators or computers to compute all definite integrals. This chapter discusses constructing antiderivatives analytically, numerically, and graphically. Substitutions and the Fundamental Theorem of Calculus are now used to evaluate indefinite and definite integrals. The last section expands on the connection between antiderivatives and the definite integral.

7.1   CONSTRUCTING ANTIDERIVATIVES ANALYTICALLY

One class.

Key Points

- Antiderivatives.
- The indefinite integral.
- Develop formulas and basic properties of antiderivatives.

Ideas for the Class

Students should be introduced to antiderivatives and their notation as indefinite integrals, since they may come across them in another course. Recognize that finding antiderivatives is the reverse process of differentiation.

Problems

Assign a good number of the three sets of routine problems. Although these problems are all simple as far as integration goes, do not overestimate the students' ability on finding antiderivatives. Antiderivatives are a notoriously weak spot in the make-up of the average student.
   Problems 1– 18: deal with finding general antiderivatives.
   Problems 19– 24, 55– 58: ask students to find particular antiderivatives.
   Problems 25– 54: involve finding indefinite integrals.

Suggested Problems

1, 4, 10, 12, 15, 18, 23, 24, 28, 35, 40, 41, 43, 45, 49, 53, 56, 57

7.2   INTEGRATION BY SUBSTITUTION

One class.

Key Points

- Use of substitution to evaluate indefinite integrals.
- Periodic functions and substitution.

Ideas for the Class

Take this material rather slowly. Students find substitution very confusing at first. First do several examples of differentiation using the chain rule, and then write the corresponding antidifferentiation formulas. Emphasize that integration by substitution is just the reversal of the chain rule; the hard part is learning to recognize to which integrals we can apply this "reverse chain rule."

Then do some more examples, starting with the integral first. Possibly introduce the integral $\int e^{x^2}\,dx$ and have students try to find the antiderivative using the guess-and-check method. Show how the obvious guess e^{x^2} does not work because of the extra term $2x$ that appears when we apply the chain rule. Then point out if the $2x$ from the chain rule had been under the integral sign in the first place, we could have easily done this integral. This gets the point across that integrands have to be in a special form to be integrated using substitution—a point you will need to come back to again and again. Then go on to do several more examples like $\int t^2 e^{t^3}\,dt$ and $\int x^4\sqrt{x^5-4}\,dx$ by the guess-and-check method. Working through several examples this way leads naturally to the quest for a more systematic method, namely w-substitution. (We use w instead of the more traditional u to distinguish it from the u, v of integration by parts.)

Once you have introduced the w-substitution, go back and repeat the same examples from before. Remind students that this method grew out of looking for the end products of the chain rule; therefore you are looking for an "inside" function whose derivative is somewhere on the "outside." And when they have this function, set it equal to w. This makes the tricky ones easier, such as $\int x^4\sqrt{x^5-4}\,dx$. Since once they have set $w=x^5-4$, the rest is mechanical. Do at least one example where the outside and inside functions are hard to recognize, for example $\int x/(1+x^2)\,dx$ (let $w=1+x^2$). Be prepared for questions about what dx means or where dw comes from.

Encourage your students not to rely too much on the mechanics of substitution; do examples to convince them that they really can guess-and-check in simple cases. Encourage them to look for patterns. For example, $\int e^{2x}\,dx=e^{2x}/2$ and $\int \sin(3x)\,dx=-\cos(3x)/3$ can be done without substitution. See if the students can see the patterns. Keep on emphasizing that integration is the reverse of differentiation. They are familiar with the constant coming out front with differentiating, so when it is absent you must divide by the constant.

An interesting example is $\int 1/(2x)\,dx$. The answer is either $\frac{1}{2}\ln x + C$ or $\frac{1}{2}\ln(2x)+C$ (using $w=2x$). Show that these two antiderivatives differ only by a constant.

Problems

Problems 1–39: students need plenty of practice on the techniques of this section. Assign as many of these problems as you feel are appropriate. Some students may need to do more than others–maybe let them judge this on their own. There are a couple of problems that do not work with the techniques of this section (e.g., 9, 16). It is a good idea to mix these in.

Problem 40: a problem where the antiderivatives are different—with a chance to require an explanation.

Suggested Problems

1, 2, 3, 4–38 evens, 39, 40

7.3   USING THE FUNDAMENTAL THEOREM TO FIND DEFINITE INTEGRALS

One-half to one class.

Key Points

- Use of antiderivatives and the Fundamental Theorem to evaluate definite integrals.
- Definite integrals by substitution
- Improper integrals.

Ideas for the Class

We return to the Fundamental Theorem to show students how to use antiderivatives to find the exact value of definite integrals. Make sure your students understand that this is the same definite integral they have been working with. They should realize, for example, that they can check their work using technology. Point out to them that the Fundamental Theorem gives an exact value while technology gives an approximate value. Point out to them also that the Fundamental Theorem is only helpful when we know the antiderivative of the integrand.

Try to keep it simple. We introduce basic integration formulas here. Students will be able to use technology to evaluate the definite integrals in the applications in the rest of the text.

Improper integrals are introduced as a limit of proper definite integrals. Include this material if you plan on covering Chapter 8, on probability.

Problems

Problems 1–20: routine problems asking the students to use the Fundamental Theorem to evaluate a definite integral.

Problems 21–23, 29: involve integration by substitution.

Problems 24–28, 30–33, 40: use the Fundamental Theorem in applications of the definite integral.

Problems 34–39: involve improper integrals.

Suggested Problems

#3, 6, 7, 9, 13, 17, 22–24, 27, 31, 35, 39, 40

7.4   ANALYZING ANTIDERIVATIVES GRAPHICALLY AND NUMERICALLY

One class.

Key Points

- Use the Fundamental Theorem of Calculus to evaluate antiderivatives.
- Graph a function given a graph of its derivative.

Ideas for the Class

To appreciate the power and beauty of the Fundamental Theorem of Calculus and the graphical interpretations of definite integrals, it is a good idea to work through some examples such as the following to generate an interactive class discussion.

Example   Figure 7.4.1 is the graph[1] of the rate r (in arrivals per hour) at which patrons arrive at the theater in order to get rush seats for the evening performance. The first people arrive at 8 am and the ticket windows open at 9 am. Suppose that once the windows open, people can be served at an (average) rate of 200 per hour.

Use the graph to find or provide an estimate of:

(a) The length of the line at 9 am when the windows open.

(b) The length of the line at 10 am.

(c) The length of the line at 11 am.

(d) The rate at which the line is growing in length at 10 am.

---

[1]From Calculus: The Analysis of Functions, by Peter D. Taylor (Toronto: Wall & Emerson, Inc., 1992). Reprinted with permission of the publisher.

(e) The time at which the length of the line is maximum.
(f) The length of time a person who arrives at 9 am has to stand in line.
(g) The time at which the line disappears.
(h) Suppose you were given a formula for r in terms of t. Explain how you would answer the above questions.

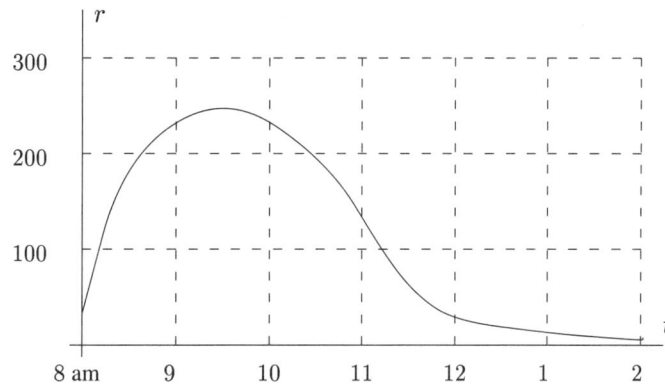

Figure 7.4.1: Rate of arrival at theater

Problems

Problems 1– 3, 7– 14, 16, 18, 19: involve evaluating, analyzing or sketching the graph of f if f' and the initial value of f are given.

Problems 4– 6, 17: give a function and ask for a sketch or as questions about antiderivatives.
Problem 15: involves recognizing the graph of a function vs. the graph of its antiderivative.
Problems 20– 25: involve graphical interpretations of the Fundamental Theorem of Calculus.

Suggested Problems

1, 3, 8, 11, 12, 15, 17, 19, 22– 25

CHAPTER EIGHT

Three classes.

Overview

8.1   DENSITY FUNCTIONS

One class.

Key Points

- The notion of density function as a smoothed-out histogram.
- General density functions.

Ideas for the Class

Almost all of the ideas presented in this section and the next section will be new to the students. Move slowly. Start with the idea of a histogram as a way of representing how some quantity is distributed through a population. The book uses age of US population, but other quantities might be incomes, grades, number of words on a printed page, etc. Students have trouble deciding what variables go on the axes; and they expect quantities to be represented by the heights of the bars rather than their areas. Smoothing out the histogram can also present difficulties; just emphasize the finer mesh. Students also want to assign meaning to the value of the density function rather than to area under the density function. For example, many students believe that $p(40) = 0.15$ means that 15% of the population has $x = 40$.

You may find it useful to have students construct their own histograms using data they collect from the class (for example, the heights of the students). Have them plot the density function.

Problems

These problems focus on interpreting density functions, and most of them are graphical or numerical in nature.

Problems 1– 8: involve understanding density functions from a given graph.

Problems 9– 11: ask students to graph density functions.

Problem 12: interpretation of a density function at a particular point.

Problems 14– 17: applied problems relying on graphs. Any of these are good for class discussion.

Suggested Problems

1– 4, 9, 11, 13– 16

8.2   CUMULATIVE DISTRIBUTION FUNCTIONS AND PROBABILITY

One class.

Key Points

- Cumulative distribution functions and probability.

Ideas for the Class

Be sure the students understand the difference between a density function and a cumulative distribution function. You may want to start class by putting a table such as Table 8.2.1 on the board. (Use the corresponding breakdown of grades for the most recent exam in your course.)

Table 8.2.1 *Grades on an exam*

Grade	50-59	60-69	70-79	80-89	90-99
Proportion of grades	0.05	0.12	0.20	0.45	0.18

Have the students tell you how to fill in Table 8.2.2, where $P(x)$ represents the fraction of grades less than x. Ask them, for example, "What proportion of the students received a grade less than 70?" and make sure they understand that this is what $P(70)$ represents.

Table 8.2.2 $P(x) = $ *proportion of grades less than x*

X	50	60	70	80	90	100
$P(x)$	0	0.05	0.17	0.37	0.82	1.00

When you have finished with this example, tell them that the first table corresponds to a density function and the second to a cumulative distribution function. Remind them repeatedly that the value of a cumulative distribution function represents the *total* (or cumulative) amount up to that point.

Plot the points for $P(x)$ given in Table 8.2.2, and sketch a graph of this function. Then ask the students to suggest properties that every graph of a cumulative distribution function will have: it will be increasing, it will "start" at a height of 0 on the left, and it will "end" at a height of 1 on the right. Have the students work on Problems 10 and 11 in groups in class. Both problems are excellent at developing understanding of the graphs of these functions.

Probability is discussed as it relates to density functions and cumulative distribution functions. Resist the urge to go into more detail unless you have lots of time.

Problems

Problem 1: reinforces the definition of a density function.

Problems 2, 7–9, 11, 16: examine just the cumulative distribution function, either graphically, numerically, or in practical terms.

Problems 3, 12: give students both the density function and the cumulative distribution function and ask them to determine which is which. These are good problems to work on in groups or for class discussion.

Problems 4–6, 10, 13, 15: investigate both the density function and the cumulative distribution function graphically.

Problem 14: involves a function defined by a formula

Problems 17 and 18 are the only problems explicitly about probability.

Suggested Problems

2–7, 10–12, 14, 16, 17

8.3 PROBABILITY AND MORE ON DISTRIBUTIONS

One class.

Key Points

- Definitions of median and mean.
- Normal Distributions.

Ideas for the Class

The material of this section is a continuation of the material of the previous section, so cover Sections 8.1 and 8.2 before covering this section.

As in the previous section, these ideas are new to the student; it is important to go slowly and give lots of basic examples. A good class problem is to hand out a probability density function for the grades in a hypothetical class. Make the curve skewed in some way, and graph it on paper with grid squares so that areas can be estimated. Then ask various questions about the class: How many students are getting a B or better? How many are failing? what are the median and mean grades? (Note that the mean is harder to estimate from the graph.) Tie this exercise to the discussion of probability density functions.

To start your discussion about normal distributions, you may want to ask the students about how they think exam grades are typically distributed. Many of them will respond with the "bell-shaped curve." Explain how this shape is often used to model real world penomena, including class grades, to heights of students in your college, to the amount of snow Denver, CO gets on a yearly basis. It is a good idea to examine the graph of a normal distribution and explain the significance of μ (the center of distributions) and σ (the horizontal distance from the center of the distibution to the change in concavity on the curve). Students like to see the graphical significance of parameters in an equation. Mention the 68-95-99.7% "rule of thumb": For any normal distribution, about 68% of the observations fall within one standard deviation of the mean, about 95% of the observations fall within two standardard deviations of the mean, and about 99.7% of observations fall withing three standard deviations of the mean.

Problems

Problems 1, 3: are graphical in nature.

Problems 2, 4–8: ask students to find the mean and median, given a formula for the density function.

Problem 9: is numerical and involves the use of a table.

Problems 10, 11: are the only questions involving normal distributions.

Suggested Problems

2–5, 8–10

CHAPTER NINE

Six classes.

Overview

Quantities of serious interest in the real world usually depend upon more than one variable. To model them mathematically we need functions of two or more variables. This chapter is a brief introduction to such functions and their use.

The Rule of Four still applies. Two-variable functions can be described numerically, graphically, algebraically, and verbally. Numerical representation is by rectangular tables of values. Graphical representation is by contour diagrams. After discussing these representations, partial derivatives are defined and interpreted as rate of change in a variety of contexts. The chapter closes with a brief introduction to constrained optimization and the method of Lagrange multipliers.

9.1 UNDERSTANDING FUNCTIONS OF MANY VARIABLES

One class.

Key Points

- Representing functions of two variables by tables of values, formulas, and words.

Ideas for the Class

Prior student experience with functions of two or more variables is likely to be almost entirely verbal, so it is good idea to begin class in the verbal mode. The opening theme is that many-variable functions are all around us. Even functions previously studied as one-variable functions can be taken as examples. For instance, the balance in a bank account earning interest continuously depends on the time it has been earning according to the formula $B = Pe^{rt}$, where we regarded initial investment P and interest rate r as constants to make this a function of the single variable t. In actuality, r and P are also variables, so we have a function of three variables $B = f(P, r, t) = Pe^{rt}$.

Some more examples of functions of several variables:

- The chirping activity C of crickets depends on the time of day t and the temperature T, so that $C = f(t, T)$.

- The temperature T can be expressed in terms of location with latitude a and longitude b, so $T = h(a, b)$.

- Your income tax due S depends on your income I, your exemptions e, and your deductions D, so $S = g(I, e, D)$.

You can get many other examples from the students. After they suggest a few, you might ask them if they can think of a quantity that depends on only a single variable, and they will not find it easy to do so.

In reality, most quantities depend on many variables. For example, the number N of years it takes to complete a college degree depends on (among other things) the major M, the choice of school s, monetary support m, academic background A, the time taken to choose a major t, intelligence I, and family support F, so $N = f(M, s, m, A, t, I, F)$. The more variables that are involved, the more difficult it is to understand the function's behaviour completely.

You might select one of the functions for money M in a bank account as a function of initial balance and time and have groups of students compute values, eventually putting together a table of values as a class. Again ask, "What observations can you make?" They should see this time that as either variable increases, M increases. Ask whether they could have predicted it in advance.

Now ask whether the function $f(x, y) = x - 2y$ is increasing or decreasing. A discussion will ensue, and students will begin to see that the language of one variable functions will sometimes have to be refined for it to make any sense at all in a two-variable setting. In this case, they might say that f is increasing with respect to x and decreasing with respect to y.

Don't take tables of values completely for granted. If you ask for a table of values for $f(x, y) = x - 2y$, you are likely to find a few of the students using the two column format they are familiar with — in the first column a list of points (x, y) and in the second column the values of the function. Rectangular tables are so natural to us that we tend to forget that it is a step requiring some real insight to appreciate the advantages of arranging the table of values in rectangular form. This can be brought home to the students if you wish by handing them a two-column table of values of a two variable function and asking them to use it to answer some simple questions, even as simple as evaluating the function or determining how the value changes when one of the independent variables increases. They will all express a preference for the rectangular format. You might ask your students to prepare a table of values both ways and comment on the differences.

Problems

Problems 1– 6: similar to examples in the section and could be done in small groups in class. These problems are all intended to make the student comfortable with the notion of a function of two variables. Interpretations and visualizations are extremely important.

Problems 7– 9: on increasing and decreasing functions in one variable.

Problems 10– 17: relatively simple but aim to increase understanding.

Problems 18– 21: form a unit and could be given as a group project.

Suggested Problems

#8, 9, 11, 13, 14, 18, 21

9.2 CONTOUR DIAGRAMS

One class.

Key Points

- Reading and interpreting contour diagrams for functions of two variables.
- Contour diagrams and tables of values.
- Finding contours algebraically.

Ideas for the Class

Because contour diagrams are widely used by the disciplines requiring this course, functions of two variables are represented graphically in this text by contour diagrams, not by surface graphs. Contour diagrams are not yet drawn well by calculators or many computer software packages, so you will want to come to class with either overhead transparancies to display or handouts to distribute. (See Part IV for overhead transparency masters.)

The first example, the weather map in Figure 9.5, will be familiar to students, though they will not have associated the word 'function' with it. Ask students some questions about this weather map; the main goal of this section is to get students comfortable reading contour diagrams. Problem 3 would be a good one to do in class. Problems 6 and 7 also work very well as class discussion problems. An additional example of interpreting contour diagrams is given in Example 5. This would work well as a class handout to work through together.

The last 15-20 minutes of class should be spent on finding contours algebraically. For many students, this is the most difficult part of the section. Emphasize that a contour line for the function $z = f(x, y)$ represents the points where z is a constant. (For example, $z = 0$ represents one contour line.) Therefore, to find the contours, we plug in fixed values for z and sketch the resulting function of two variables. Start with an easy example, such as $z = x - y$. The contour for $z = 0$ is $0 = x - y$, or the line $y = x$. The contour for $z = 1$ is $1 = x - y$, or the line $y = x - 1$. Don't bother with complicated examples; in this text and in the user disciplines, the most important thing is to be able to read and interpret a contour diagram, not to create one.

Problem 1 in the Projects for Chapter 9 makes an excellent group discussion project.

Problems

Problems 1–3, 5–8, 11–14, 23–25, 29–31: ask students to interpret and use contour diagrams. These problems are not hard, and you should assign several of them. Students will particularly enjoy Problem 23 on the species density of birds. Assign this one! Problem 31 is also particularly interesting, especially for students in life sciences.

Problems 4, 9, 10, 15–20, 32: ask the students to draw a contour diagram, given a formula or a table of values. Assign only a couple of these.

Problems 21, 22, 26–28: ask the students to match verbal descriptions or tables of values with contour diagrams.

Suggested Problems

#5, 6, 8, 9, 13, 20, 23, 26, 27, 31

9.3 PARTIAL DERIVATIVES

One class.

Key Points

- Definition of partial derivatives.
- Approximating partial derivatives with difference quotients given a contour diagram or table of values.
- Using units to interpret partial derivatives in applied contexts.

Ideas for the Class

Begin the class by reminding students of the intepretation and numerical approximation of the derivative of a function of a single variable. The following numerical example is a good one to start with.

Example   Table 9.3.1 gives H = temperature in °F at time t minutes since noon on a certain day. If $H = f(t)$, estimate $f'(10)$. What are the units of this derivative? What is its meaning?

Table 9.3.1

t (min)	0	10	20	30
H (°F)	60	65	68	70

When students have had this quick review of Chapter 2, you are ready to extend this concept to functions of two (and more) variables. Tell them that temperature is a function of elevation as well as time of day, so we can think of H as a function of two variables, elevation x (in feet above sea level) and time t (in minutes since noon). Table 9.3.2 gives values of $H = g(x, t)$. (Point out to them that Table 9.3.1, a one-variable table, represents the point at 300 feet above sea level.)

Table 9.3.2

	x			
	0	10	20	30
100	72	75	80	84
200	67	72	76	80
300	60	65	68	70
400	53	59	64	66

x (ft)

Ask the students what they think the "derivative" should mean at a point, such as the point $(300, 10)$. With luck, they will suggest the idea of "two derivatives" on their own. Work with them to estimate the two partial derivatives at the point $(300, 10)$. Emphasize the fact that when we find the partial derivative with respect to t, we hold the x variable fixed, and vice versa. When you have estimated the two partial derivatives at this point, be sure to have the students explain what the units are and to interpret the meaning. Discuss whether this function is an increasing or decreasing function of x or t, and relate this to the sign of the partial derivatives.

Do at least one additional example of interpreting partial derivatives, such as in Problem 3 or 4 in the book. Finish by estimating derivatives from a contour diagram. Remember to bring a handout or an overhead transparency for the contour diagram.

Problems

The concept of the partial derivative will be new to the students, so they will need to spend time on the interpretation. These problems require hardly any calculations, mostly interpretations.

Problems 1–3, 5, 7, 8, 17, 18, 27: are approached from a broad interpretation aspect.
Problems 4, 10–12, 15, 16, 28–31: take the approach from the graphical side.
Problems 6, 9, 13, 19–26: are numerical in nature
Problem 14: only problem that asks the students to draw contour diagrams.

Suggested Problems

#1, 4, 6, 7, 9, 12, 14, 17, 29

9.4 COMPUTING PARTIAL DERIVATIVES ALGEBRAICALLY

One class.

Key Points

- Algebraic computation of partial derivatives.
- The Cobb-Douglas production model.
- Second-order partial derivatives.

Ideas for the Class

Begin the class with some straighforward drill problems on computing partial derivatives algebraically. This is a good way to reinforce the derivative formulas learned in Chapter 3! Be sure to have the students do some of the examples at their seats.

If your students are primarily in economics and business, and in particular if you plan to cover Lagrange multipliers, you will want to spend some time on the Cobb-Douglas production functions. Example 6 is a good example to do in class.

The last part of the section is on second-order partial derivatives. This can be omitted or covered briefly.

Problems

Problems 1– 18: are practice exercises; assign a number of these.
　　Problems 19– 23: combine technical skills with interpretations.
　　Problems 24– 28: concern Cobb-Douglas functions.
　　Problems 29– 41: ask the student to calculate second-order partial derivatives.

Suggested Problems

1, 5, 8, 9, 17, 19, 20, 22, 27, 28, 39

9.5   CRITICAL POINTS AND OPTIMIZATION

One class or less.

Key Points

- Local maxima and minima of functions of two variables.
- Critical points.
- Second Derivative Test.

Ideas for the Class

Students used to reading topographical maps will know that local maxima and minima show up on contour diagrams surrounded by closed contours. That these local extrema can only occur where both partial derivatives equal zero is an easy analog of the one variable criterion for local extrema.

Finding critical points algebraically from a formula for a two-variable function is not necessarily easy or even possible. It requires solving a system of (often nonlinear) equations. You may not want to ask your students to do much of this by hand.

Saddle points are not discussed in the text, but they are critical points. Ask your students for a contour diagram representing a mountain pass. You may want to show them a simple example such as the function $f(x,y) = x^2 - y^2$ at the origin. Then introduce the Second Derivative Test as an analytic tool to determine optimization points.

Problems

Problem 1: is a good problem to do in class.
　　Problems 2– 11: are necessary practice problems. Encourage students to do as many as it takes for them to feel comfortable with them.
　　Problems 12– 16: are graphical problems which the students will enjoy doing. Assign at least a couple.
　　Problem 17: involves calculation and a little more thought.
　　Problems 18– 20: involve cost, demand, and profit. Problem 18 requires careful thinking and is suitable for group work.
　　Problem 21: a neat applied problem. It is suggested that you assign this one.

Suggested Problems

#3, 5, 9, 13–15, 18, 21

9.6   CONSTRAINED OPTIMIZATION

One class.

Key Points

- Extrema of functions of two variables subject to a constraint.
- Method of Lagrange multipliers and the meaning of the multiplier.

Ideas for the Class

Constrained optimization is a multivariable concept. It has no analogue in one variable mathematics, so it is wise to spend some time on the difference between unconstrained and constrained optimization of a two variable function. You might show a huge table of values or a contour diagram and ask for the point corresponding to the greatest value of the function. Then introduce a constraint by drawing a curve right across the table or diagram and ask for the point on the curve corresponding to the greatest value of the function. Find the highest point anywhere in North America, or constrain to the United States-Canada border. Place a mound of playdough on a table and find the highest point, then make a vertical cut with dental floss and find the highest point on the cut. Consider Example 1, a linear budget constraint, for the analytic view.

An in-class discovery activity suitable for working in groups goes as follows. Distribute contour diagrams for a number of functions. Students draw some constraint curves of their own choosing on the diagrams, and search for geometric properties of the maximal points on the constraints. They will find that the maximal points can be only where the constraint curve is tangent to a level curve of the diagram. Now go back to the playdough, and draw the level curve round the playdough that goes through the maximal point on the cut. Tangency again, this time of the cut and the level curve.

Students know that tangents are related to derivatives, and so will be ready for the statement and use of the method of Lagrange multipliers. There is no formal derivation of the method in the text.

Note that the Lagrange multiplier λ has a practical interpretation, which is explained in the text. It tells how the maximal value changes if the constraint is changed slightly. Use of this interpretation of λ is illustrated in Examples 3 and 4.

Problems

Problems 1–10: These practice problems will help students familiarize themselves with the terminology and the technique of Lagrange Multipliers.

Problem 11: A valuable graphical problem.

Problems 12, 13, 15, 16, 18, 22: Students should now be familiar with Cobb-Douglas functions and ready to work on these applied problems.

Problems 14, 17, 19: have business applications involving cost and production functions.

Problem 21: is a numerical problem, and it is suggested that this be assigned.

Suggested Problems

#1, 8, 11, 18, 21, 22

CHAPTER TEN

Six classes.

Overview

In this chapter, the emphasis should be on mathematical modeling as much as on the details of the differential equations. When students write differential equations to model real world situations, the concept of the derivative as a rate of change is reinforced and the students see some of the most important applications of calculus. Focus your attention on developing a qualitative understanding of how differential equations are used to model real-world applications, and this chapter will pull it all together for your students.

What a differential equation is and what it means for a function to be a solution of a differential equation are introduced in Sections 10.1 and 10.2. Section 10.3 presents slope fields as a graphical representation of a differential equation. Differential equations leading to exponential growth and decay are presented in Section 10.4. An introduction to modeling in such settings as Newton's law of cooling and net worth of a company is discussed in Section 10.5. Modeling the interaction of two populations and the spread of disease are introduced in Sections 10.6 and 10.7. The last two sections are particularly appropriate for students in the life sciences.

10.1 MATHEMATICAL MODELING: SETTING UP A DIFFERENTIAL EQUATION

One class.

Key Points

- How to recognize a differential equation.
- Writing a different equation from a verbal description.
- The logistic model.

Ideas for the Class

Begin the class by discussing the Marine Harvesting example on page 376. Emphasize that differential equations are equations involving a *derivative*, or rate of change. If we have information about a derivative, this gives rise to a *differential equation*. Follow this with a couple more examples of differential equations, as in the following examples. Have the students try these at their seats.

Example   Write a differential equation for each of the following:
(a) The amount of money in a bank, B, is increasing at a rate of 8% times the amount of money in the bank.
(b) Radioctive carbon (carbon-14) decays at a rate proportional to the amount of carbon-14 present.
(c) A yam is placed inside a 200°C oven. The temperature, Y, of the yam increases at a rate proportional to the difference between the oven temperature and its temperature.

Solution   (a) $\dfrac{dB}{dt} = 0.08B$.

(b) $\dfrac{dP}{dt} = -kP$. Ask them why the constant is negative.

(c) $\dfrac{dY}{dt} = k(200 - Y)$.

We will spend more time in Sections 10.2, 10.4, and 10.5 showing students how to write differential equations given information about the rate of change. For now, you just want to give them an understanding of where the differential equations come from.

Problems

Problems 1, 14–18: involve matching verbal descriptions or equations with graphs.
Problems 2–4, 6–9, 11: ask students to write the differential equation.
Problems 5, 10: have students find and use a differential equation to make estimates.
Problems 12, 13: give the differential equation and ask questions about the solutions.

Suggested Problems

1, 3, 5, 7, 9, 10, 12, 15, 16

10.2 SOLUTIONS OF DIFFERENTIAL EQUATIONS

One class.

Key Points

- Numerical solutions to differential equations.
- General and particular solutions, initial conditions.
- How to tell whether a proposed solution is a solution.

Ideas for the Class

The purpose of this section is to discuss the *solution* of a differential equation. Students have trouble with the idea that the solution to a differential equation is a *function* that satisfies the differential equation. One way to help illustrate this is by analogy with algebraic equations. Ask the students: "How can you prove that $x = 2$ is a solution to the equation $5x - x^3 = 2$?" This will help to remind them about substituting to verify a solution—the technique that they will need to use with differential equations. The difference is that differential equations have functions as solutions rather than numbers. Algebraic equations model simple problems where the solution is a number; differential equations model more complex problems where the solution is a function.

Tell the students that they will see in Sections 10.4 and 10.5 how to find the solution to a differential equation. In this section, they will only be asked to verify that a given function is a solution. Do several examples at the board—the students have a surprisingly hard time with this. For example, verify that $y = 2x - 4$ is a solution to the differential equation $\dfrac{dy}{dx} = x - \dfrac{1}{2}y$, and that $y = x^2$ is a solution to the differential equation $\dfrac{dy}{dx}x - 2y = 0$.

You can extend the yam in the oven example from Section 10.1. Have the students check that the function $Y = 200 - Ce^{-kt}$ is a solution to the differential equation $\dfrac{dY}{dt} = k(200 - Y)$. Discuss the idea of arbitrary constants and initial conditions. Ask them what they learn about the solution if they are told that the yam is 20°C when $t = 0$. (This *initial condition* allows them to determine that $C = 180$.) Then ask the students what they learn about the solution from the information that the temperature of the yam is increasing at 2° per minute when the temperature is 120°. (This information allows them to determine that the constant of proportionality $k = 0.025$.) If we put together all of this information, we see that the solution is $Y = 200 - 180e^{-0.025t}$. If there is time, include Examples 1, 2, and 3 from the text and an expanded discussion of the arbitrary constant and particular solutions given by initial conditions.

You can include the discussion on solving a differential equation numerically given on page 376, or you can omit this. It is not essential to the rest of the chapter.

Problems

Problems 1–3, 11, 14, 16, 17: check proposed solutions.
    Problems 4–8: give practice in estimating a solution numerically.
    Problems 9, 10: match the description of a rate of change with a graph.
    Problems 12, 13: ask for conclusions about constants that appear in proposed solutions.
    Problem 15: involves using antiderivatives to find a solution.

Suggested Problems

1–4, 9, 10, 12, 13, 17

10.3 SLOPE FIELDS

One class.

Key Points

- Slope fields.
- Graphical solutions of differential equations.

Ideas for the Class

It's best for this section (and others that follow) if you have available a computer program that draws slope fields and can project them on a screen, such as the University of Arizona program SLOPES. Slope field programs for the graphing calculator are included in this manual. Graphing calculator slope fields can be used with an overhead projector, though they are generally not as good as those generated by computer. If no technology is available, you can photocopy some of the prepared slope fields in this manual, or make overhead transparencies of various slope fields and trace the solutions in pen. In any event, do at least one example by hand, just to make sure that the students see how a slope field is constructed. In particular, make sure that students remember what various slopes look like: a large positive slope, a small positive slope, a large negative slope, a slope of 1/2, 1 or 2. You can explain slope fields as a set of sign posts. There's one at each point, and wherever you are, it tells you in what direction to move. You move a little, and there's the next sign post, etc. Students are often uncomfortable with eyeballing slope fields at first. However, it is surprising how accurate you can be if you draw the solution curves carefully. Try drawing the solution to $dy/dx = 1/x$ from $x = 1$, $y = 0$ to $x = 2$, and see how close to $\ln 2 \approx 0.7$ you get (or get the students to do this on work sheets). If you are projecting slope fields onto a white board, get students to come up and draw solutions directly on the board, over the projected slope field, and then ask them to criticize each other's efforts. Make sure that they understand that the solutions they draw on the slope field are the same as the solutions given in the previous class. For example, if you gave the example $dy/dx = x - (1/2)y$ with solution $y = 2x - 4$ in the previous class, show in this class how the graph of $y = 2x - 4$ fits into the slope field of that differential equation.

One good problem that can be done either in the way described above or on worksheets that you hand out, is to give students the slope field of the yam equation from the previous class (or, if you gave them another example, give the slope field from that example). You can see a lot more about the general behavior of solutions from the slope field than from the specific solution. Illustrate how the general behavior of the solution depends on the initial conditions by getting the students to draw three solutions, one starting at the equilibrium solution ($T = 200$), one starting above it, and one starting below it. It is a good test of their understanding whether they will cross the equilibrium solution, or whether starting on it, they will stay on it. Ask them why a solution can never cross the equilibrium solution. (Once you are on the line $T = 200$ you can't leave it because the signposts don't let you.) Also, point out how to read the arbitrary constant in the previously derived analytic solution from the graph of the solution on the slope field (it is the difference between 200 and the point where the solution crosses the y-axis).

Another useful class example is to give a slope field and ask about the "long-run" behavior of the solution (i.e. what happens to y as x or $t \to \infty$.)

For the homework problems you might want to hand out worksheets which are photocopies of the problems in the book. Students can hand these in instead of ripping out pages from their books.

Problems

Problems 1– 5: give students practice in sketching and analyzing slope fields.
Problems 6, 7, 14, 15: match the differential equations with their slope fields.
Problems 8– 13: ask about long-term behavior.
Problem 16: has students analyze the properties of the solutions to two different differential equations.

Suggested Problems

1, 3, 4, 5, 7, 8, 12, 15, 16

10.4 EXPONENTIAL GROWTH AND DECAY

One class.

Key Points

- The differential equation of exponential growth and decay.
- Applications in modeling pollution and quantity of a drug in the body.

Ideas for the Class

Many of the ideas and applications of this section have been seen before, but there is a shift in emphasis to modeling in terms of a differential equation. Emphasize the translation of verbal statements into differential equations.

For example, ask the students how many rabbits there are in a population that starts at 100 and grows at the continuous rate of 3% per year. They may know that $P = 100e^{0.03t}$, but they may not yet understand that $dP/dt = 0.03P$. That is the point in going over such an example in detail. Help them get to the idea of expressing the growth rate in terms of the derivative.

When it comes to continuously compounded interest, they may remember the formula $B = Pe^{rt}$ from Chapter 1, but they may need help in seeing that $dB/dt = rB$. Some students may want to write something like $dB/dt = e^{rB}$. (Don't bring this up unless they do; just be ready for it.) Learning how to set up the differential equation will help immensely in the next section, where they will meet situations for which they do not know formulas for the solutions. Knowing how to think in terms of derivatives will enable them to solve the problems.

Problems

Problems 1– 6: are straightforward.
Problems 7– 19: are applications.

Suggested Problems

1, 3, 4, 8, 11, 13, 19

10.5 APPLICATIONS AND MODELING

One or two classes.

Key Points

- Setting up differential equations from verbal descriptions.
- Equilibrium solutions and stability.

Ideas for the Class

This section deals primarily with differential equations of the form

$$\frac{dy}{dt} = k(y - A).$$

Start by explaining that the solution to this differential equation is $y = A + Ce^{kt}$. Have the students get started by doing some straightforward examples as in Example 1. As soon as they understand the mechanics, move straight to the applications.

Differential equations cover a broad range of situations. You can include examples such as the following. In each case, have the students write the differential equation, solve the differential equation, and (if possible) find the arbitrary constant.

Example

(a) Money in a bank account earns 5% annual interest compounded continuously, and is paid out of the account at a continuous rate of $5000 a year. The initial amount of money in the account is $25,000.

(b) A drug is administered intravenously at a continuous rate of 25 mg per hour. The drug is metabolized and excreted from the body at a rate proportional to the amount there, with proportionality constant 0.05.

(c) A hot cup of coffee, at temperature $140°$F, is placed in a $65°$ room. The temperature, H, of the coffee changes at a rate proportional to the difference between the temperature of the room and the temperature of the coffee. (This is called *Newton's Law of Heating and Cooling*.)

Take your time setting up each differential equation and solving it. Graph the solution to the money example. If possible, draw a slope field for the solution to the drug example. In the coffee example, point out the fact that the initial condition does not enter into the writing of the differential equation, but comes into play only in finding the arbitrary constant in the solution.

Students have a tendency to assume that in the solution $y = A + Ce^{kt}$ to the differential equation $dy/dt = k(y - A)$, the constant C of integration is going to take on the initial value of y when the particular solution is found. Point out that this is not the case, and re-emphasize it with each example.

Problems 23 and 24 make excellent group discussion problems.

Discuss how to recognize an equilibrium solution by looking at the differential equation of the form $dy/dx = f(y)$. [Set $dy/dx = 0$.] Show how to tell if an equilibrium solution is stable or unstable by looking at a slope field.

Problems

Problems 1– 8: ask for formulas for solutions.
   Problem 9: has students check the solution of a differential equation.
   Problems 10– 12, 14, 19– 21, 23– 25: are applications.
   Problems 13, 15- 18, 22: deal with graphical solutions and equilibrium solutions.

Suggested Problems

1, 4, 7, 13– 17, 21, 24

10.6 MODELING THE INTERACTION OF TWO POPULATIONS

One class.

Key Points

- Systems of differential equations.
- Predator-prey model.

Ideas for the Class

Have fun with this section! You want to leave the students with the idea that mathematics is both interesting and applicable. Make the point early that we can use systems of differential equations to model the interaction of any two species (or businesses or chemicals...). Work through the top half of page 405. Emphasize two things: that we need to understand what would happen to the population in the *absence* of the other species and then that we need to understand whether the interaction with the other species has a positive or negative effect on the population. Be sure the students understand why the terms are positive or negative in the following two differential equations:

$$\frac{dw}{dt} = w - wr \qquad \text{and} \qquad \frac{dr}{dt} = -r + wr.$$

To check the students' understanding, do an example similar to Example 2: put each system of differential equations on the board and have the students describe the interaction between the species. See if they can think of species which interact in each of the ways described.

Once the students understand where the differential equations come from, the next step is to investigate the solution to the system of differential equations. With a single differential equation, we look at a slope field to visualize the solution. That is not possible here because each equation in the system has three variables (r, w, and t) in it. Instead, with a system of differential equations, we eliminate the t variable and create a single differential equation involving $\frac{dr}{dw}$. The slope field for this differential equation is called a *phase plane*. Be sure to emphasize to students that the two axes represent the populations of robins and worms. Draw a trajectory and have your students discuss how the two populations change over time. Start at the point $r = 3, w = 2$, for example. As you trace out the trajectory counterclockwise, have the students discuss what is happening: There are many worms—this is good for the robins so the robin population increases. But as the robin population gets bigger, they eat up all the worms and the worm population shrinks. When the worm population gets too small, however, there are two few worms to sustain the robins and we see a sharp dropoff in the robin population. When the robin population gets very small, the worm population starts to recover, and we start all over again.

Tell your students about the study of the lynx and hare by the Hudson Bay Company. The data matches the behavior predicted by this mathematical model.

Problems

Problems 1– 4, 16– 23: involve the interaction of two species: reading a trajectory and interpreting or creating differential equations.

Problems 5– 15: refer explicitly to the robins and worms example. Problem 11 raises some interesting questions and is good for class discussion.

Suggested Problems

1– 4, 8, 9, 11, 14, 16, 17, 19– 21

10.7   MODELING THE SPREAD OF DISEASE

One class.

Key Points

- The *S-I-R* model.
- Threshold value for an epidemic.

Ideas for the Class

This section extends the ideas developed in the last section to a very different kind of interaction: the interaction between infected people and healthy people. We use a mathematical model to determine how many people should be vaccinated to avoid an epidemic. Work through the example given in the text of flu in a British boarding school. Hopefully, your students will be impressed at how much information the mathematical model gives us. This is why the Center for Disease Control in Atlanta uses mathematical models to monitor many different diseases.

Problems

Problems 1– 7: involve interpretations of the S-I-R model for the spread of a disease and epidemic. Assign as many as you can and encourage discussions.

Problems 8– 10: involve different types of flu models. A good discussion is to compare the effects of each term and each coefficient in the differential equation in these problems.

Suggested Problems

3, 4, 6, 8, 9, 10

CHAPTER ELEVEN

Three classes.

Overview

11.1 GEOMETRIC SERIES

One class.

Key Points

- An introduction to finite and infinite geometric series.
- Sums of geometric series.

Ideas for the Class

Some students may have been introduced to geometric series in high school, but do not assume that most have. Use an example like that of repeated drug dosage given in this section to develop the general form for a geometric series. Talk about the idea of the long-term effect of the drug. Derive the formulas for the sums of both finite and infinite series. Have students find the sum of at least one finite geometric series both ways, by adding it term by term and by using the formula for a finite sum.

Problems

Problems 1– 13: are straightforward problems involving the sums of geometric series.

Problems 14, 15: bank account problems. Assign these, especially if you plan to cover Section 11.2.

Problems 16, 17, 20: are more problems on repeated drug dosage. Problem 20 requires more thought; a good problem to assign for group work.

Problems 18, 19: ask students to further investigate an example from the section. Problem 18 enforces the concept of the sum of an infinite number of terms can in fact be finite. Problem 19 shows an infinite sum that is infinite.

Suggested Problems

1, 2, 6, 8, 11, 12, 15, 17– 19

11.2 APPLICATIONS TO BUSINESS AND ECONOMICS

One-hald to one class.

Key Points

- Annuities; present and future values.
- The multiplier effect.
- Market stabilization.

Ideas for the Class

A review of continuously compounded interest is a good place to start. Carefully discuss annuities; finding present values tends to be confusing. A good example to use is one pertaining to a lottery. Ask the student if they were to win $1,000,000 in a lottery, would they prefer a lump sum (all $1,000,000 given to them immediately) or 20 equal annual payments of $50,000. Which would the government (who runs the lottery) rather do? Ask them if they were to choose the 20 payments, how much money the government would have to put in an account now that earns 5% annual interest to cover the 20 future payments.

Students find market stabilization interesting. Example 5 pertaining to the amount of pennies in circulation is a good example for class discussion.

George Bush's recent rebate on 2000 taxes can generate a good discussion on the multiplier effect.

Problems

Problems 1, 2: compound interest problems.

Problems 3– 7: deal with annuities. Problems 3– 6 ask students to find the present value of specific annuities while Problem 7 asks how long an annuity will last.

Problems 8, 9: investigate the multiplier effect. Problem 9 is a good problem to do together in class or have the students work in together in groups.

Problems 10, 13, 14: are market stabilization problems.

Problems 11, 12: involve salaries and finding the sums of finite geometric series. Problem 11 is a well-known question that asks how much would ones total earnings be over a certain amount of time if they were paid a penny the first day working and their salary doubled every day. Most students are fascinated and surprised by this problem. It should be assigned or worked together in class.

Suggested Problems

2, 4, 7, 8, 10, 11, 13 # 2, 4, 7, 8, 10, 11, 13

11.3 APPLICATIONS TO LIFE SCIENCES

One-half to one class.

Key Points

- Applications in life sciences: drug dosage, toxin accumulation, and natural resource depletion.
- Geometric series vs. differential equations.

Ideas for the Class

Students have already seen what happens with repeated drug doses in Section 11.1. Revisit a simple example to show how much of a drug is in the body at a certain time and when the amount of a drug in the body tends to stabilize. Half-life is used in the problems, so we suggest you do a class example mentioning half-life as a reminder to the students.

Depletion of our natural resources is probably a concern to most of your students, if not all. Work out Problems 13– 16 with the students, either together as a class or let them work on them in groups first, then discuss as a class. They will appreciate the real world applications of this material.

Problems

Problems 1– 8, 12: deal with repeated drug dosage. Problems 1, 2, 4, and 5 are straightforward. Problems 3, 7 and 12 involve variables and students may need some guidance. Problem 6 investigates how much of a drug remains in the system at various times after the dosages are stopped.

Problems 9, 11: are problems dealing with toxin accumulation and steady state amounts.

Problems 10, 13– 19: investigate the depletion of natural resources. Problems 13– 16 are related and can be assigned as a group exercise or used for class discussion.

Suggested Problems

2, 5, 7, 11, 13– 16

PART III

SAMPLE SYLLABI

Several sample syllabi appear on the following pages, and are summarized below. Many other variations are also possible. We encourage you to see what works best for your students.

1. **Sample Syllabus #1**
   Standard one-semester course.

   - **Assumes a 3-hour course and a 14-week semester.**

2. **Sample Syllabus #2**
   Two-semester course.

   - **Assumes a 3-hour course and a 14-week semester.**

3. **Sample Syllabus #3**
   Quarter system.

   - **Assumes a quarter system, with three 4-hour courses over 10 weeks each.**

4. **Sample Syllabus #4**
   One-semester emphasis on business applications

   - **Assumes a 3-hour course and a 14-week semester.**

5. **Sample Syllabus #5**
   One-semester emphasis on life sciences applications.

   - **Assumes a 3-hour course and a 14-week semester.**

Sample Syllabus #1: Standard one-semester course

(Assumes a 3-hour course and a 14-week semester.)

Week	Monday	Wednesday	Friday
1	1.1	1.2	1.3
2	1.4	1.5	1.6
3	1.7	1.8	1.9
4	1.10	Review	First Exam
5	2.1	2.2	2.3
6	2.4	2.5	3.1
7	3.2	3.3	3.4
8	3.5	4.1	4.2
9	4.3	Review	Second Exam
10	4.4	4.5	4.6
11	4.7	4.8	5.1
12	5.2	5.3	5.4
13	5.5	Review	Third Exam
14	6.1	6.2	Review

Optional second semester would start with a review of Chapter 5 and then cover Chapters 6 through 11.

Sample Syllabus #2: Two-semester course

(Assumes a 3-hour course and a 14-week semester.)

A two-semester course allows ample time for in-class student explorations and projects. Take advantage of the projects offered in the book, both in the appendix and at the end of each chapter. Alternatively, use the time to move more slowly through the material and allow time for regular in-class group work using ideas and examples from the text or those suggested in this Instructor's Manual. We recommend the pace given by this syllabus.

First Semester:

Week	Monday	Wednesday	Friday
1	1.1	1.2	1.3
2	1.4	1.5	Focus on: compound interest
3	1.6	1.7	1.8
4	1.9	1.10	Focus on: fitting formulas to data
5	Exploration/Project	Review	First Exam
6	2.1	2.2	2.3
7	2.4	2.5	Exploration/Project
8	Ch. 2 Focus on Theory	3.1	3.2
9	3.3	Review	Second Exam
10	3.4	3.5	Ch. 3 Focus on Theory
11	4.1	4.2	4.3
12	4.4	4.5	4.6
13	Exploration/Project	Review	Third Exam
14	4.7	4.8	Review

Second Semester:

Week	Monday	Wednesday	Friday
1	5.1	5.2	5.3
2	5.4	5.5	Exploration/Project
3	6.1	6.2	6.3
4	6.4	Review	First Exam
5	7.1	7.2	7.3
6	7.4	8.1	8.2
7	8.3	9.1	9.2
8	9.3	9.4	9.5
9	9.6	Review	Second Exam
10	10.1	10.2	10.3
11	10.4	10.5	Exploration/Project
12	10.6	10.7	Exploration/Project
13	Review	Third Exam	11.1
14	11.2	11.3	Review

Sample Syllabus #3: Quarter system

(Assumes a quarter system, with three 4-hour courses over 10 weeks each.)

If courses meet **3 hours a week**, some of the exploration/projects can easily be omitted.

First Quarter:

Week	Monday	Tuesday	Thursday	Friday
1	1.1	1.2	1.3	1.4
2	1.5	Compound interest	1.6	1.7
3	Exploration/Project	Fitting formulas to data	Review	First Exam
4	1.8	1.9	1.10	Exploration/Project
5	2.1	2.2	2.3	Exploration/Project
6	2.4	2.5	Review	Second Exam
7	Ch. 2 Focus on Theory	3.1	3.2	3.3
8	Exploration/Project	3.4	3.5	Ch. 3 Focus on Theory
9	Exploration/Project	Focus on Practice	Review	Third Exam
10	Exploration/Project	Exploration/Project	Review	Review

Second Quarter:

Week	Monday	Tuesday	Thursday	Friday
1	Ch. 2 (review)	Ch. 3 (review)	4.1	4.2
2	4.3	Exploration/Project	4.4	4.5
3	Exploration/Project	4.6	Review	First Exam
4	4.7	4.8	Exploration/Project	5.1
5	5.2	5.3	5.4	Exploration/Project
6	5.5	Ch. 5 Focus on Theory	Review	Second Exam
7	6.1	6.2	6.3	Exploration/Project
8	6.4	7.1	7.2	7.3
9	Exploration/Project	7.4	Review	Third Exam
10	Focus on Practice	Exploration/Project	Review	Review

Third Quarter:

Week	Monday	Tuesday	Thursday	Friday
1	Review	8.1	8.2	8.3
2	Exploration/Project	9.1	9.2	Exploration/Project
3	9.3	Exploration/Project	Review	First Exam
4	9.4	9.5	9.6	Exploration/Project
5	Ch. 9 Focus on Theory	10.1	10.2	Exploration/Project
6	10.3	Exploration/Project	Review	Second Exam
7	10.4	10.5	Exploration/Project	10.6
8	10.7	Exploration/Project	11.1	11.2
9	11.3	Exploration/Project	Review	Third Exam
10	Exploration/Project	Exploration/Project	Review	Review

Sample Syllabus #4: Emphasis on Business Applications

(One-semester course)

(Assumes a 3-hour course and a 14-week semester.)

Trying to fit multivariable functions in a 3-hour one-semester course is difficult, but it can be done. This is a crowded course. You will have to skip some of the detail in some sections.

Option: If you hope to get to **Lagrange multipliers** in one semester, it is possible. Adjust the syllabus below by choosing any two of the four sections 4.6, 4.7, 6.2, 6.3 to omit. This will leave room for 9.5 and 9.6 at the end.

Week	Monday	Wednesday	Friday
1	1.1	1.2, 1.3	1.4
2	1.5	1.6	1.7
3	1.8	1.9	2.1
4	2.2	2.3	2.4
5	2.5	Review	First Exam
6	3.1	3.2	3.3
7	3.4	3.5	4.1
8	4.2	4.3, 4.4	4.5, 4.6
9	4.7	Review	Second Exam
10	5.1	5.2	5.3
11	5.4, 5.5	6.1	6.2
12	6.3	9.1, 9.2	9.3
13	9.4	Review	Third Exam
14	11.1	11.2	Review

Sample Syllabus #5: Emphasis on Life Sciences Applications

(One-semester course)

(Assumes a 3-hour course and a 14-week semester.)

Ideally, life sciences students should take two semesters. If that is not possible, the following syllabus enables them to see the applications of differential equations in the life sciences in just one semester.

Week	Monday	Wednesday	Friday
1	1.1	1.2	1.3
2	1.5	1.6	1.7
3	1.8	1.9	1.10
4	2.1	2.2	2.3
5	2.4	Review	First Exam
6	3.1	3.2	3.3
7	3.4	3.5	4.1
8	4.2	4.3	4.7
9	4.8	Review	Second Exam
10	5.1	5.2	5.3
11	5.4, 5.5	10.1	10.2
12	10.3	10.4, 10.5	10.6
13	10.7	Review	Third Exam
14	11.1	11.3	Review

MASTERS FOR
OVERHEAD
TRANSPARENCIES

82

Transparency Master for Figure 1.11 for Problem 12 in Section 1.1

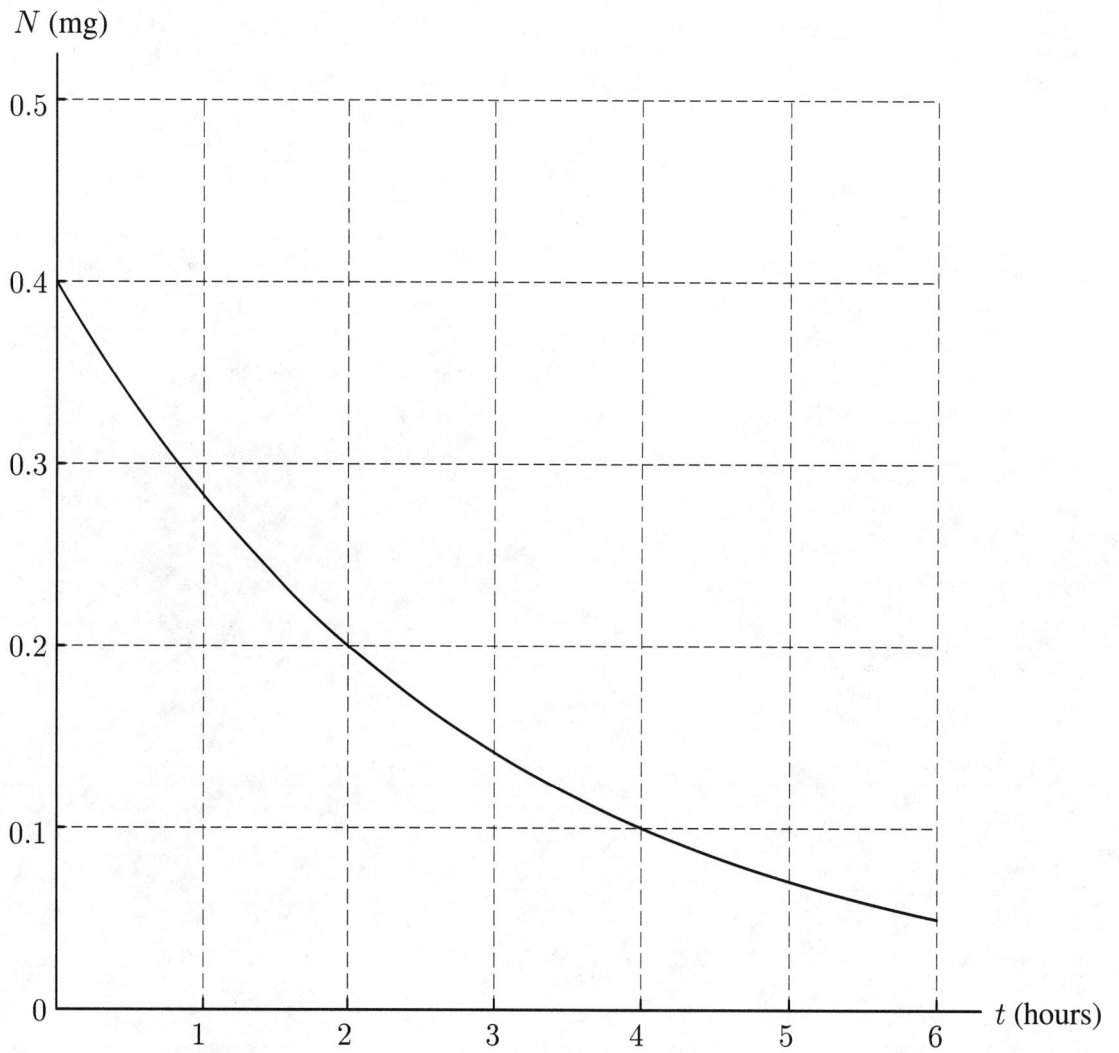

Transparency Master for Figure 1.25 for Example 2 in Section 1.3

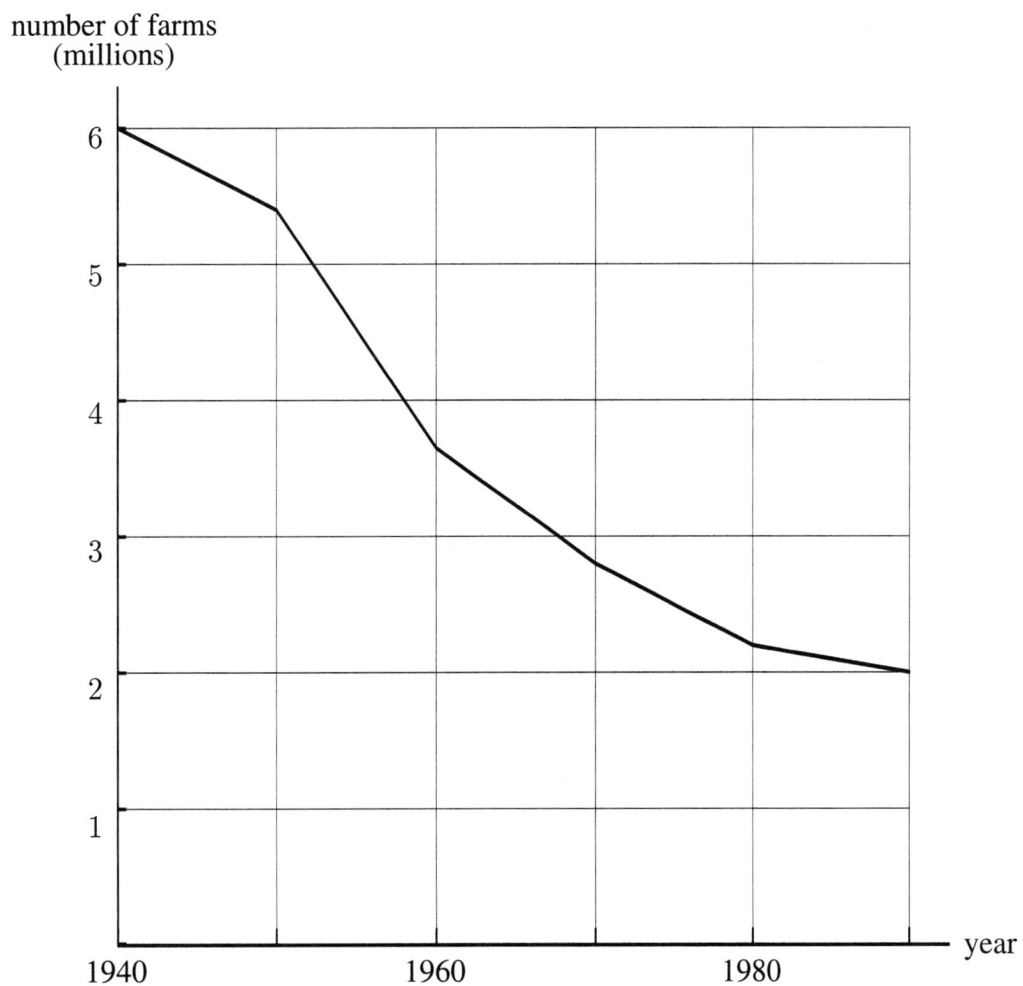

number of farms
(millions)

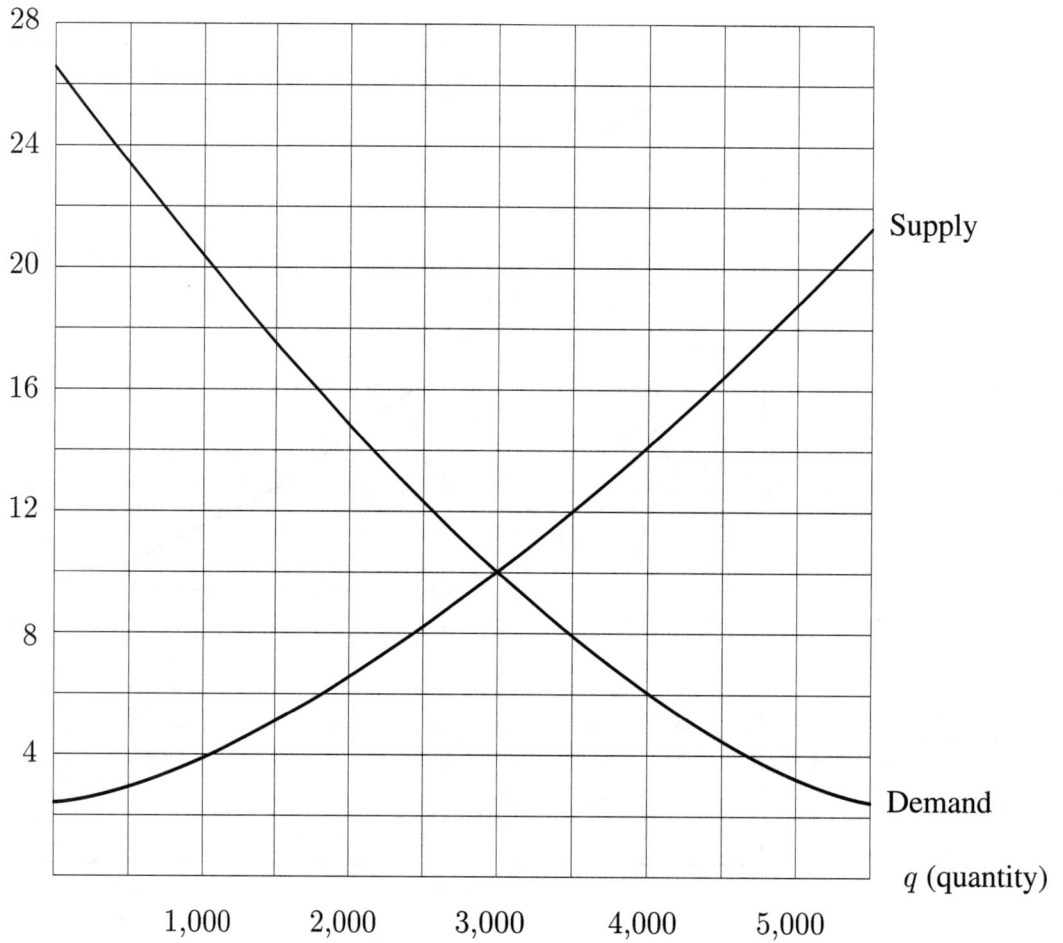

Transparency Master for Section 1.9

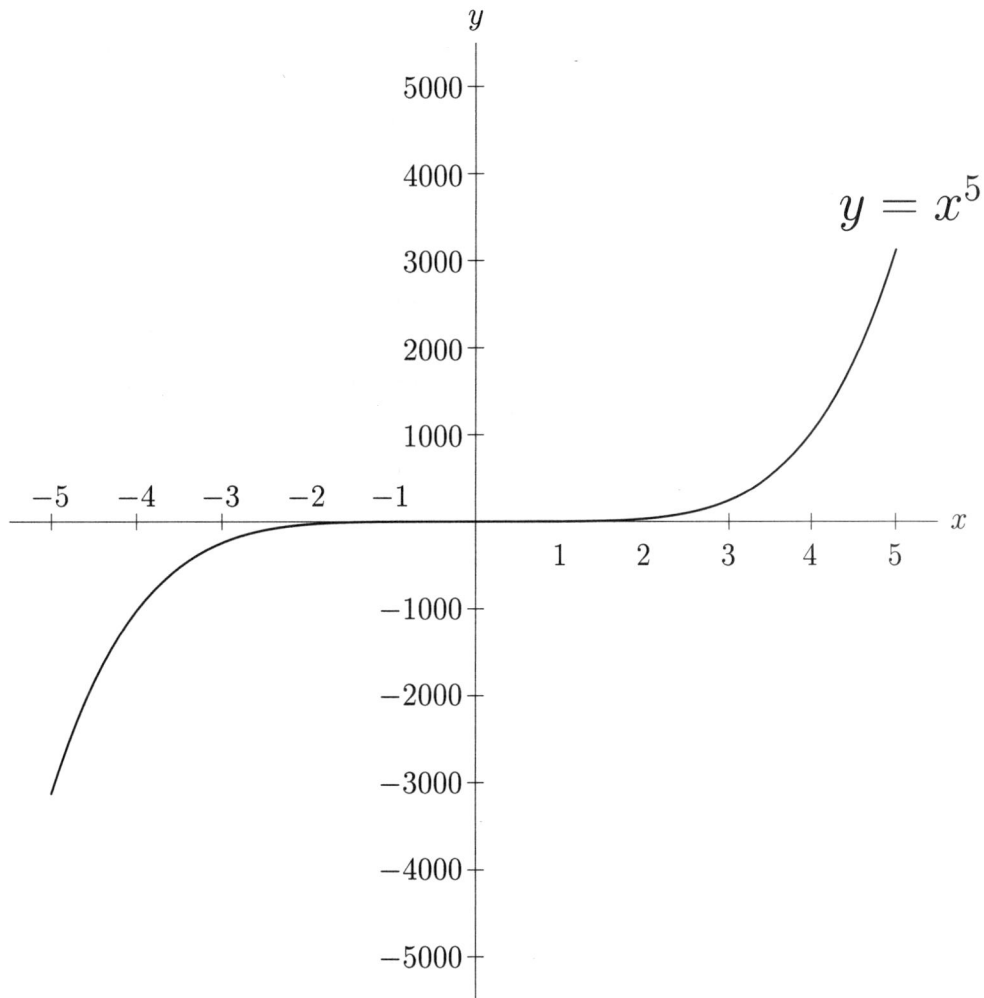

$$y = x^5$$

Transparency Master for Section 1.9

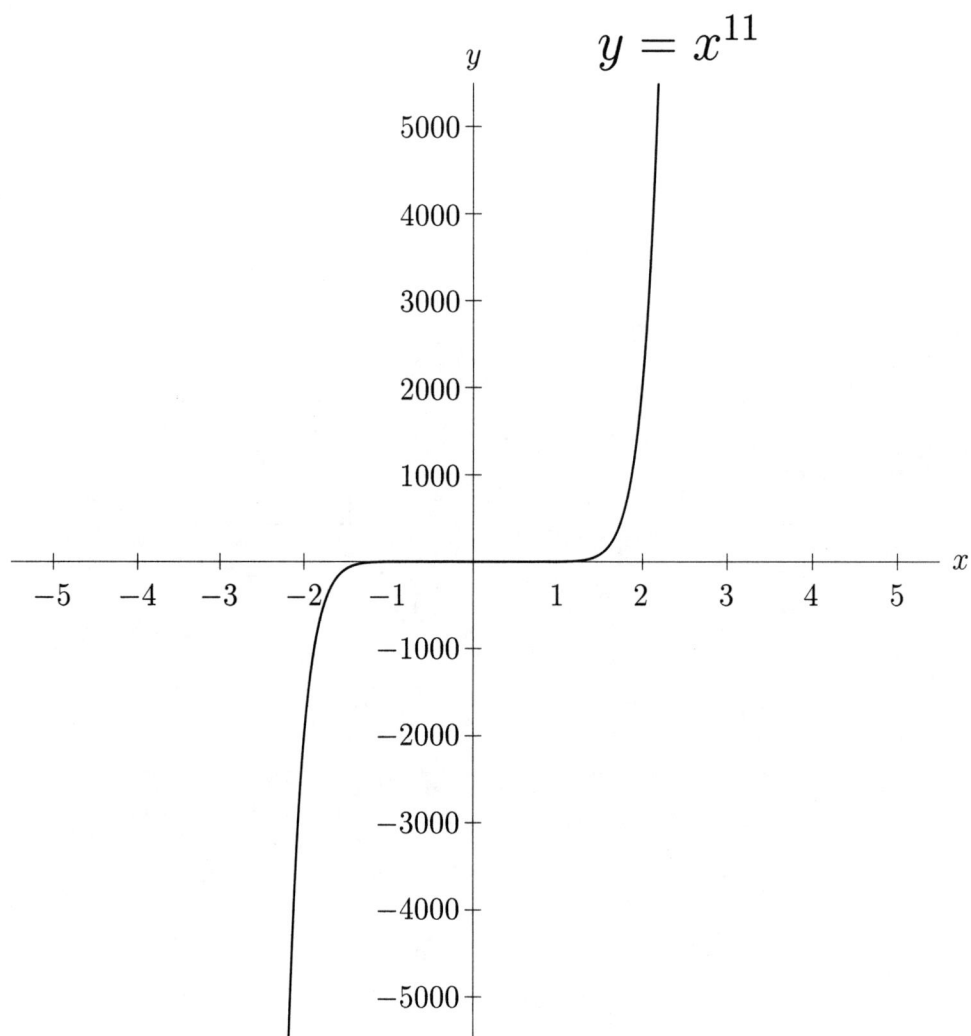

$$y = x^{11}$$

Transparency Master for Section 1.9

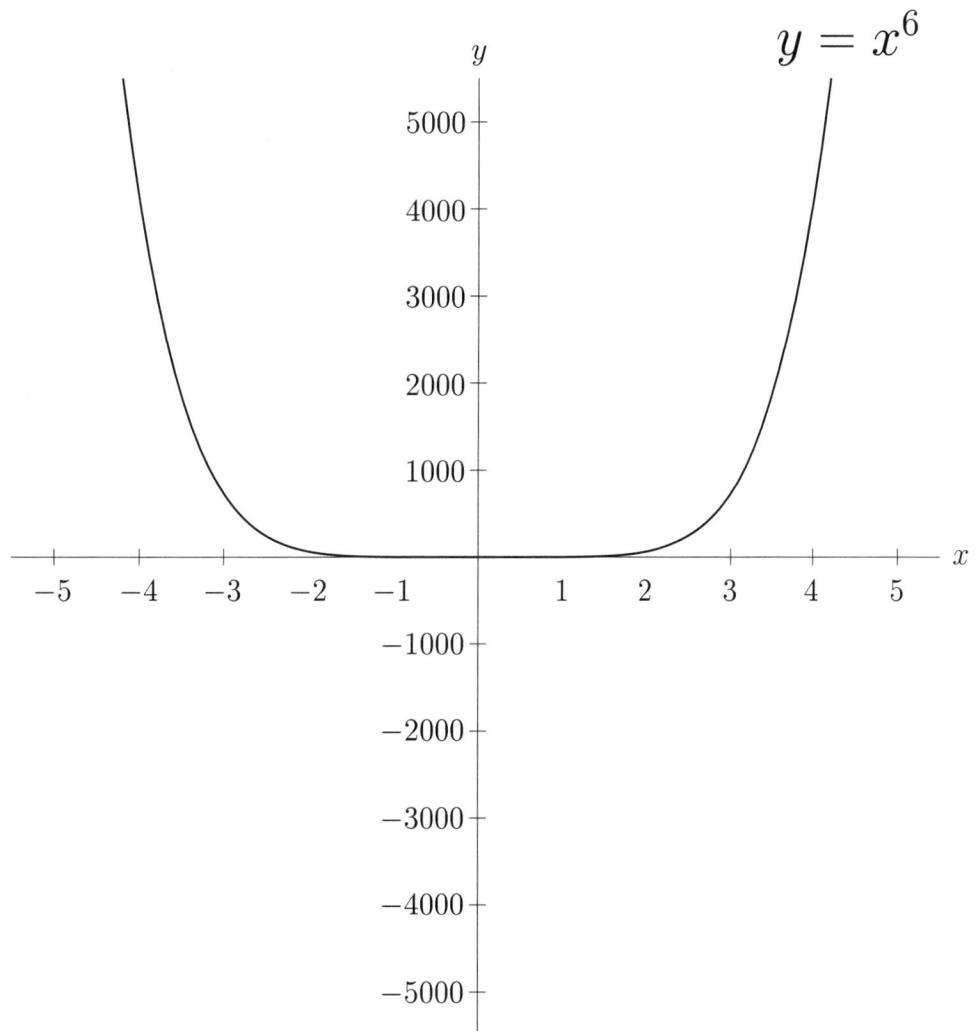

$y = x^6$

Transparency Master for Section 1.9

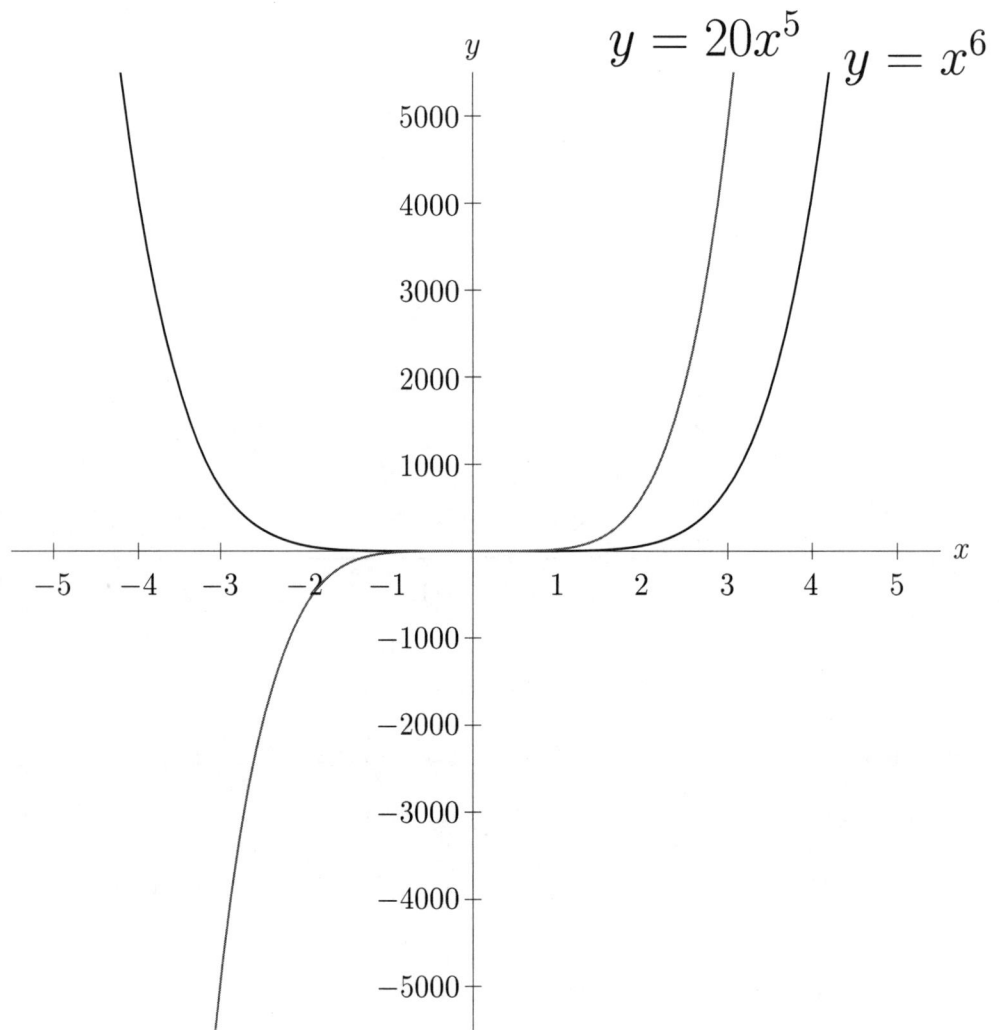

$y = 20x^5$

$y = x^6$

Transparency Master for Section 1.9

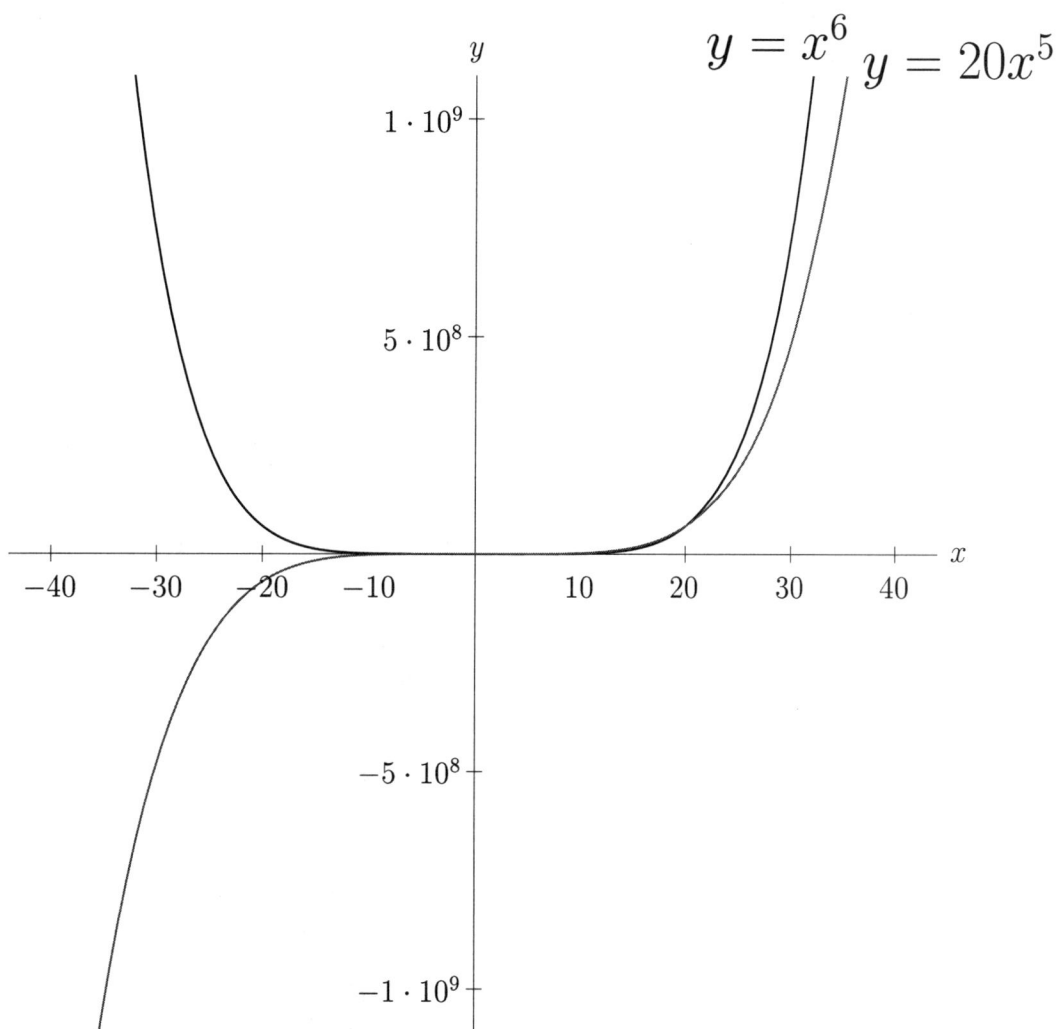

$y = x^6$ $y = 20x^5$

Transparency Master for Section 1.10

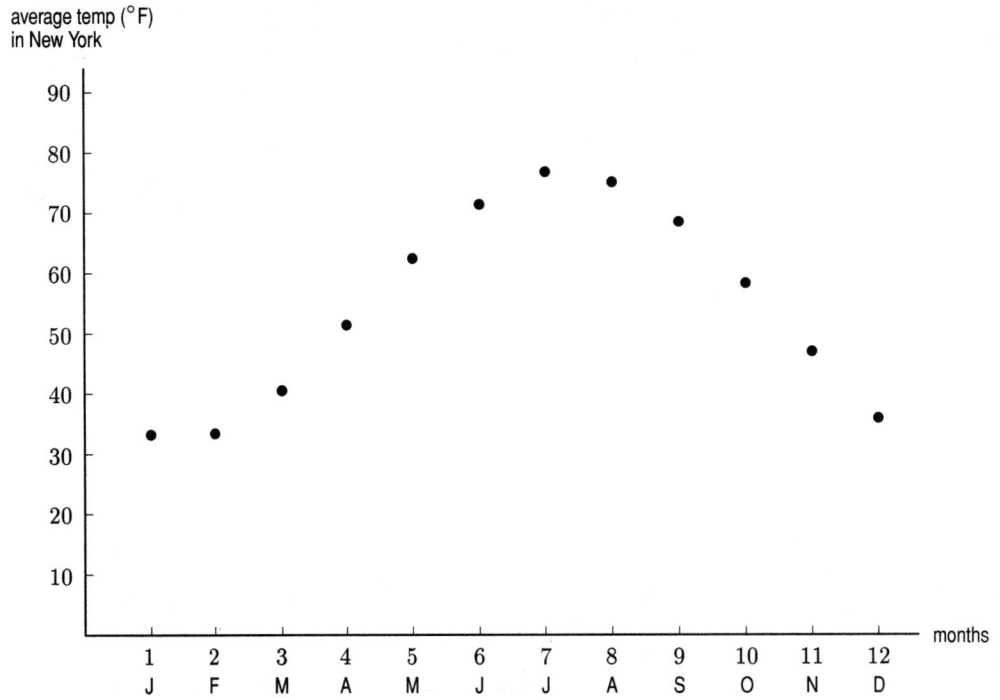

average temp (°F) in New York

Table 0.0.1
*Average
Temperature (°F)
in New York by
Month*

Jan	33.2
Feb	33.4
Mar	40.5
April	51.4
May	62.4
Jun	71.4
July	76.8
Aug	75.1
Sept	68.5
Oct	58.3
Nov	47.0
Dec	35.9

91

Transparency Master for Section 2.2

The graph of f is given. Sketch the graph of f' on the axes below.

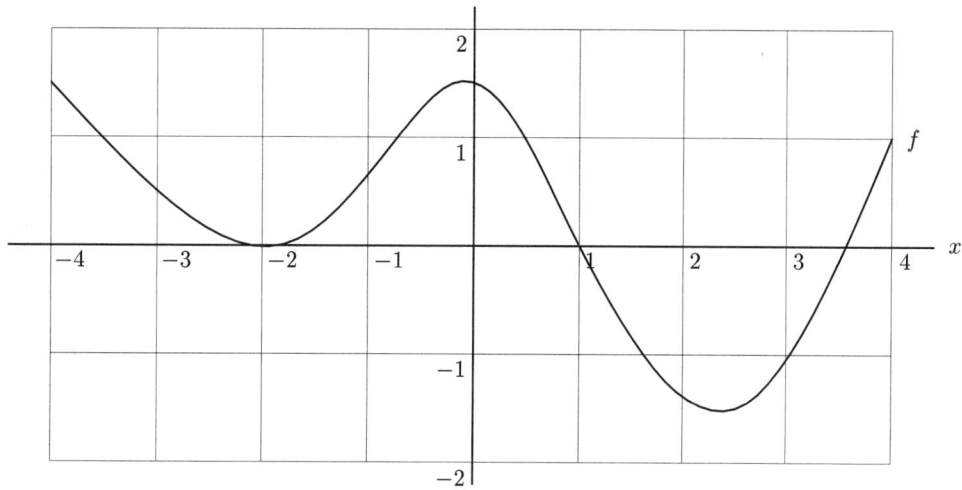

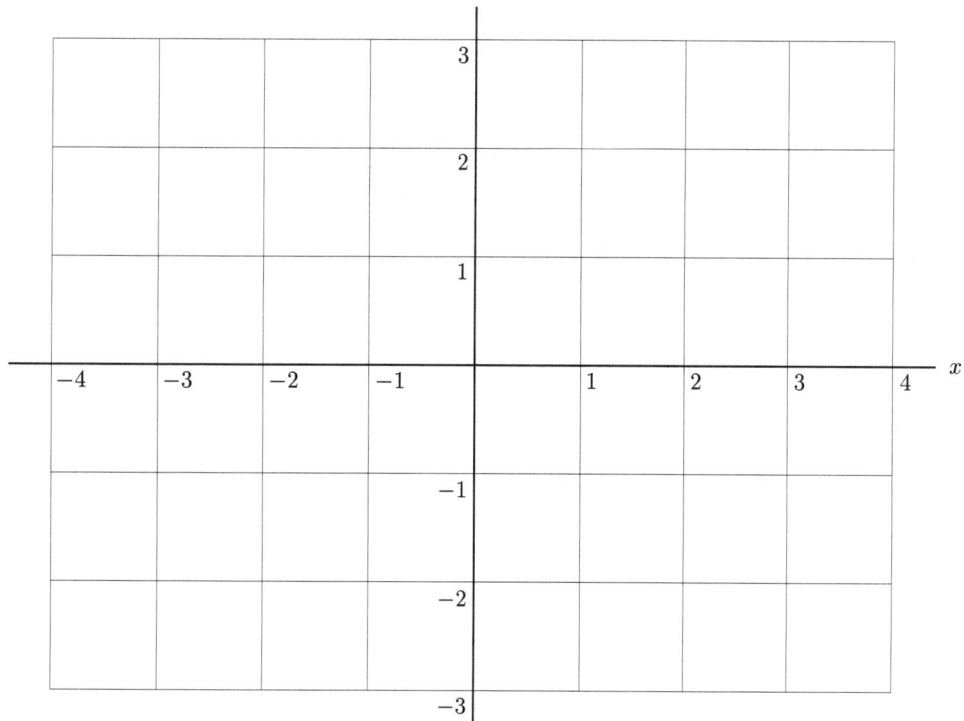

Transparency Master for Figure 2.32 for Example 2 in Section 2.4

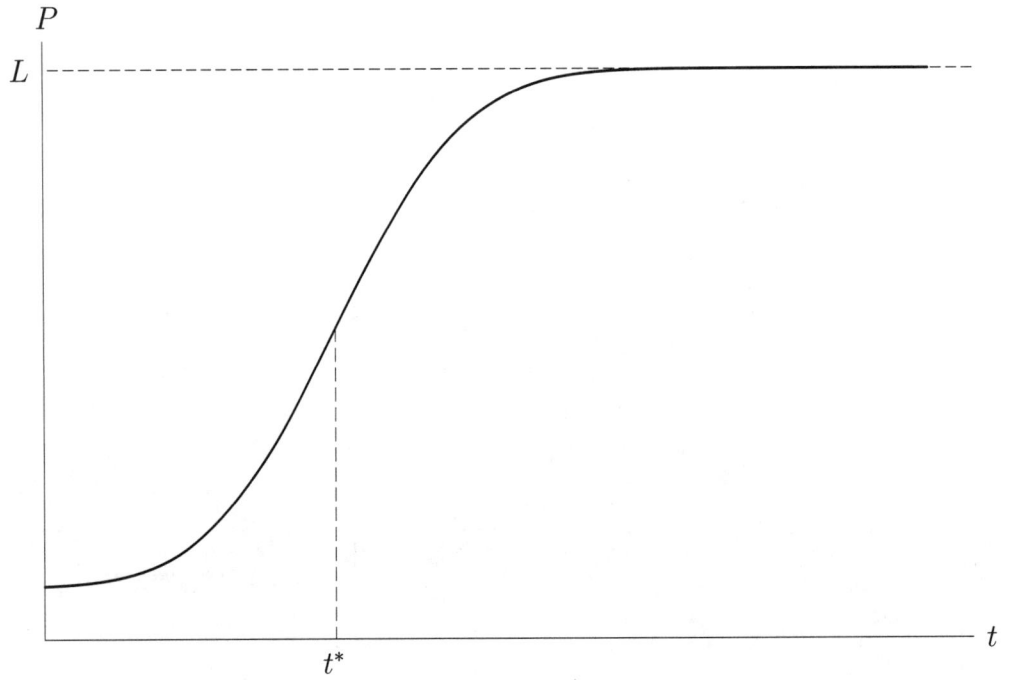

Transparency Master for Figure 2.54 for Problem 6 in Section 2.5

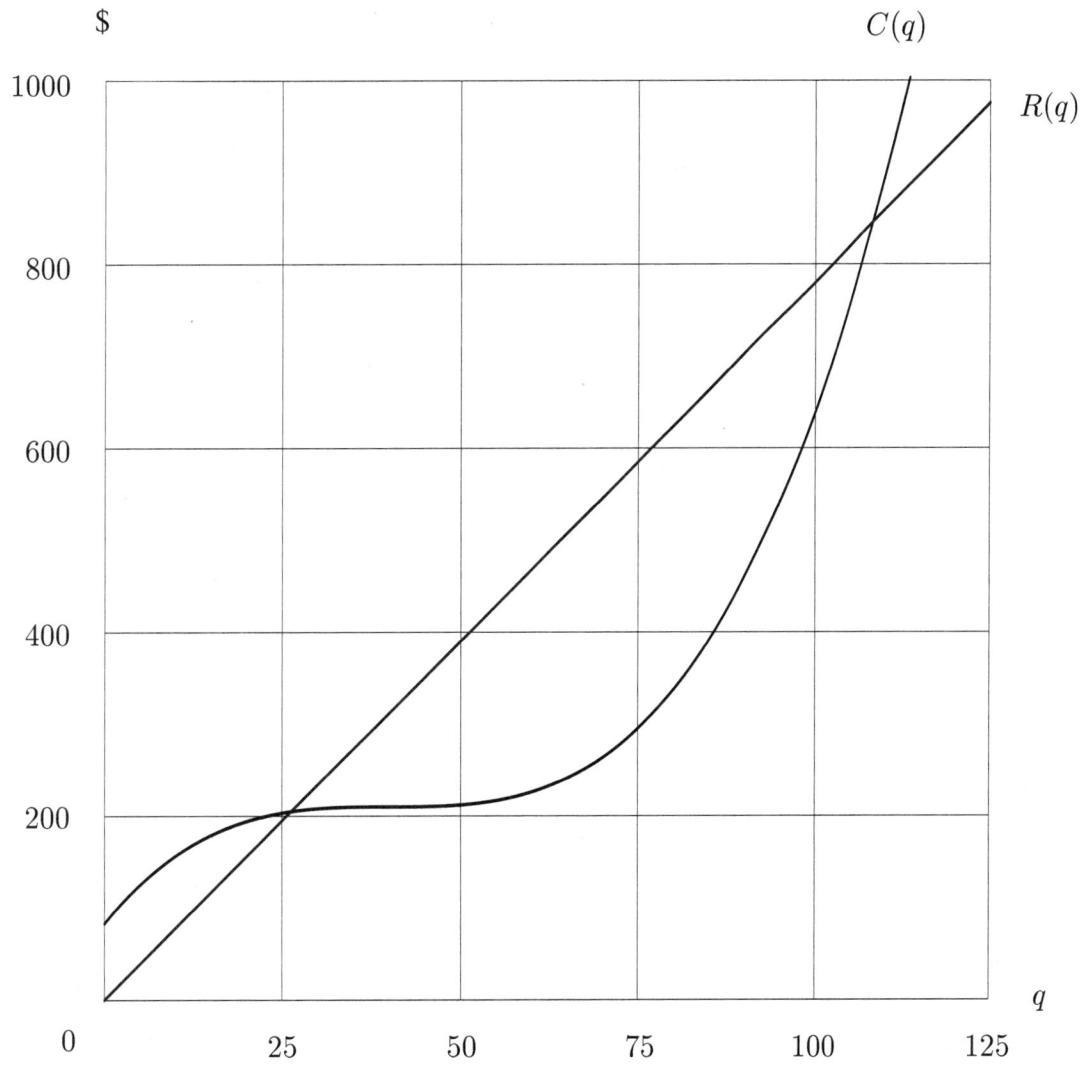

Transparency Master for Table 4.6 in Section 4.6

Table 4.6.2

Air travel	1.10	Jewelry	2.60
Automobiles	1.50	Milk	0.31
Automobile parts	0.50	Oranges	0.97
Lightbulbs	0.33	Poultry	0.27
Flour	0.79	Radios	1.50
Furniture	3.04	Sporting goods	1.20
Furs	2.30	Sugar	0.44

Elasticity of demand for selected products

Transparency Master for Figure 4.83 for Problem 3 in Section 4.8

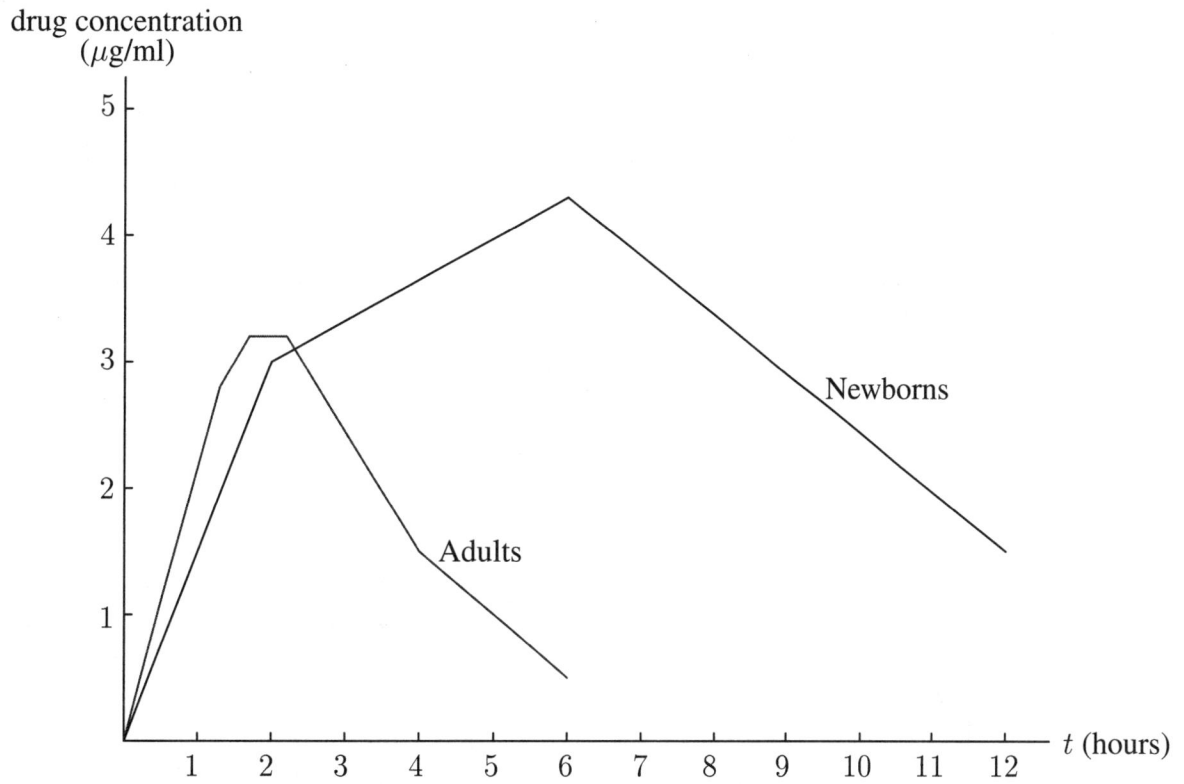

Transparency Master for Figure 5.4 for Problem 7 in Section 5.1

v (m/sec)

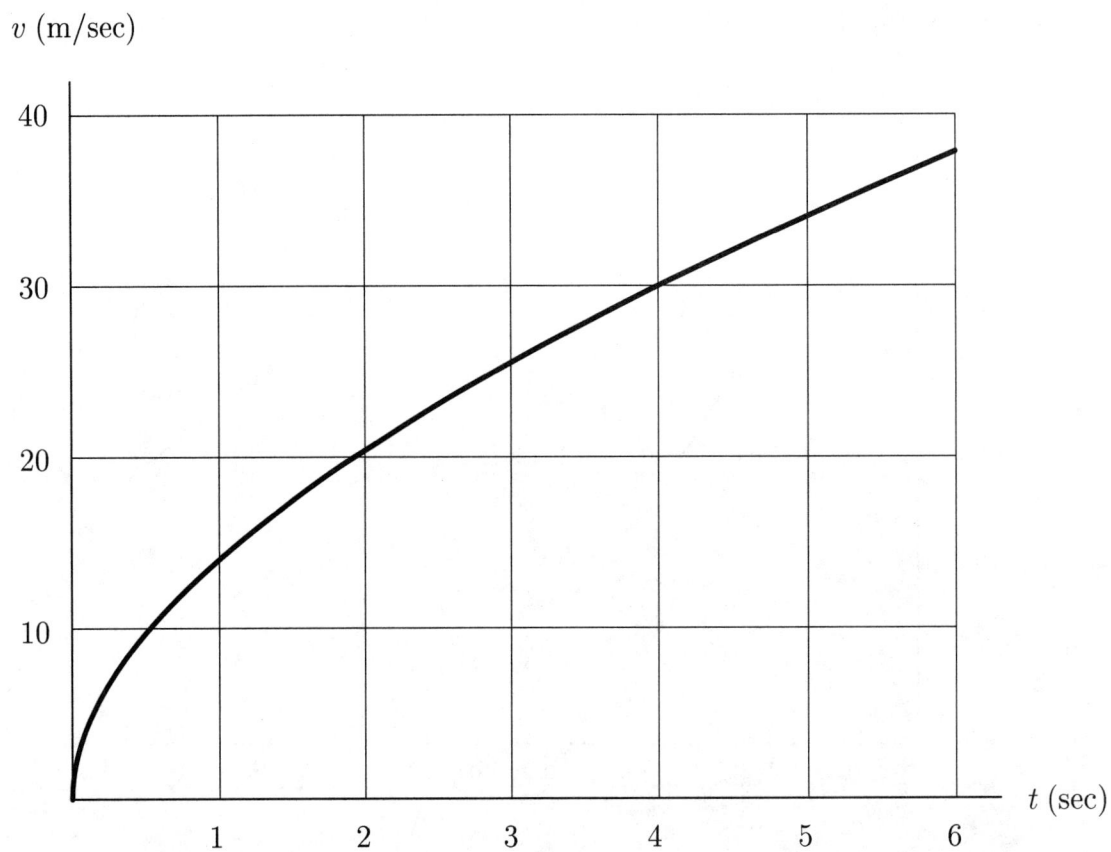

t (sec)

Transparency Master for Figure 5.37 for Example 4 in Section 5.4

new plants per year

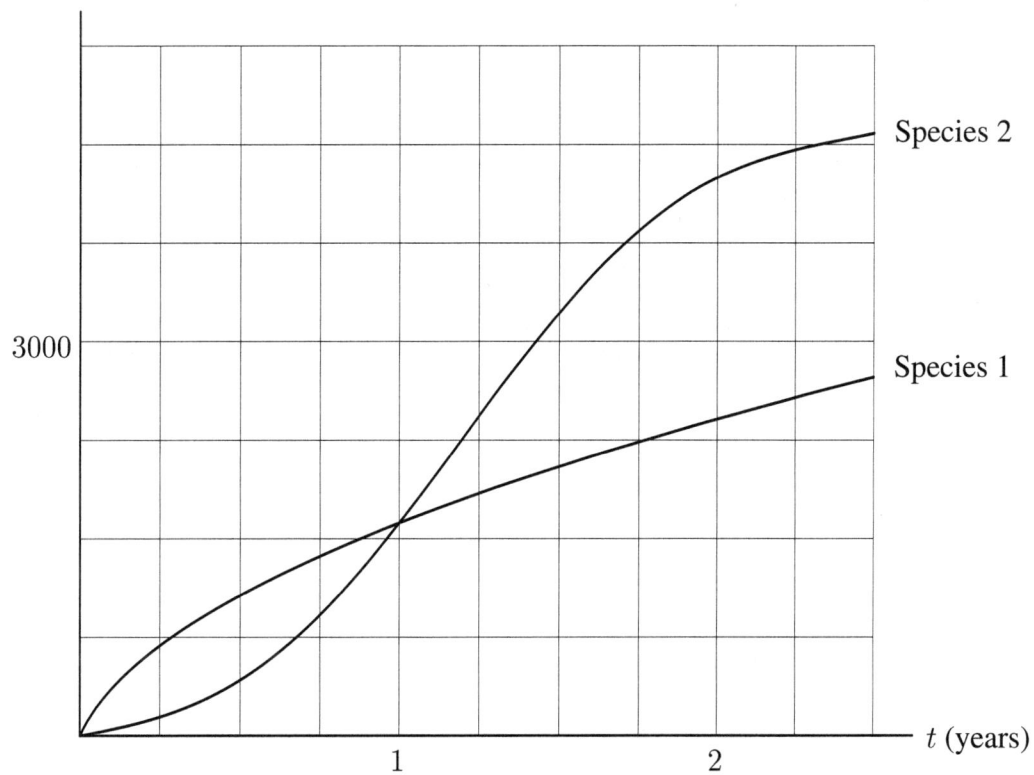

Transparency Master for Figure 5.51 for Example 2 in Section 5.5

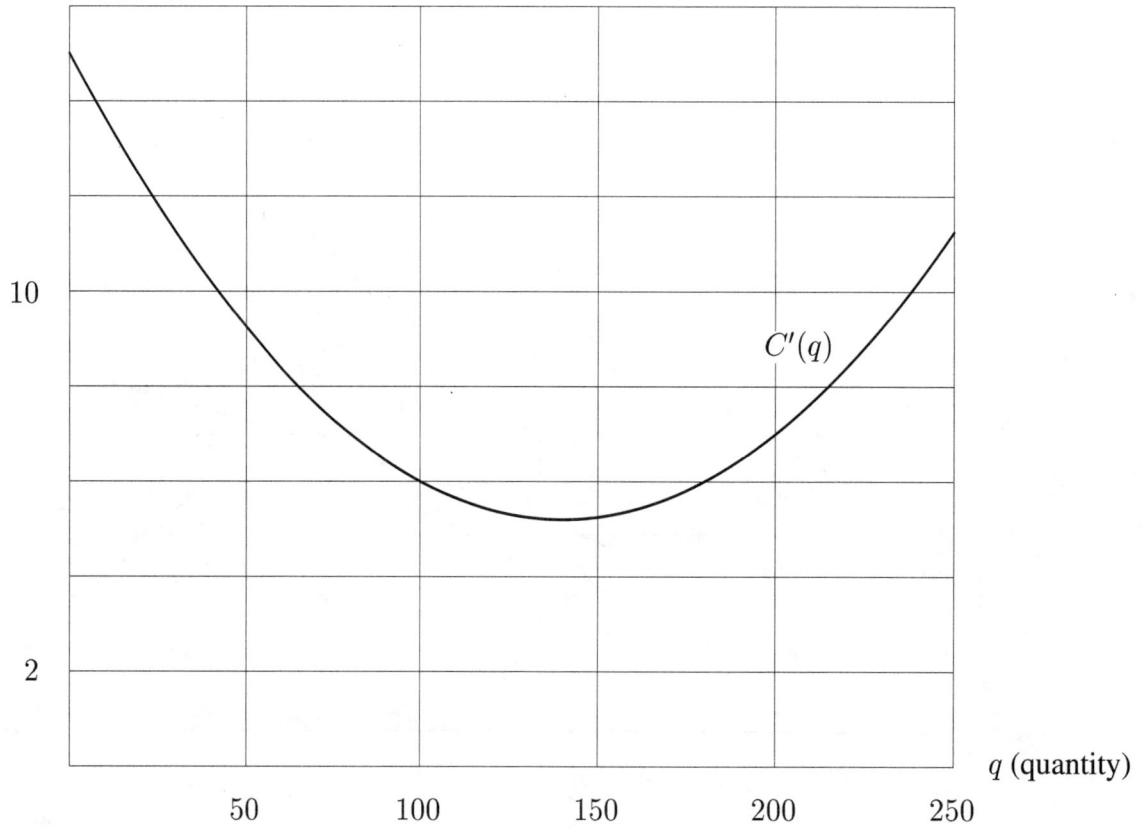

$ per item

$C'(q)$

10

2

q (quantity)

50   100   150   200   250

Transparency Master for Figure 6.5 for Problem 2 in Section 6.1

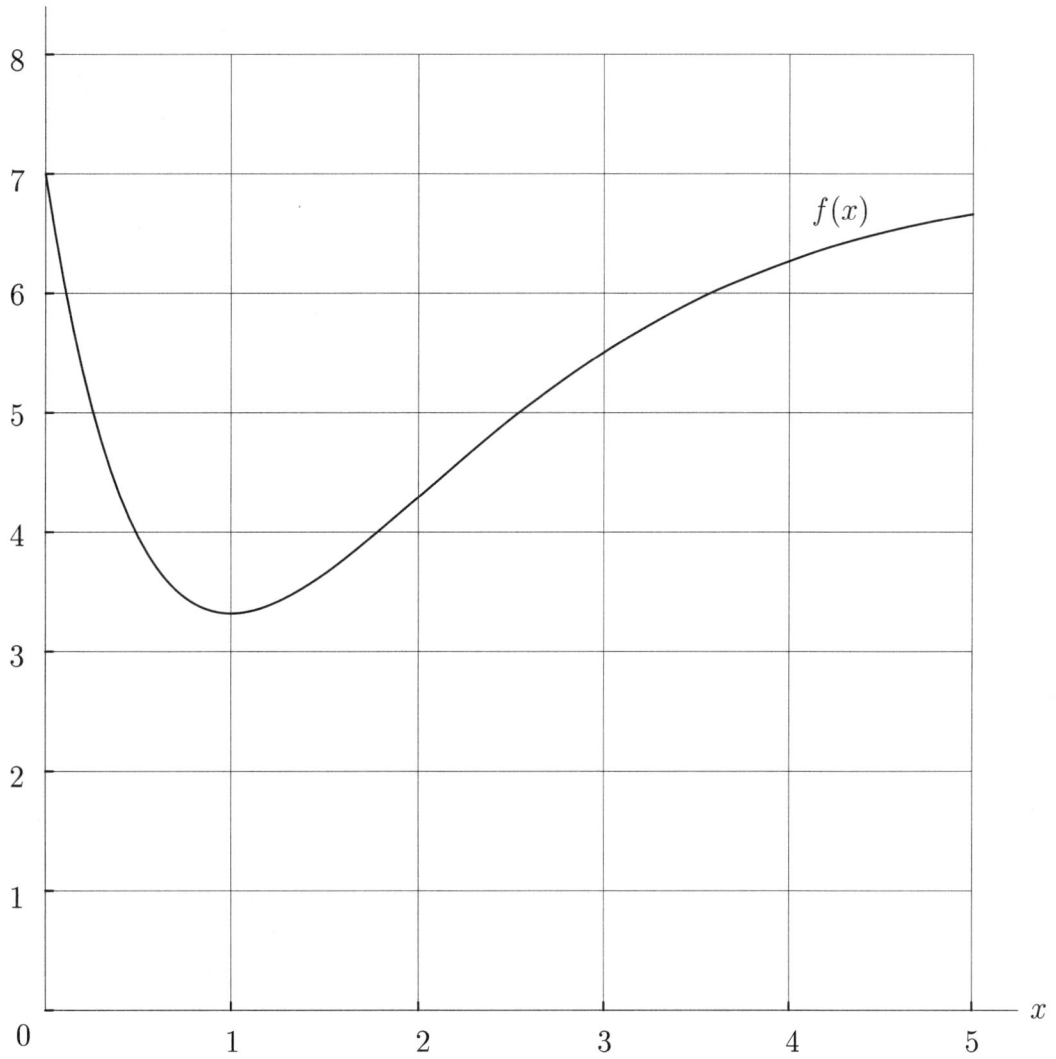

Transparency Master for Table 9.1 in Section 9.1

Table 9.1.1

		Number of full price tickets, x			
		100	200	300	400
Number of discount tickets, y	200	75,000	110,000	145,000	180,000
	400	115,000	150,000	185,000	220,000
	600	155,000	190,000	225,000	260,000
	800	195,000	230,000	265,000	300,000
	1000	235,000	270,000	305,000	340,000

Revenue from ticket sales (dollars)

Transparency Master for Figure 9.5 in Section 9.2

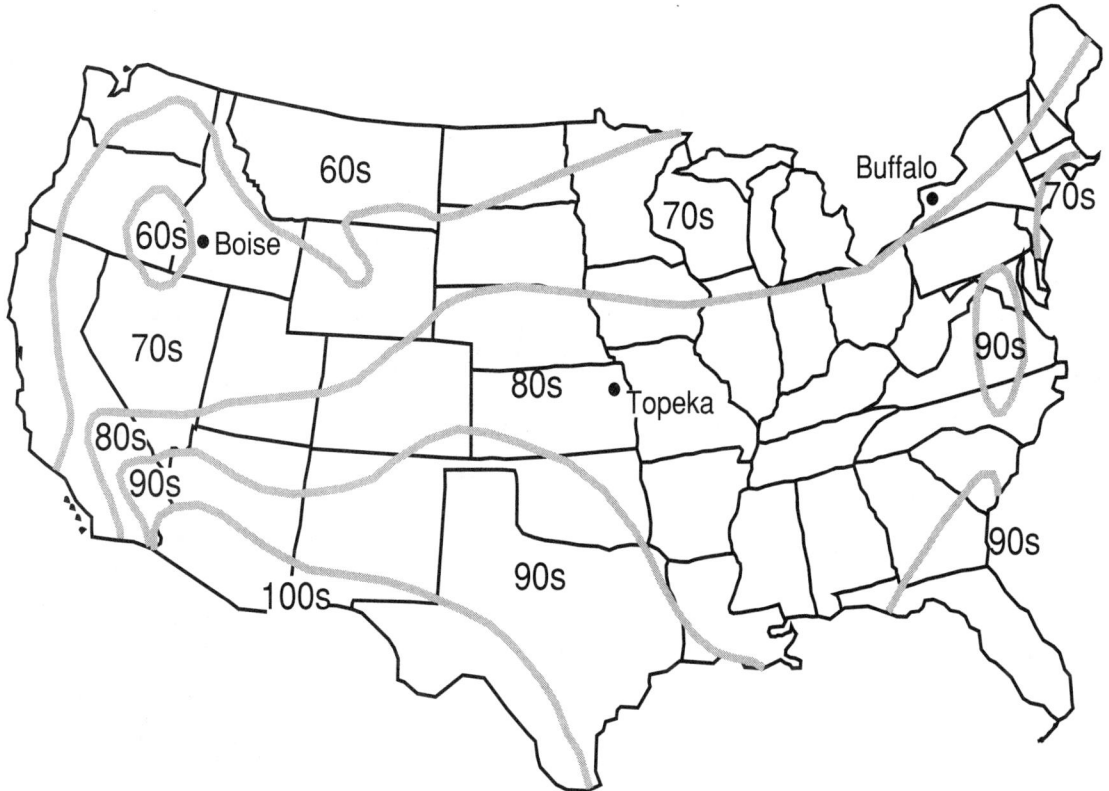

Transparency Master for Figures 9.7 and 9.8 for Example 2 in Section 9.2

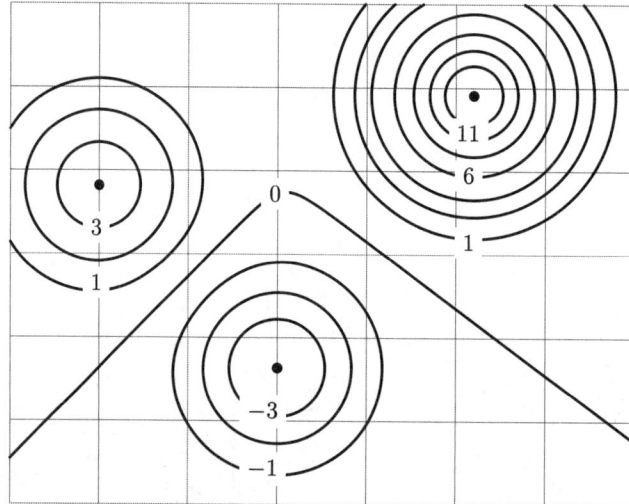

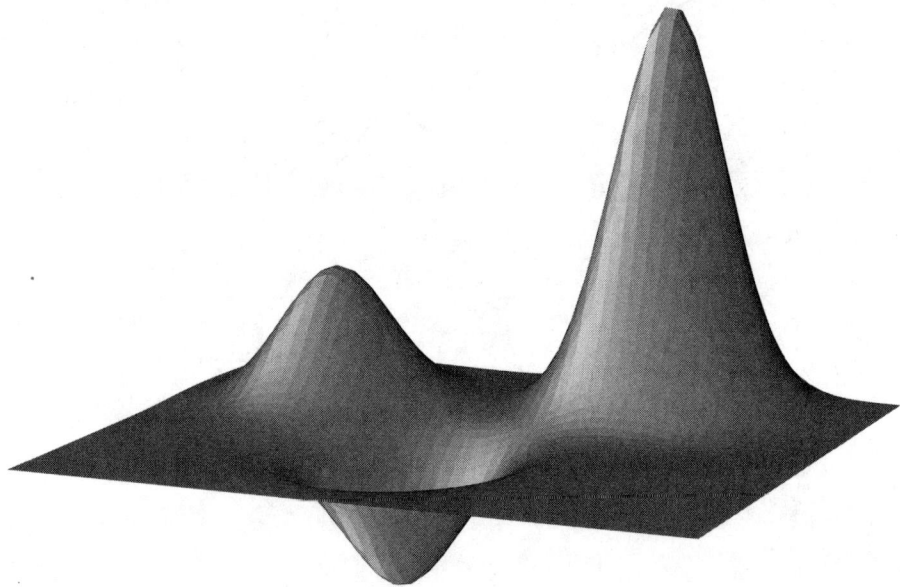

Transparency Master for Figure 9.14 in Section 9.2

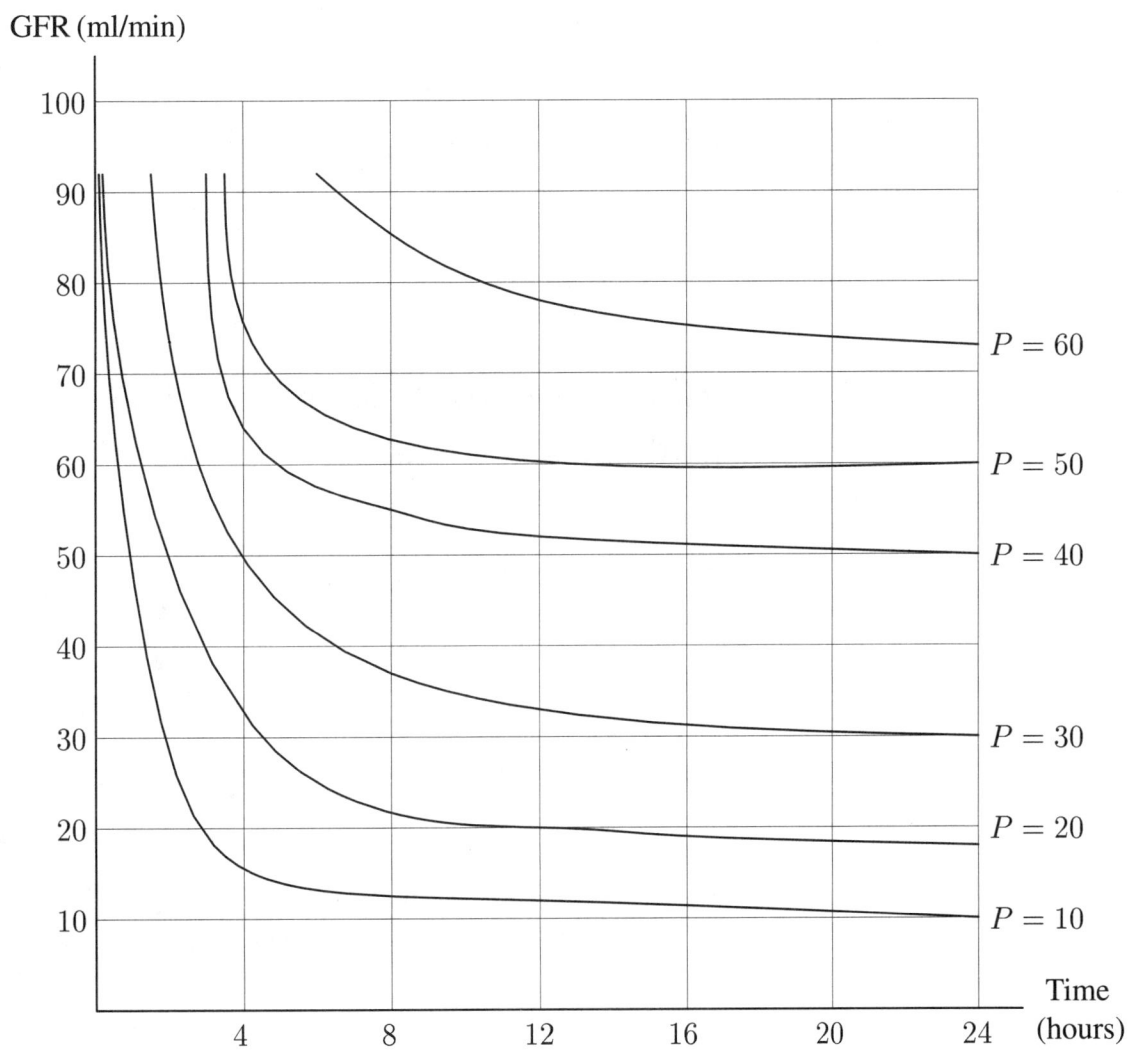

Table 9.1.2 *Quantity of beef bought (lbs/household/week)*

		p			
		3.00	3.50	4.00	4.50
	20	2.65	2.59	2.51	2.43
	40	4.14	4.05	3.94	3.88
I	60	5.11	5.00	4.97	4.84
	80	5.35	5.29	5.19	5.07
	100	5.79	5.77	5.60	5.53

Beef consumption in the US
(in pounds per household per week)

Transparency Master for Figure 9.25 for Problem 13 in Section 9.2

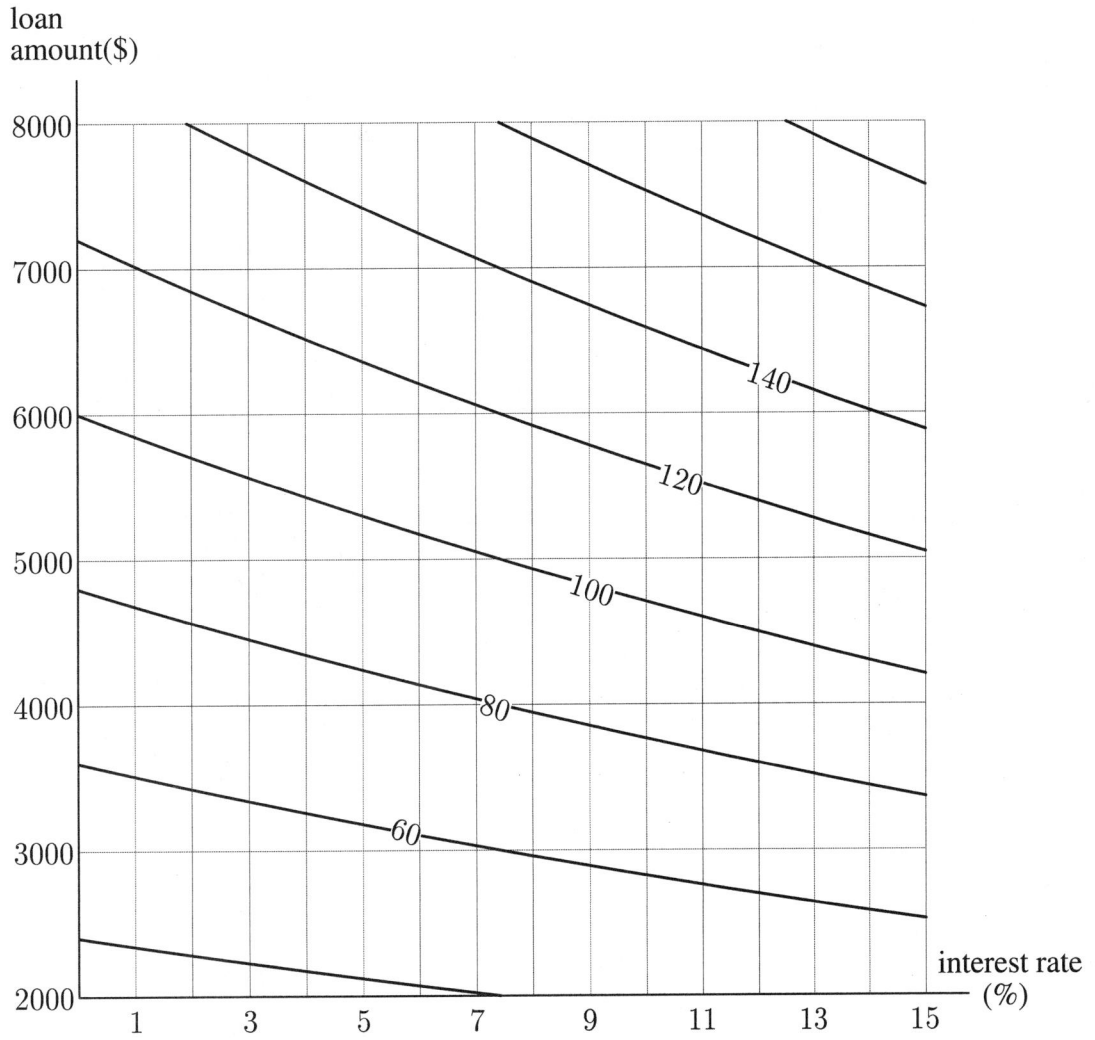

Transparency Master for Figure 9.26 for Problem 14 in Section 9.2

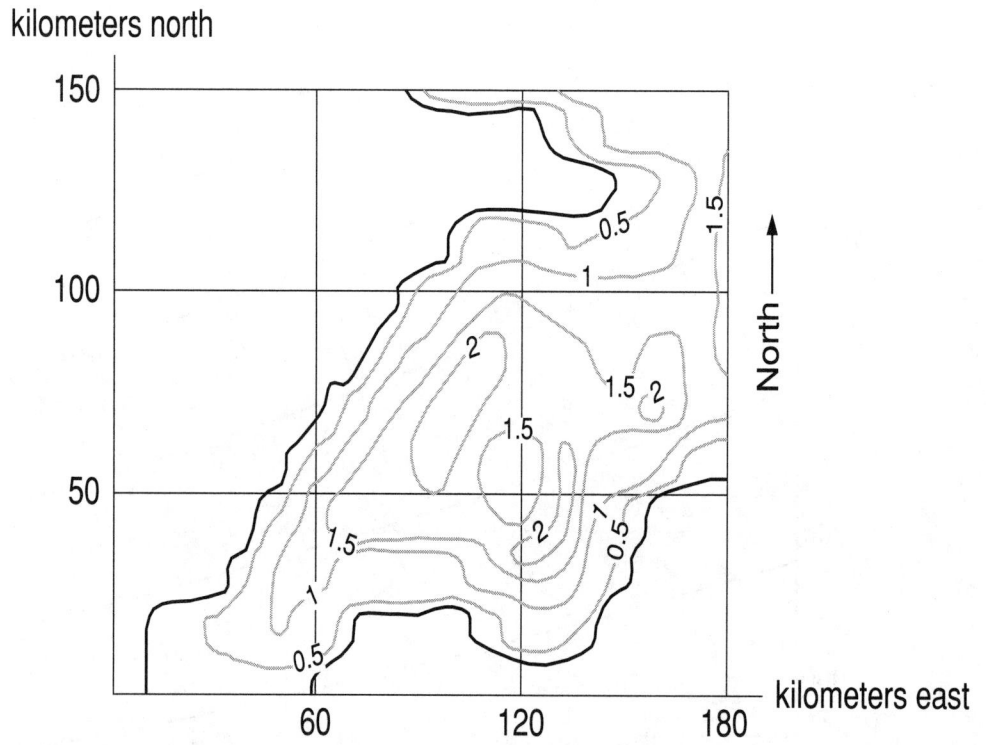

Transparency Master for Figure 9.36 for Problem 31 in Section 9.2

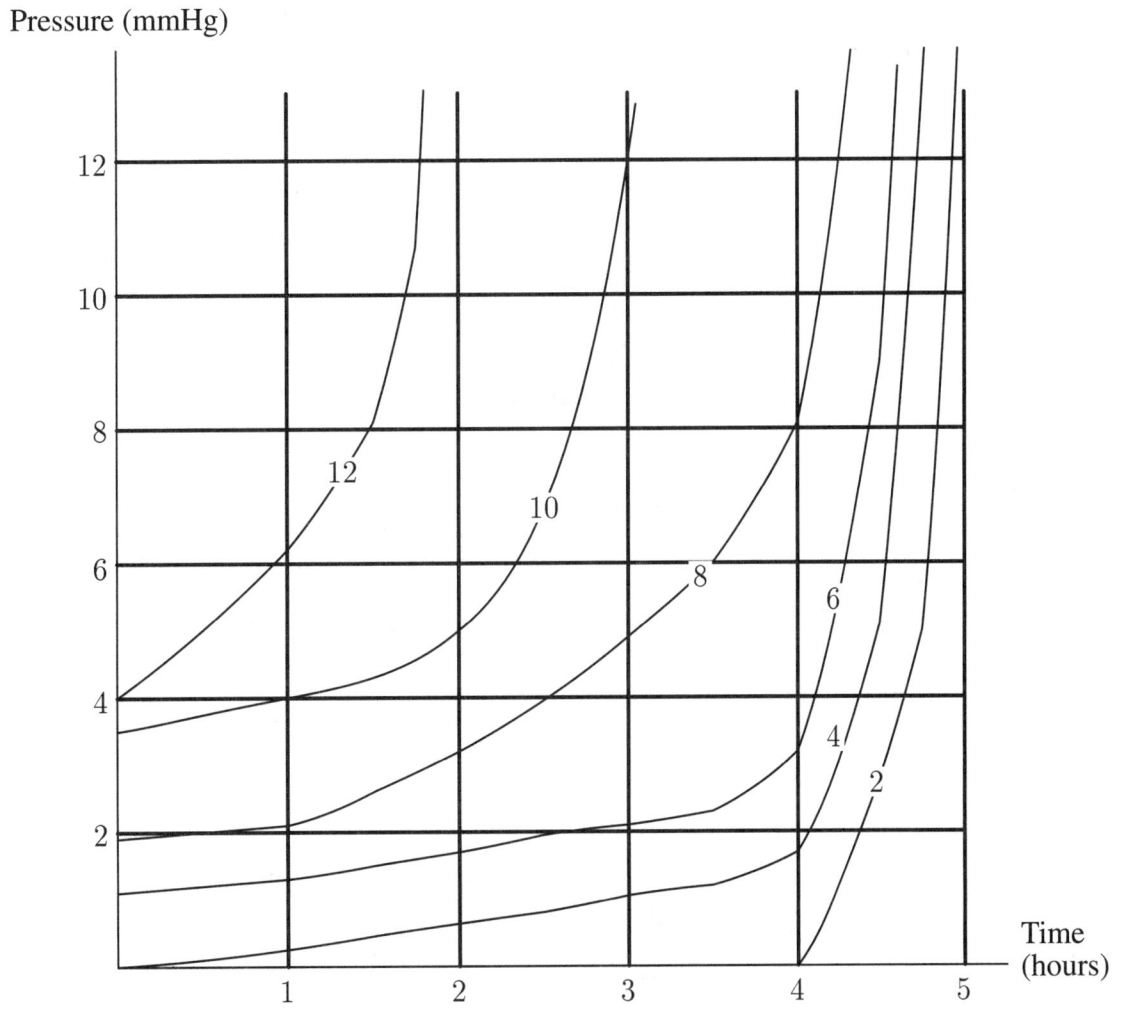

**Transparency Master for Figure 9.61 for Problem 1
in the Projects for Chapter 9**

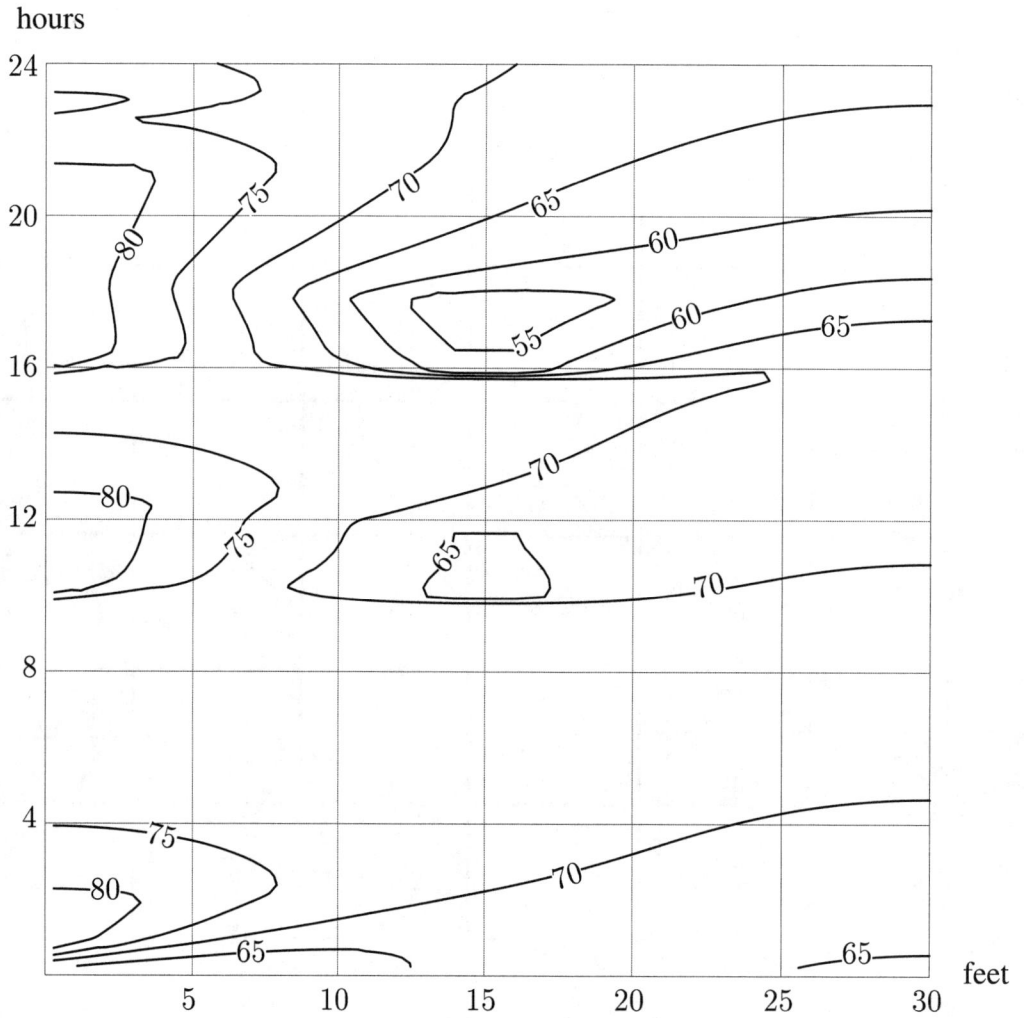

Transparency Master for Figure 10.14 for Example 3 in Section 10.3

(I)

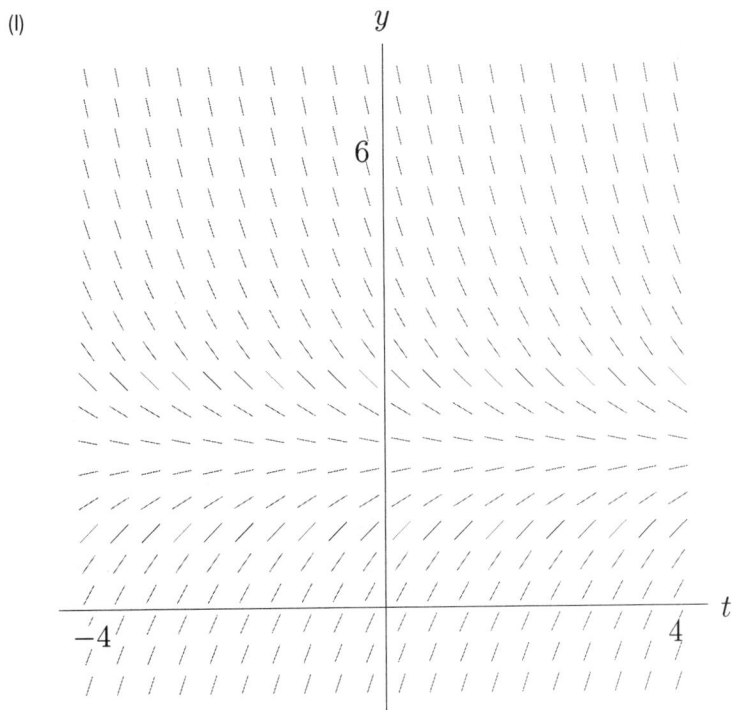

(II)

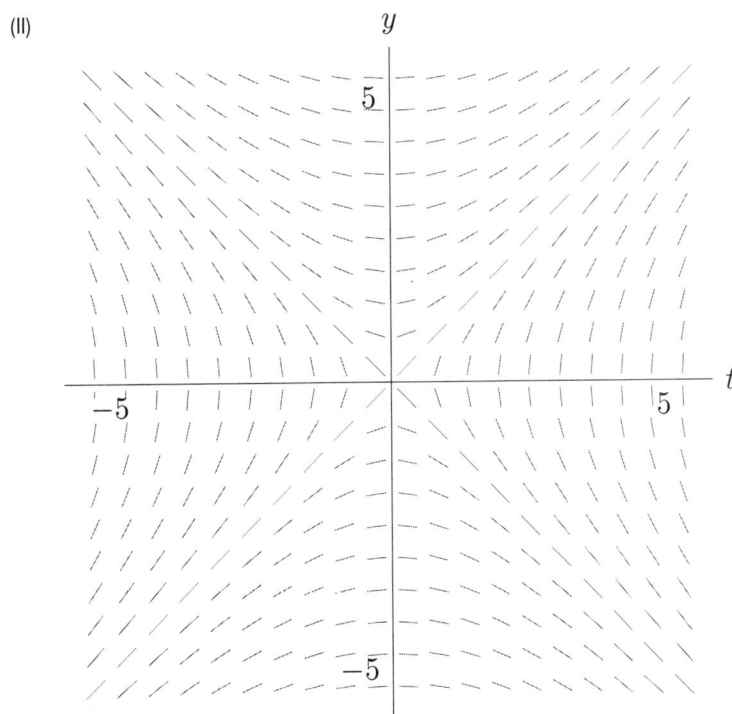

Transparency Master for Figure 10.40 in Section 10.6

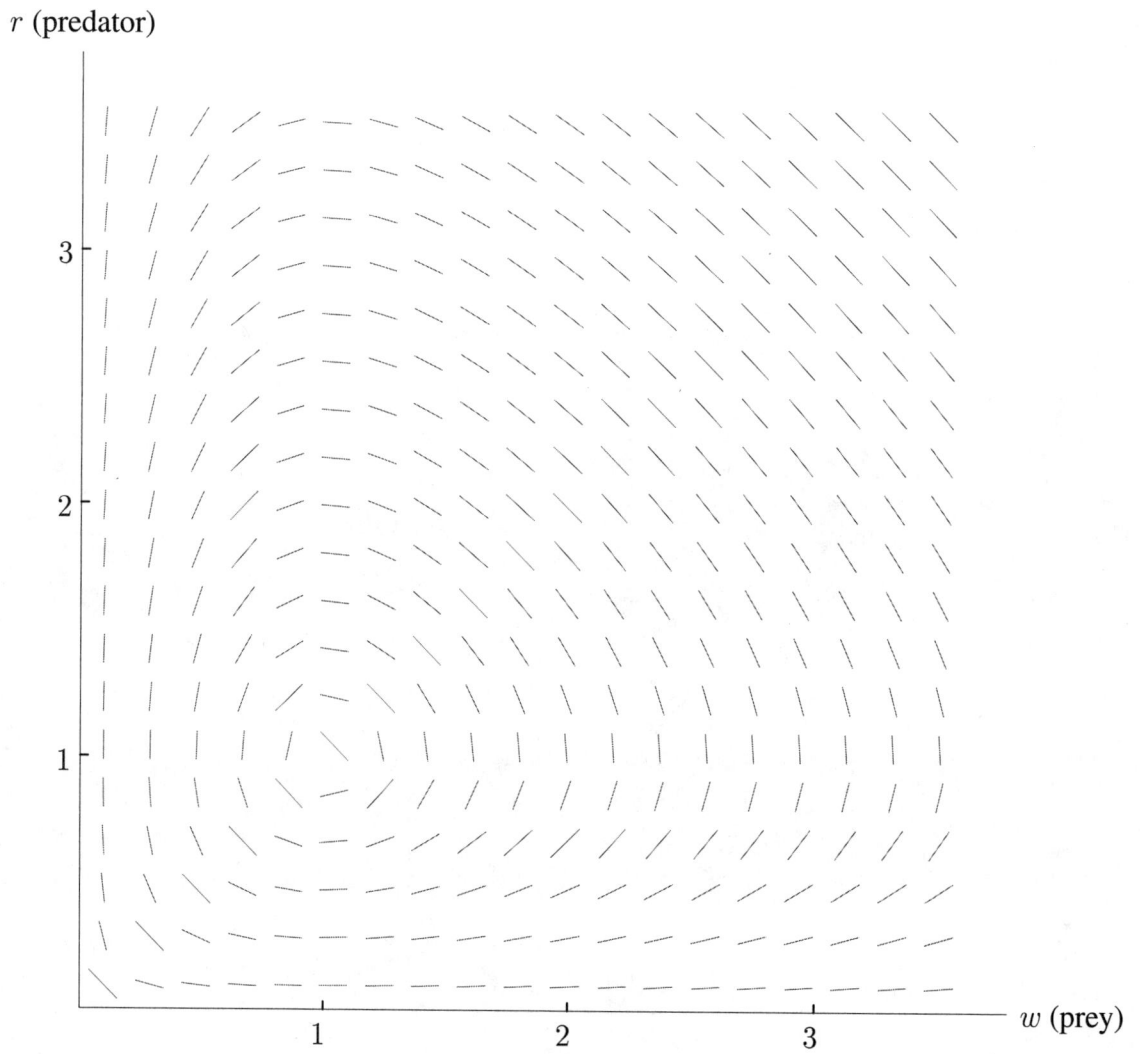

Transparency Master for Figures 10.45 and 10.46 in Section 10.7

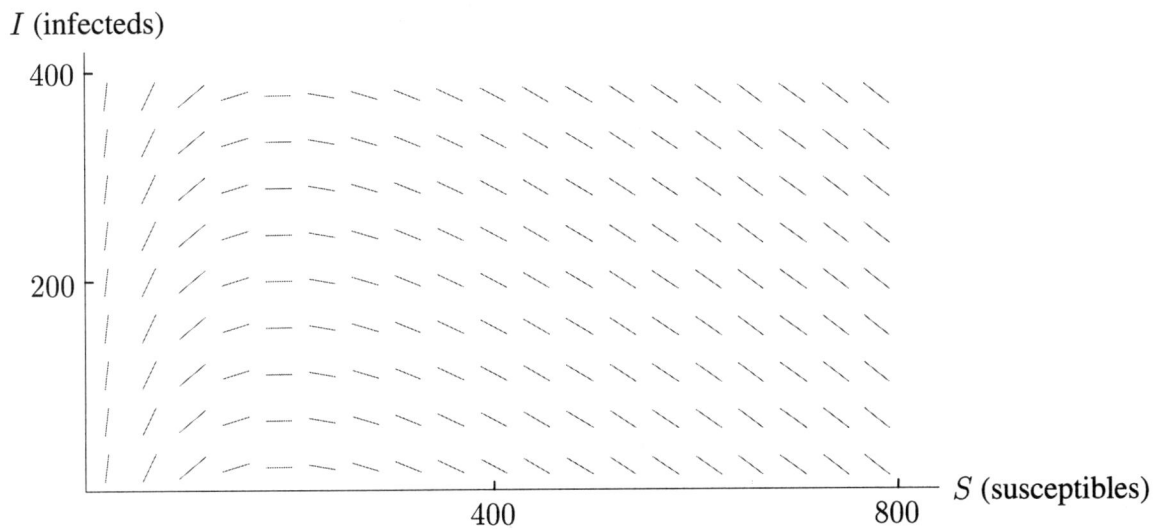

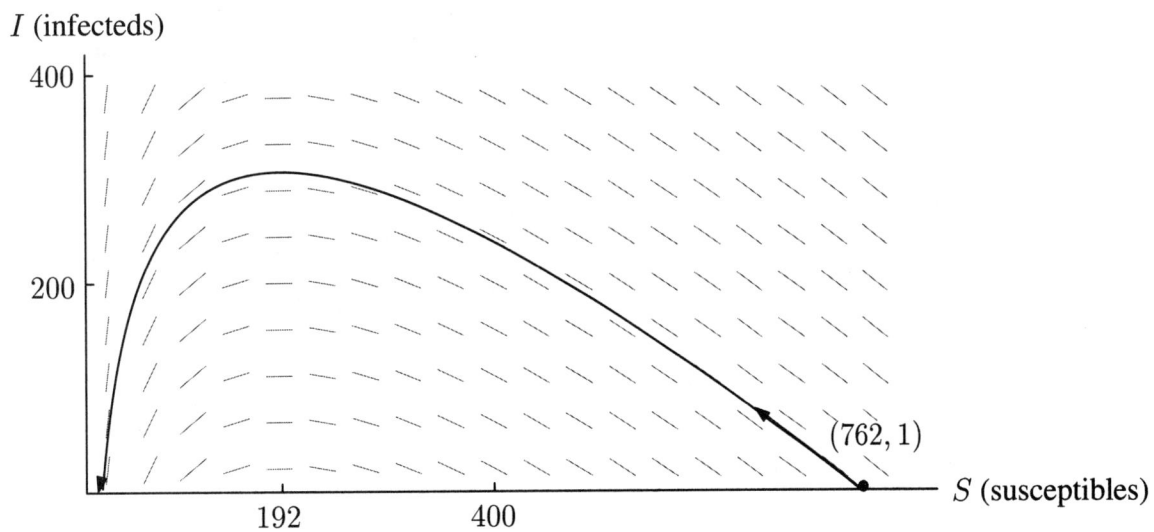

PART V

CALCULATOR PROGRAMS

CALCULATOR PROGRAMS

for the

TI - 81

Program to Calculate Riemann Sums to Evaluate a Definite Integral (TI-81)

Select 'PRGM' to get the program menu, move to 'EDIT' to enter a program. When you select a program number, you must first give it a name (for example, 'RSUMS') to the right of the program number.

Prgm: RSUMS	*Where to Find The Commands*
:Disp "LOWER LIMIT"	Disp and Input are accessed via PRGM, I/O, pressed while editing a program.
:Input A	The " " are on the $+$ key when you have pressed the ALPHA key.
:Disp "UPPER LIMIT"	
:Input B	
:Disp "DIVNS"	
:Input N	
:$A \rightarrow X$	$\rightarrow$ is on the STO key.
:$0 \rightarrow S$	0 is Zero, not oh.
:$1 \rightarrow I$	
:$(B - A)/N \rightarrow H$	
:Lbl P	Lbl is accessed via PRGM, CTL.
:$S + H * Y_1 \rightarrow S$	Y_1 is accessed via Y-VARS, i.e., 2nd VARS. (Do not use Y followed by 1, which means Y multiplied by 1.)
:$X + H \rightarrow X$	
:$IS > (I, N)$	$IS > ($ and Goto are accessed via PRGM, CTL.
:Goto P	
:Disp "LEFT SUM"	
:Disp S	
:$S + Y_1 * H \rightarrow S$	
:$A \rightarrow X$	
:$S - Y_1 * H \rightarrow S$	
:Disp "RIGHT SUM"	
:Disp S	

Things to watch for:
1. Difference between $-$, which means subtract, and (-), which means negative. For example, $-2 - 3$ must be entered as (-)2 $-$ 3.

2. Disp and Input are not typed in letter-by-letter, but are obtained by highlighting them under PRGM, I/O and hitting ENTER.

To run this program:
1. The integrand (the function you want to integrate) $f(x)$ must be entered into Y_1, with X as the independent variable.

2. Make sure the lower limit of the integral you're approximating is less than the upper limit.

3. Test on $\int_1^3 x^3 \, dx = 20$ with 100 subintervals. You should get left- and right-hand sums of 19.7408 and 20.2608, respectively.

Numerical Integration Program (TI-81)

This program calculates left- and right-hand Riemann sums, and the trapezoidal, midpoint and Simpson approximations. Since there's not room on the calculator to label each approximation separately, we use a compressed method of displaying the results. For instance, the label LEFT/RIGHT indicates that the next two numbers are the left- and right-hand Riemann sums, respectively.

Notes:

1. To enter a program, hit PRGM and select EDIT and a program number. To finish a line, hit ENTER; to finish editing a program, hit 2nd QUIT.

2. The function to be integrated must be entered as Y_1 (accessed by the "$Y =$" button). When Y_1 occurs in a program, it is evaluated at the current value of X.

3. The lower limit of integration must be less than the upper limit.

4. $IS > ($ means that the PRGM button must be pushed and then $IS > ($ selected, not that $I, S, >$ and $($ are to be entered separately. 'Disp' and 'Input' are to be found under PRGM, I/O, pushed while entering a program.

5. To run a program, select PRGM, EXEC. To stop a program while it is running, hit ON.

6. Test the program by evaluating $\int_1^3 x^3 \, dx = 20$, using 100 subdivisions. You should get left- and right-hand sums of 19.7408 and 20.2608, respectively. For the trapezoid approximation, you should get 20.0008. For the mid point approximation, you should get 19.9996. For Simpson's rule, you should get exactly 20.

Prgm: INTEGRAL	*Where to Find The Commands*
:Disp "LOWER LIMIT"	Disp and Input are accessed via PRGM, I/O
:Input A	Enter lower limit of integration.
:Disp "UPPER LIMIT"	
:Input B	Enter upper limit of integration.
:Disp "DIVNS"	
:Input N	Enter number subdivisions.
:$(B - A)/N \to H$	Stores size of one subdivision in H (Note that $\to$ means hit STO button).
:$A \to X$	Start X off at beginning of interval.
:$0 \to L$	Initialize L, which keeps track of left sums, to zero.
:$0 \to M$	Initialize M, which keeps track of midpoint sums, to zero.
:$1 \to I$	Initialize I, the counter for the loop.
:Lbl P	Label for top of loop. Lbl is accessed via PRGM, CTL.
:$L + H * Y_1 \to L$	Increment L by $Y_1 H$, the area of one more rectangle. (Y_1 is accessed via Y-VARS, or 2nd VARS.)
:$X + .5H \to X$	Move X to middle of interval.
:$M + H * Y_1 \to M$	Evaluate Y_1 at the middle of interval and increment M by rectangle of this height.
:$X + .5H \to X$	Move X to start of next interval.
:$IS > (I, N)$	$IS >$ (is accessed via PRGM, CTL. This is the most difficult step in the program: adds 1 to I and does the the next step if $I \leq N$ (i.e., if haven't gone through loop enought times); otherwise, skips next step. Thus, if $I \leq N$, goes back to Lbl P and loops through again. If $I > N$, loop is finished and goes on to print out results. Continue here if $I > N$, in which case the value of X is now B.
:Goto P	Goto is accessed via PRGM, CTL. Jumps back to Lbl P if $J \leq N$.
:Disp "LEFT/RIGHT"	
:Disp L	L now equals the left sum, so display it.
:$L + H * Y_1 \to R$	Add on area of right-most rectangle, store in R.
:$A \to X$	Reset X to A.
:$R - H * Y_1 \to R$	Subtract off area of left-most rectangle.
:Disp R	R now equals right sum, so display it.
:$(L + R)/2 \to T$	Trap approximation is average of L and R.
:Disp "TRAP/MID/SIMP"	
:Disp T	Display trap approximation.
:Disp M	Display midpoint approximation.
:$(2M + T)/3 \to S$	Simpson is weighted average of M and T.
:Disp S	Display Simpson's approximation.

CALCULATOR PROGRAMS

for the

TI - 82 and TI - 83

Introduction to the TI-82 and TI-83

The following are programs for the TI-82 calculators to do numerical integration, graph slope fields, demonstrate solutions to differential equations using Euler's method, and plot trajectories and solutions to systems of differential equations. Although these programs were originally written for the TI-82, they will work on the TI-83 as well.

Notes on Using the TI-82: The commands on the TI-82 are basically the same as on the TI-81, though they may be in different menus. Select 'PRGM' to get the program menu, move to 'NEW' to enter a new program. You must first give it a name (for example, 'RSUMS') when prompted; to finish entering/editing a program, hit 2nd QUIT. To delete a program, press 2nd MEM and select [2:Delete]. The quantity AB represents the product of A and B. Writing $Y_3 = Y_1 Y_2$ defines Y_3 as the product Y_1 and Y_2. Writing $Y_3 = Y_1(Y_2)$ defines Y_3 as the composition of Y_1 and Y_2.

Program to Calculate Riemann Sums to Evaluate a Definite Integral (TI-82)

Select 'PRGM' to get the program menu, move to 'NEW' to enter a new program. You must first give it a name (for example, 'RSUMS') when prompted; to finish entering/editing a program, hit 2nd QUIT.

Name= RSUMS	*Where to Find The Commands*
:Disp "LOWER LIMIT"	Disp and Input are accessed via PRGM, I/O, pressed
	while in the middle of a program.
:Input A	The " " are on the + key when you have
	pressed the ALPHA key.
:Disp "UPPER LIMIT"	
:Input B	
:Disp "DIVNS"	
:Input N	
:$A \rightarrow X$	$\rightarrow$ is on the STO key.
:$0 \rightarrow S$	0 is Zero, not oh.
:$1 \rightarrow I$	
:$(B - A)/N \rightarrow H$	
:Lbl P	Lbl is accessed via PRGM, CTL.
:$S + H * Y_1 \rightarrow S$	Y_1 is accessed via Y-VARS, i.e., 2nd VARS.
	(Do not use Y followed by 1, which means
	Y multiplied by 1.)
:$X + H \rightarrow X$	
:$IS > (I, N)$	$IS > ($ and Goto are accessed via PRGM, CTL.
:Goto P	
:Disp "LEFT SUM"	
:Disp S	
:$S + Y_1 * H \rightarrow S$	
:$A \rightarrow X$	
:$S - Y_1 * H \rightarrow S$	
:Disp "RIGHT SUM"	
:Disp S	

Things to watch for:
1. Difference between $-$, which means subtract, and (-), which means negative. For example, $-2 - 3$ must be entered as (-)2 $- 3$.

2. Disp and Input are not typed in letter-by-letter, but are obtained by highlighting them under PRGM, I/O and hitting ENTER.

To run this program:
1. The integrand (the function you want to integrate) $f(x)$ must be entered into Y_1, with X as the independent variable.

2. Make sure the lower limit of the integral you're approximating is less than the upper limit.

3. Test on $\int_1^3 x^3 \, dx = 20$ with 100 subintervals. You should get left- and right-hand sums of 19.7408 and 20.2608, respectively.

Numerical Integration Program (TI-82)

This program calculates left- and right-hand Riemann sums, and the trapezoidal, midpoint and Simpson approximations. Since there's not room on the calculator to label each approximation separately, we use a compressed method of displaying the results. For instance, the label LEFT/RIGHT indicates that the next two numbers are the left- and right-hand Riemann sums, respectively.

Notes:
1. Select 'PRGM' to get the program menu, move to 'NEW' to enter a new program. You must first give it a name (for example, 'INTEGRAL') when prompted; to finish entering/editing a program, hit 2nd QUIT.
2. The function to be integrated must be entered as Y_1 (accessed by the "$Y =$" button). When Y_1 occurs in a program, it is evaluated at the current value of X.
3. The lower limit of integration must be less than the upper limit.
4. $IS > ($ means that the PRGM button must be pushed and then $IS > ($ selected, not that $I, S, >$ and $($ are to be entered separately. 'Disp' and 'Input' are to be found under PRGM, I/O, pushed while entering a program.
5. To run a program, select PRGM, EXEC. To stop a program while it is running, hit ON.
6. Test the program by evaluating $\int_1^3 x^3\, dx = 20,$ using 100 subdivisions. You should get left- and right-hand sums of 19.7408 and 20.2608, respectively. For the trapezoid approximation, you should get 20.0008. For the mid point approximation, you should get 19.9996. For Simpson's rule, you should get exactly 20.

Name= INTEGRAL	*Where to Find The Commands*
:Disp "LOWER LIMIT"	Disp and Input are accessed via PRGM, I/O
:Input A	Enter lower limit of integration.
:Disp "UPPER LIMIT"	
:Input B	Enter upper limit of integration.
:Disp "DIVNS"	
:Input N	Enter number subdivisions.
:$(B - A)/N \to H$	Stores size of one subdivision in H (Note that $\to$ means hit STO button).
:$A \to X$	Start X off at beginning of interval.
:$0 \to L$	Initialize L, which keeps track of left sums, to zero.
:$0 \to M$	Initialize M, which keeps track of midpoint sums, to zero.
:$1 \to I$	Initialize I, the counter for the loop.
:Lbl P	Label for top of loop. Lbl is accessed via PRGM, CTL.
:$L + H * Y_1 \to L$	Increment L by $Y_1 H$, the area of one more rectangle. (Y_1 is accessed via Y-VARS, or 2nd VARS.)
:$X + .5H \to X$	Move X to middle of interval.
:$M + H * Y_1 \to M$	Evaluate Y_1 at the middle of interval and increment M by rectangle of this height.
:$X + .5H \to X$	Move X to start of next interval.
:$IS > (I, N)$	$IS > ($ is accessed via PRGM, CTL. This is the most difficult step in the program: adds 1 to I and does the the next step if $I \le N$ (i.e., if haven't gone through loop enought times); otherwise, skips next step. Thus, if $I \le N$, goes back to Lbl P and loops through again. If $I > N$, loop is finished and goes on to print out results. Continue here if $I > N$, in which case the value of X is now B.
:Goto P	Goto is accessed via PRGM, CTL. Jumps back to Lbl P if $J \le N$.
:Disp "LEFT/RIGHT"	
:Disp L	L now equals the left sum, so display it.
:$L + H * Y_1 \to R$	Add on area of right-most rectangle, store in R.
:$A \to X$	Reset X to A.
:$R - H * Y_1 \to R$	Subtract off area of left-most rectangle.
:Disp R	R now equals right sum, so display it.
:$(L + R)/2 \to T$	Trap approximation is average of L and R.
:Disp "TRAP/MID/SIMP"	
:Disp T	Display trap approximation.
:Disp M	Display midpoint approximation.
:$(2M + T)/3 \to S$	Simpson is weighted average of M and T.
:Disp S	Display Simpson's approximation.

CALCULATOR PROGRAMS

for the

TI - 85

Introduction to the TI-85

The following are programs for the TI-85 calculator to do numerical integration, graph slope fields, demonstrate solutions to differential equations using Euler's method, and plot trajectories and solutions to systems of differential equations.

Notes on Using the TI-85: When you want to multiply two quantities, say A and B, on the TI-85 you must use "*". On a TI-81, AB means $A * B$, whereas on a TI-85 AB is a single variable. If you can't find a command, try 2nd, CATALOG, which gives a list of all the commands. Press a letter to go quickly to the commands beginning with that letter. Press ENTER to select. Letters can be both upper and lower case. Upper case letters are obtained by hitting ALPHA; lower case by hitting 2nd ALPHA. To enter a function (for example, $y1 =$) under the GRAPH menu, lower case variables should be used.

Program to Calculate Riemann Sums to Evaluate a Definite Integral (TI-85)

Select 'PRGM' to get the program menu, then select 'EDIT' to enter a program. When you enter a program, you must first give it a name (for example, 'RSUMS'). To finish editing, hit EXIT.

Name=RSUMS	*Where to Find The Commands*
:Disp "LOWER LIMIT"	Disp, " ", and Input are accessed via I/O, pressed while in
	the middle of a program.
:Input A	
:Disp "UPPER LIMIT"	
:Input B	
:Disp "DIVNS"	
:Input N	
:$A \rightarrow x$	$\rightarrow$ is on the STO key; use x-VAR key for x.
:$0 \rightarrow S$	0 is Zero, not "oh".
:$1 \rightarrow I$	
:$(B-A)/N \rightarrow H$	
:Lbl P	Lbl is accessed via CTL or CATALOG.
:$S + y1 * H \rightarrow S$	$y1$ is typed in by entering 2nd ALPHA Y and then 1.
:$x + H \rightarrow x$	
:$IS > (I, N)$	$IS >$ (and Goto are accessed via CTL or CATALOG.)
:Goto P	
:Disp "LEFT SUM"	
:Disp S	
:$S + y1 * H \rightarrow S$	
:$A \rightarrow x$	
:$S - y1 * H \rightarrow S$	
:Disp "RIGHT SUM"	
:Disp S	

Things to watch for:
1. Difference between $-$, which means subtract, and (-), which means negative. For example, $-2 - 3$ must be entered as (-)2 $- 3$.
2. Disp and Input can also be typed in letter-by-letter.

To run this program:
1. The integrand (the function you want to integrate) $f(x)$ must be entered into y_1, under the GRAPH menu, with x as the independent variable.
2. Make sure the lower limit of the integral you're approximating is less than the upper limit.
3. Test on $\int_1^3 x^3 \, dx = 20$ with 100 subintervals. You should get left- and right-hand sums of 19.7408 and 20.2608, respectively.

Numerical Integration Program (TI-85)

This program calculates left and right Riemann sums, and the trapezoidal and midpoint approximations. Since there's no room on the calculator for a separate labeling of each approximation, we use a compressed method of displaying the results. For instance, the label LEFT/RIGHT indicates that the next two numbers are the left- and right-hand Riemann sums, respectively.

Notes:
1. Select 'PRGM' to get the program menu, then select 'EDIT' to enter a program. When you enter a program, you must first give it a name (for example, 'INTEG'). To finish editing, hit EXIT.

2. The function to be integrated must be entered as $y1$ (accessed by GRAPH, followed by $y(x) =$). When $y1$ occurs in a program, it is evaluated at the current value of x.

3. The lower limit of integration must be less than the upper limit.

4. $IS > ($ is selected from under CTL, while enter entering the program. "Disp" and "Input" are selected from under I/O.

5. Use MORE to see items on a menu which are currently off the screen.

6. To run a program, select PRGM, NAMES. To stop a program while it is running, hit ON.

7. Test the program by evaluating $\int_1^3 x^3 \, dx = 20,$ using 100 subdivisions. You should get left- and right-hand sums of 19.7408 and 20.2608, respectively. For the trapezoid approximation, you should get 20.0008. For the mid point approximation, you should get 19.9996. For Simpson's rule, you should get exactly 20.

Name=INTEG	*Where to Find The Commands*
:Disp "LOWER LIMIT"	Disp, Input, and " " are accessed via PRGM, I/O.
:Input A	Enter lower limit of integration.
:Disp "UPPER LIMIT"	
:Input B	Enter upper limit of integration.
:Disp "DIVNS"	
:Input N	Enter number subdivisions.
:$(B - A)/N \to H$	Stores size of one subdivision in H. (Note that $\to$ means hit STO button).
:$A \to x$	Start x off at beginning of interval.
:$0 \to L$	Initialize L, which keeps track of left sums, to zero.
:$0 \to M$	Initialize M, which keeps track of right sums, to zero.
:$1 \to I$	Initialize I, the counter for the loop.
:Lbl P	Label for top of loop. Lbl is accessed via CTL
:$L + H * y1 \to L$	Increment L by $H * y1$, the area of one more rectangle. ($y1$ is typed in as y and then 1.)
:$x + 0.5H \to x$	Move x to middle of interval.
:$M + H * y1 \to M$	Evaluate $y1$ at the middle of interval and increment M by a rectangle of this height.
:$x + 0.5H \to x$	Move x to start of next interval.
:$IS > (I, N)$	Access $IS > ($ from under CTL. This is the most difficult step in the program: adds 1 to I and does the next step if $I \leq N$ (i.e., if haven't gone through loop enought times); otherwise, skips next step. Thus, if $I \leq N$, goes back to Lbl 1 and loops through again. If $I > N$, loop is finished and goes on to print out results.
:Goto P	Goto is accessed via CTL. Jumps back to Lbl P if $I \leq N$.
:Disp "LEFT/RIGHT"	
:Disp L	L now equals the left sum, so display it.
:$L + H * y1 \to R$	Add on area of right-most rectangle, store in R.
:$A \to x$	Reset x to A.
:$R - H * y1 \to R$	Subtract off area of left-most rectangle.
:Disp R	R now equals right sum, so display it.
:$(L + R)/2 \to T$	Trap approximation is average of L and R.
:Disp "TRAP/MID/SIMP"	
:Disp T	Display trap approximation.
:Disp M	Display midpoint approximation.
:$(2 * M + T)/3 \to S$	Simpson is weighted average of M and T.
:Disp S	Display Simpson's approximation.

CALCULATOR PROGRAMS

for the

CASIO fx-7700GB

Introduction to the CASIO fx-7700GB

The following are programs for the CASIO fx-7700GB calculator to do numerical integration, graph slope fields, demonstrate solutions to differential equations using Euler's method, and plot trajectories and solutions to systems of differential equations.

Notes on Using the CASIO fx-7700GB:

- All the Casio integration programs call on the function you put in f_1. To store a function in f_1, first select the Function Memory Menu by pressing SHIFT then F·MEM. Clear the screen by pressing AC if needed. Type out the function, then press STO (F1 key) followed by 1. Press LIST (F4 key) to see the list of stored functions.

- You can put either a colon (:) or a carriage return (EXE) after each instruction (to separate them), except after display sign ▶, which provides its own carriage return.

Riemann Sums Program (CASIO)

To enter the program, press 'MODE' then '2' to select 'WRT' mode. Move the cursor to an empty program number, then press 'EXE'. You'll see a blank screen with the blinking cursor at the upper left corner. Now you can proceed to the beginning of the program. When finished, press 'MODE' then '1' to get back to 'RUN' mode.

Program	*Where to Find The Commands*
"RSUMS"	
"L-LIM" ?$\to$ A	'?' is accessed by SHIFT, PRGM
"R-LIM" ?$\to$ B	
"DIVNS" ?$\to$ N	
A $\to$ X	
0 $\to$ S	
$f_1 \to$ Y	'f_1' is accessed by SHIFT, $\boxed{F}$MEM, f_n, 1
(B$-$A)div N $\to$ H	
Lbl 1	'Lbl' is in JMP menu
X + H $\to$ X	
f_1 + S $\to$ S	
Dsz N	'Dsz' is in JMP menu
Goto 1	'Goto is also in JMP menu
"L-SUM="	
$(S - f_1 + Y)$H$\blacktriangleright$	
"R-SUM="	
HS$\blacktriangleright$	

To run this program:
1. The integrand (the function you want to integrate) $f(x)$ must be entered into f$_1$, with X as the independent variable.

2. Make sure the lower limit of the integral you're approximating is less than the upper limit.

3. Test on $\int_1^3 x^3\,dx = 20$ with 100 subintervals. You should get left- and right-hand sums of 19.7408 and 20.2608, respectively.

Numerical Integration Program (CASIO)

This is a **Casio fx series** calculator program for various numerical integrals. It will display the Left and Right Riemann Sums and the Trapezoid Rule, Midpoint Rule, and Simpson's Rule approximations all at once. The way this program evaluates integrals is by keeping a running total of function values on n subintervals, and then multiplying by the width of the rectangles to obtain the area at the very end. At the end of the program, hitting EXE will let you reevaluate the integral with a different number of subdivisions, and hitting AC will let you out of the program.

Program	Comments
"INTEGRAL"	
"L-LIM"?$\rightarrow$A	Integrate from X=a
"U-LIM"?$\rightarrow$B	to X=b
"DIVNS"?$\rightarrow$N	over N subdivisions.
(B$-$A)div (2N)$\rightarrow$H	calculates half the width of a subdivision.
0$\rightarrow$L	initialize L, which will keep track of the left sums.
0$\rightarrow$M	initialize M, the midpoint sum.
A$\rightarrow$X	place X at A, the beginning of the interval.
Lbl 1	top of loop:
f_1+L$\rightarrow$L	evaluate the function at the left edge of the interval add the result to the left-hand sum running total.
X+H$\rightarrow$X	move X to the middle of the interval.
f_1+M$\rightarrow$M	evaluate the function at the middle of the interval add the result to the midpoint running total.
X+H$\rightarrow$X	move X to the beginning of the next interval.
Dsz N	decrease N by 1; if N=0, skip the next step and go on.
Goto 1	bottom of loop.
"LEFT,RIGHT,TRAP"	
2HL$\rightarrow$L	multiply the sum of the left-hand function values by width 2H.
L$\blacktriangleright$	display the left-hand sum.
L+2H$f_1$$\rightarrow$T	evaluate the function at X=b: the rightmost function value add the area of the rightmost rectangle with the left-hand sum.
A$\rightarrow$X	put X back at A.
T$-$2H$f_1$$\rightarrow$T	evaluate the function at the left-most edge of the interval take the area of the leftmost rectangle out of T.
T$\blacktriangleright$	display what is now the right-hand sum.
(L+T)div 2$\rightarrow$T	average the left- and right-hand sums.
T$\blacktriangleright$	display the trapezoid approximation.
"MID,SIMP"	
2HM$\rightarrow$M	multiply the midpoint values sum by the interval width.
M$\blacktriangleright$	display the midpoint approximation.
(2M+T)div 3$\rightarrow$M	calculate Simpson's Rule by weighted averaging.
M	display Simpson's Rule approximation.

CALCULATOR PROGRAMS

for the

SHARP EL-9200 and EL-9300

Introduction to the Sharp EL-9200 and EL-9300

The following programs are for the Sharp EL-9200 and Sharp EL-9300 calculators. Enclosed you will find programs to do numerical integration, graph slope fields, demonstrate solutions to differential equations using Euler's method, and plot trajectories and solutions to systems of differential equations.

Notes:

1. All the following programs are in REAL mode.

2. To access the commands, first press 2ndF COMMAND, then select them via the appropriate menus.

3. After editing a line in a program, make sure to press ENTER or the button with downward pointing triangle before you quit; otherwise the changes will not be saved.

4. Since the user-defined functions (Y_1 ... etc.) are not shared by different programs, the formula for a function has to be given when it is used in a program. Thus, each programs has a subroutine that starts with the line "Label eqn" and ends with "Return", where the desired function should be entered.

5. Variables are case sensitive. Single uppercase letters (A to Z) are global variables, i.e., the values stored in memories designated by single letter A to Z can be shared by different programs. Lowercase letters and lowercase words are local variables, i.e., the values of local variables are specific to the program in which they're used. You can use a string of up to 12 lowercase letters to designate a local variable. Note also that you can not mix uppercase and lowercase letters to form a variable.

Riemann Sums Program (SHARP)

Select NEW in the program menu. Then select REAL in the MODE menu. When prompted for a title, use "riemann." The "..." in the following program is where the integrand–the function to be integrated–should be entered.

Program	Where the Commands are
Goto start	Goto is in BRANCH menu
Label eqn	Label is in BRANCH menu
f=...	replace "..." by the integrand $f(x)$
Return	Return is in BRANCH menu
Label start	
Print "l-limit	Print and " are in PROG menu
Input a	Input is in PROG menu
Print "u-limit	
Input b	
Print "divns	
Input n	
x=a	= is also in INEQ menu
s=0	reset memory s to zero
i=1	
h=(b−a)/n	
Label 1	
Gosub eqn	Gosub is in BRANCH menu
s=s+f*h	
x=x+h	
i=i+1	
If i ¡= n Goto 1	If and Goto are in BRANCH menu; ¡= is in INEQ menu
Print "left sum	
Print s	
Gosub eqn	
s=s+f*h	
x=a	
Gosub eqn	
s=s−f*h	
Print "right sum	
Print s	
End	End is in PROG menu

To run this program:

1. The integrand (the function you want to integrate) $f(x)$ must be entered into where "..." is, with x as the independent variable.

2. Make sure the lower limit of the integral you're approximating is less than the upper limit.

3. Test on $\int_1^3 x^3\, dx = 20$ with 100 subintervals. You should get left- and right-hand sums of 19.7408 and 20.2608, respectively.

Numerical Integration Program (SHARP)

This program calculates left- and right-hand Riemann sums, and the trapezoidal, midpoint and Simpson approximations. Since there's not room on the calculator to label each approximation separately, we use a compressed method of displaying the results. For instance, the label "left/right" indicates that the next two numbers are the left- and right-hand Riemann sums, respectively.

To enter the program select NEW in the program menu. Then select REAL in the MODE menu. When prompted for a title, use "integral."

Program	Where the Commands are
Goto start	Goto is in BRANCH menu
Label eqn	Label is in BRANCH menu
f=...	replace "..." by the integrand $f(x)$
Return	Return is in BRANCH menu
Label start	
Print "l-limit	Print and " are in PROG menu
Input a	Input is in PROG menu
Print "u-limit	
Input b	
Print "divns	
Input n	
x=a	= is also in INEQ menu
s=0	reset memory s to zero
m=0	reset memory m to zero
i=1	
h=(b−a)/n	
Label 1	
Gosub eqn	Gosub is in BRANCH menu
s=s+f*h	
x=x+ .5h	
Gosub eqn	
m=m+f*h	
x=x+ .5h	
i=i+1	
If i ¡= n Goto 1	If and Goto are in BRANCH menu; ¡= is in INEQ menu
Print "left/right	
Print s	
Gosub eqn	
r=s+f*h	
x=a	
Gosub eqn	
r=r−f*h	
Print r	
Wait	Wait is in PROG menu
Print "trap/mid/simp	
t=.5(s+r)	
Print t	
Print m	
s=(2m+t)/3	
Print s	
End	End is in PROG menu

To run this program:

1. The integrand (the function you want to integrate) $f(x)$ must be entered into where "..." is, with x as the independent variable.

2. Make sure the lower limit of the integral you're approximating is less than the upper limit.

3. Test on $\int_1^3 x^3\,dx = 20$ with 100 subintervals. You should get left- and right-hand sums of 19.7408 and 20.2608, respectively.

CALCULATOR PROGRAMS

for the

HP-48S, HP-48G and HP-38G

Riemann Sums for the HP-48S/G

The following is a directory for experimenting with different Riemann sums. To use a directory press VAR to get the user's menu and then press the name of directory on menu keys, in this case RSUM. To leave the directory when you are finished, press (left shift) UP. To create a directory, type the name, say 'RSUM' followed by CRDIR on the MEMORY menu.

RSUM Directory

These short programs can be entered by hand or transferred via infrared from another HP-48. Programs are given in the order you will find most convenient to use on the menu. The name of each program is given before the program. As you enter each program, store it under the given name. Thus for first program, type ¡¡ ANS − ¿¿ followed by ENTER and then type 'ERR' and press STO.

ERR ¡¡ ANS − ¿¿
LFT ¡¡ A SUM ¿¿
RGT ¡¡ A H + SUM ¿¿
MID ¡¡ A H 2 / + SUM ¿¿
TRP ¡¡ LFT B F A F − H * 2 / + ¿¿
SMP ¡¡ MID 2 * TRP + 3 / ¿¿
NSTO ¡¡ 'N' STO B A − N / 'H' STO ¿¿
ABSTO ¡¡ 'B' STO 'A' STO ¿¿
FSTO ¡¡ 'F(X)' SWAP = DEFINE ¿¿
SUM ¡¡ → X 'H*Σ(I=0, N−1,F(X+I*H))' ¿¿

The other variables used by these programs will also be on your menu: A, B, F, N, H, ANS. To make sure your menu is in the most convenient order, just use FSTO, ABSTO, and NSTO once, which will creat variables A, B, F, N, and H. Also do 0 'ANS' STO to create a variable called ANS. Then press (left shift) {} and press menu buttons to get this:

{ERR LFT RGT MID TRP SMP NSTO ABSTO FSTO ANS}

Press ENTER. Then type in ORDER followed by ENTER (or find ORDER on the MEMORY menu and press the menu button).

How to Use Directory RSUM

Here is an example. If you want to experiment with $\int_0^1 \frac{1}{1+x^2}\,dx$ with $N = 10$ subdivisions do this:

Enter '1/(1 + X^2)' and press FSTO.
Enter 0. Enter 1. Press ABSTO.
Enter 10 and press NSTO.

Now press LFT, RGT, MID, TRP, SMP to get the left, right, midpoint, trapezoid, and Simpson approximations for the given integral. If you know the actual value of the integral, store that value in ANS. Then when you press ERR the value of ANS will be subtracted from whatever is at level one of the stack. Thus, if you press MID followed by ERR, the error for the midpoint approximation will appear on the stack. This is especially useful if you are trying to study how the errors of the different methods are related to each other and to N.

Riemann Integration on the HP 38G

Enter the programming environment on the HP 38G by pressing (shift) PROGRAM. There you will see a list of all programs in the machine, plus the name Editline. To enter a new program, press **[NEW]** on the menu bar. You will be prompted for a name for the program; we call this one R.INT. Type in R.INT and press **[ENTER]**. A screen titled R.INT PROGRAM will appear. Type in the following program; you may look for commands on the menus, but it is far easier to type the letters one by one. To type a string of capital letters, hold down the [**A...Z**] button and type letters; lower case letters must be typed with (shift) a...z before each one. The symbol ▶ is typed by pressing [**STO** ▶] on the menu. Other special symbols (", =, ?) are typed from the Character Browser, accessed by typing (shift) CHARS. To get a new line, press [**ENTER**]. Indentation is done here for ease in reading; on the machine, it is optional.

Program **R.INT**:

```
SELECT Function:              0 ▶ L:
2 ▶ Format:                   FOR I=1 TO N STEP 1;
9 ▶ Digits:                       L+F1(X) ▶ L:
MSGBOX                            X+H ▶ X:
    "STORE INTEGRAND IN F1":  END:
INPUT A;                      L+F1(X) ▶ R:
    "LOWER LIMIT";            L*H ▶ L:
    "A";                      A ▶ X:
    "ENTER LOWER LIMIT";      R-F1(X) ▶ R:
    0:                        R*H ▶ R:
INPUT B;                      MSGBOX
    "UPPER LIMIT;                 "LHS=" L
    "B";                          "RHS=" R
    "ENTER UPPER LIMIT";          "AVE=" (L+R)/2:
    1:                        1 ▶ C:
DO                            CHOOSE C;
    INPUT N;                      "DO IT AGAIN?";
        "SUBINTERVALS";           "YES";
        "N;                       "NO":
        "ENTER NUMBER OF      UNTIL
SUBINTERVALS";                    C==2
    10:                       END:
    (B-A)/N ▶ H:              1 ▶ Format:
    A ▶ X:
```

The program is saved automatically. To put the integrand in F1, press [**LIB**] and highlight Function; then press [**START**] on the menu bar, highlight F1, and press [**EDIT**], then type in the formula. To run the program, press (shift) PROGRAM and press [**RUN**].

PART VI

SAMPLE EXAM QUESTIONS

Chapter 1 Exam Questions

Problems and Solutions   for Section 1.1

1. $f(x)$ is the age of antarctic ice (in hundreds of years) at a depth of x meters below the surface.
   (a) Give in words the practical meaning of the equation $f(10) = 15$.
   (b) Is f increasing or decreasing, and why?

   ANSWER:

   (a) The antarctic ice at a depth of 10 meters below the surface is 1500 years old.
   (b) f is increasing, since the deeper the ice is, the older it is.

2. The empirical function $W = f(t)$, given in the graph below, comes from the <u>Wall Street Journal</u>, September 4, 1992. From the graph, describe the domain of this function and the range of this function. In a sentence, apply the general definition of the word "function" to explain why you think that the given curve is in fact a function.

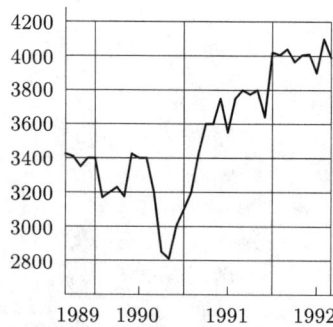

   ANSWER:

   Domain: September 1989 to August 1992. Range: 2810 (approx.) $\leq W \leq$ 4090 (approx.). For every date in the domain, there is a unique value of W.

3. Consider a ten-story building with a single elevator. From the point of view of a person on the sixth floor, sketch a graph indicating the height of the elevator as a function of time as it travels. Remember to indicate when it stops. Try to take into account all *types* of cases that can happen, but do not worry about *every* possible situation. (There are many different possible graphs that could be drawn for this.)

   ANSWER:

   A possible diagram: An elevator first goes from the ground floor to the third floor, then to the eighth floor, and finally back to the ground floor.

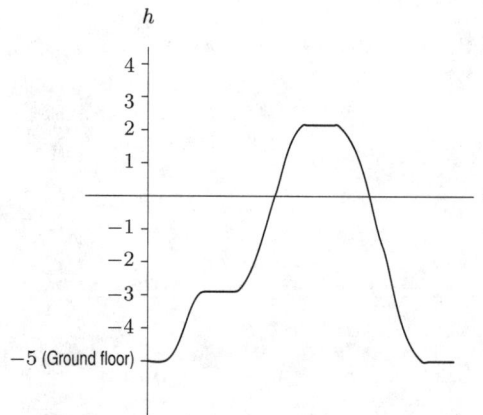

4. Draw a graph which accurately represents the temperature of the contents of a cup left overnight in a room. Assume the room is at $70°$ and the cup is originally filled with water slightly above the freezing point.

ANSWER:

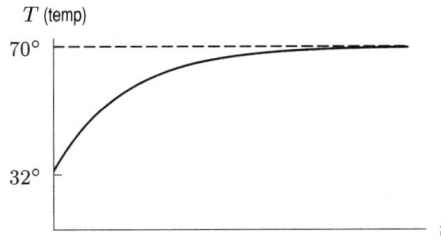

5. Suppose the Long Island Railroad train from Easthampton to Manhattan leaves at 4:30 pm and takes two hours to reach Manhattan, waits two hours at the station and then returns, arriving back in Easthampton at 10:30 pm. Draw a graph representing the distance of the train from the Farmingdale station in Easthampton as a function of time from 4:30 pm to 10:30 pm. The distance from Easthampton to Manhattan is 150 miles.

ANSWER:

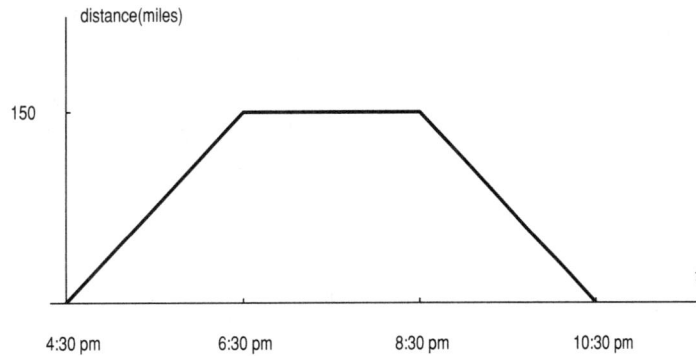

6. A graph of $y = f(x)$ is given in Figure 1.1.1.

   (a) What is $f(4)$?
   (b) What is the domain of this function?
   (c) What is the range of this function?

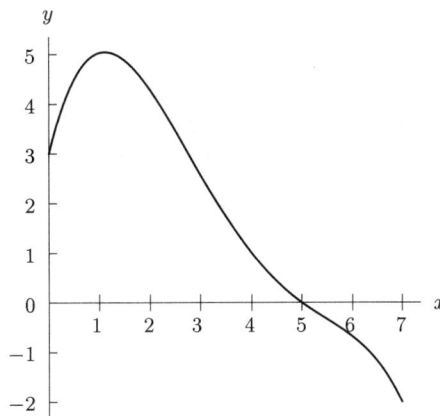

Figure 1.1.1

ANSWER:

   (a) $f(4) = 1$.
   (b) The domain is $0 \le x \le 7$.
   (c) The range is $-2 \le y \le 5$.

7. An object is put outside on a cold day and its temperature, H, in $°C$ is a function of the time, t, in minutes since it was put outside. A graph of the function is given in Figure 1.1.2.

   (a) What does the statement $f(30) = 10$ mean in terms of temperature? Include units for 30 and for 10 in your answer.

   (b) Explain in terms of temperature of the object and the time outside what each of the following represents: (i) the vertical intercept a (ii) the horizontal intercept b

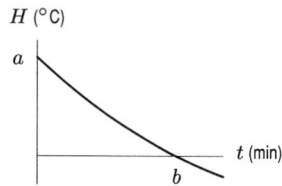

Figure 1.1.2

ANSWER:

(a) $f(30) = 10$ means that 30 minutes after the object is first put outside, its temperature is $10°C$.

(b)   (i) The vertical intercept a represents the value of H when $t = 0$, in other words, the temperature of the object when it is first put outside.

    (ii) The horizontal intercept b represents the value of t when $H = 0$, in other words, the number of minutes that it takes (after the object is first put outside) for its temperature to drop to $0°C$.

8. The empirical function $P = g(t)$ graphed below represents the population P of a city (in thousands of people) at time t. Describe the domain and range of this function.

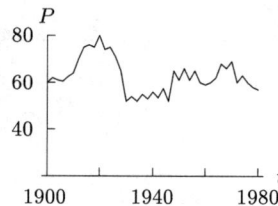

Figure 1.1.3

ANSWER:

Domain: 1900 to 1980. Range: $80,000 \le P \le 48,000$ (approximately).

9. A pond has a population of 500 frogs. Over a ten-year period of time the number of frogs drops quickly by 50%, then increases slowly for 5 years before dropping to almost zero. Sketch a graph to represent the number of frogs in the pond over the ten-year period of time.

   ANSWER:

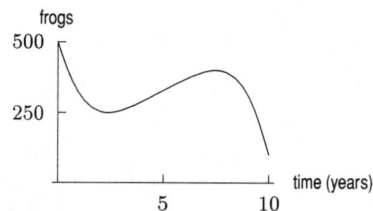

Figure 1.1.4

Problems and Solutions   for Section 1.2

1. A function is linear for $x \leq 2$ and also linear for $x \geq 2$. This function has the following values: $f(-4) = 3$; $f(2) = 0$; $f(4) = 6$. (a) Find formula(s) (or equation(s)) which describe this function. (b) Graph the function.

   ANSWER:

   (a) For $x \leq 2$, the slope is $\dfrac{-3}{6} = \dfrac{-1}{2}$ and the y-intercept is 1; thus $y = \dfrac{-1}{2}x + 1$. For $x \geq 2$, the slope is $\dfrac{6}{2} = 3$. To find the y-intercept we substitute: $6 = 3(4) + b$, $6 - 12 = b$, so that $b = -6$. Hence, $y = 3x - 6$.

   This is an example of a *piece-wise function*: $f(x) = \begin{cases} -\dfrac{1}{2}x + 1 & \text{when } x \leq 2 \\ 3x - 6 & \text{when } x \geq 2 \end{cases}$

   (b)

   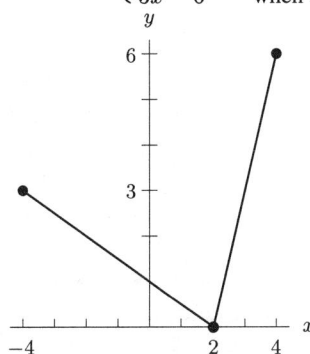

2. The average weight in pounds of American men in their sixties (in 1979) as a function of their heights in inches is given in Table 1.2.1.

   Table 1.2.1

height (h)	68	69	70	71	72	73
weight (w)	166	171	176	181	186	191

   (a) How do you know from the table that weight w is a linear function of height h?
   (b) What is the slope of the function? (Be sure to give the units of measurement as part of your answer.)
   (c) Explain the meaning of the slope in terms of heights and weights.
   (d) Find a formula that expresses the weight w in terms of the height h.

   ANSWER:

   (a) When height goes up by 1 inch, weight always goes up by the same amount, namely 5 pounds.
   (b) 5 pounds per inch.
   (c) For American men in their sixties in 1979, each extra inch of height corresponds to about 5 extra pounds of weight.
   (d) $w = 166 + 5(h - 68) = -174 + 5h$

3. Suppose that $y = f(t)$ is the distance in miles traveled in t hours by a car moving at 50 miles per hour.

   (a) Give a table of values for the function $f(t)$.
   (b) Give a formula for the function $f(t)$.
   (c) Draw the graph of the funtion $f(t)$. Label the axes and include a scale.

   ANSWER:

   (a)

t time in hours	0	1	2	3	4	5
$f(t)$ distance in miles	0	50	100	150	200	250

   (b) $f(t) = 50t$

(c)

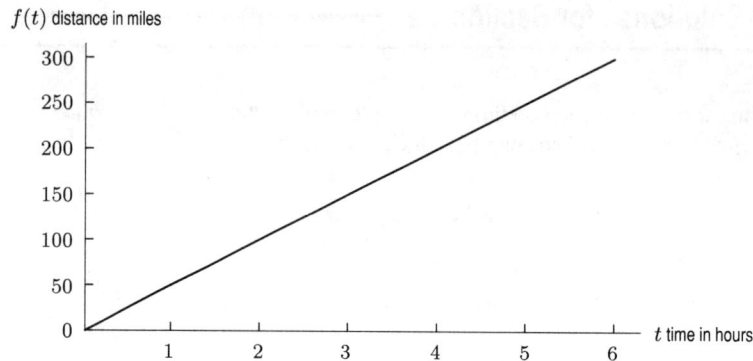

4. Complete the tables of values for the three linear functions f, g, and h:

(a)

x	0	1	2	3	4
$f(x)$	10	20	?	?	?

(b)

x	0	5	10	15	20
$g(x)$	10	20	?	?	?

(c)

x	0	100	200	300	400
$h(x)$	10	20	?	?	?

ANSWER:

(a)

x	0	1	2	3	4
$f(x)$	10	20	30	40	50

(b)

x	0	5	10	15	20
$g(x)$	10	20	30	40	50

(c)

x	0	100	200	300	400
$h(x)$	10	20	30	40	50

5. Find formulas for the three linear functions in Problem 4.

ANSWER:

(a) $f(x) = 10x + 10$
(b) $g(x) = 2x + 10$
(c) $h(x) = 0.1x + 10$

6. A car is worth \$14,000 when it is one year old, and it is worth \$9000 when it is three years old.

(a) Write the value of the car, V (in dollars), as a linear function of the age of the car, a (in years).
(b) What is the slope of your line? Interpret it in terms of the value of the car.
(c) What is the vertical intercept of your line? Interpret it in terms of the value of the car.

ANSWER:

(a) The function will be of the form $V = b + ma$, since it is linear. To find m, we use the formula $m = \dfrac{y - y_0}{x - x_0}$:

$$m = \frac{9000 - 14{,}000}{3 - 1} = \frac{-5000}{2} = -2500.$$

Thus we have $V = b - 2500a$. To find b, we substitute in the known point $(1, 14{,}000)$ and solve the resulting equation, $14{,}000 = b - (2500)(1)$, to get $b = 16{,}500$. So the function is $V = 16{,}500 - 2500a$.

(b) The slope of the line is -2500. This represents the amount in dollars that the car depreciates in value each year.
(c) The vertical intercept of the line is 16,500. This represents the value of the car in dollars when it is first purchased.

7. Find the equation of the line through the points $(3, 2)$ and $(5, -4)$.

ANSWER:

The equation will be of the form $y = m + bx$. We use the equation $m = \dfrac{y - y_0}{x - x_0}$ to find m:

$$m = \frac{-4 - 2}{5 - 3} = \frac{-6}{2} = -3.$$

Substituting in the known point $(3, 2)$ gives the equation:

$$y - 2 = -3(x - 3)$$
$$y - 2 = -3x + 9$$
$$y = 11 - 3x.$$

8. The bill for electricity is \$150 when 40 kilowatt hours are used and is \$225 when 70 kilowatt hours are used.

(a) Find a linear function for the electricity bill as a function of the number of kilowatt hours.
(b) What is the slope? Give units with your answer and interpret it in terms of the cost of electricity.
(c) What is the vertical intercept? Give units with your answer and interpret it in terms of the cost of electricity.

ANSWER:

(a) The equation will be of the form $E = m + bk$, where E is the electricity bill and k is the number of kilowatt hours used. We use the equation $m = \dfrac{E - E_0}{k - k_0}$ to find m:

$$m = \frac{225 - 150}{70 - 40} = \frac{75}{30} = 2.5.$$

Substituting in the known point $(40, 150)$ gives the equation:

$$E - 150 = 2.5(k - 40)$$
$$E - 150 = 2.5x - 100$$
$$E = 50 + 2.5k.$$

(b) The slope is 2.5 \$/kilowatt hour. This means that each additional kilowatt hour used costs \$2.50.
(c) The vertical intercept is \$50. This means that the base cost (without using any electricity) is \$50.

9. A school library opened in 1980. In January, 2000 they had 30,000 books. One year later, they had 30,480 books. Assuming they acquire the same number of books at the start of each month:

(a) How many books did they have in January, 2003?
(b) How many books did they have in July of 1980?
(c) Find a linear formula for the number of books, N, in the library as a function of the number of years t the library has been open.
(d) If you graph the function with domain 1980-2010, describe in words what the y-intercept of the graph means.

ANSWER:

(a) They acquire $30,480 - 30,000 = 480$ books per year. In January, 2003 the library will have 3×480 more books than they did in January, 2000 for a total of $31,440$ books.
(b) From part (a), the number of books the library acquires each year is 480. January, 1980 was 20 years before January, 2000, therefore the number of books the library had in January, 1980 was $30,000 - (20 \times 480) = 20,400$. From part (a), the number of books acquired each month is $480/12 = 40$. By July, 1980 the library acquired $6 \times 40 = 240$ books therefore the total number of books in the library will be $20,400 + 240 = 20,640$.
(c) We find the slope m and the intercept b in the linear equation $N = b + mt$. From part (a), we use $m = 480$. We substitute to find b:

$$30,000 = b + (480)(20)$$

$$b = 20,400$$

The linear formula is $N = 20,400 + 480t$.
(d) The y-intercept is the number of books the library had in 1980.

10. A furniture moving company charges a fixed amount plus a charge for each pound that they move. A person who shipped 50 lbs of furniture was charged $300 while someone else was charged $780 to move 210 lbs.

   (a) Write a function that represents the moving cost, C, in terms of pounds, x and fixed cost.
   (b) What does the y-intercept represent?
   (c) The company changes their rates. They increase the per pound charge by $2 but cut the fixed amount they charge by half. What is the new function that represents the new moving cost D?
   (d) Will someone who ships 210 lbs pay more or less with the new rates than they would have with the original rates?

   ANSWER:

   (a) Charge per pound = Δ dollars/Δ lbs = (780 - 300)/(210 - 50) = $ 3.00/ lb.

$$C = 3x + b$$

   Substitute one pair of values

$$300 = 3(50) + b$$

$$b = \$150$$

   So,
$$C = 3x + 150$$

   (b) The y-intercept represents the fixed cost charged by the moving company.
   (c) $D = 5x + 75$.
   (d) With the new rates,

$$D = 5(210) + 75$$

$$D = \$1125$$

   So it is more expensive to ship 210 lbs with the new rates than it was with the old rates.

Problems and Solutions   for Section 1.3

1. The population of Los Angeles, California was 2,811,801 in 1970 and was 3,448,613 in 1994.[1] Give the change and the average rate of change in the population of Los Angeles between 1970 and 1994. Include units with both your answers and write a sentence explaining the meaning of each answer.

   ANSWER:

   The change is $3,448,613 - 2,811,801 = 636,812$ people. The average rate of change is found by dividing the change by the number of years in which it occured:

$$\frac{3,448,613 - 2,811,801}{1994 - 1970} = \frac{636,812}{24} = 26,534 \text{ people per year.}$$

So between 1970 and 1994 the population of Los Angeles increased by 636,812 people, which represents an average annual increase of 26,534 people per year.

2. The number of reported offenses of violent crime in the US between 1983 and 1996 is given in Table 1.3.2.[2]

   (a) Find the average rate of change between 1983 and 1992.
   (b) Find the average rate of change between 1992 and 1996.
   (c) Write a sentence explaining what the two numbers tell you about the rate of change in reported offenses of violent crime in the US between 1983 and 1996.

   ---

   [1] *The World Almanac and Book of Facts 1998*, p. 385 (Mahwah, NJ: K-111 Reference Corporation).
   [2] *The World Almanac and Book of Facts 1998*, p. 958 (Mahwah, NJ: K-111 Reference Corporation).

Table 1.3.2

Year	Reported offenses
1983	1,258,090
1984	1,273,280
1985	1,328,800
1986	1,489,170
1987	1,484,000
1988	1,566,220
1989	1,646,040
1990	1,820,130
1991	1,911,770
1992	1,932,270
1993	1,926,020
1994	1,857,670
1995	1,798,790
1996	1,682,280

ANSWER:

(a) The average rate of change between 1983 and 1992 is $\dfrac{1{,}932{,}270 - 1{,}258{,}090}{1992 - 1983} = \dfrac{674{,}180}{9} = 74{,}909$ reported offenses per year.

(b) The average rate of change between 1992 and 1996 is $\dfrac{1{,}682{,}280 - 1{,}932{,}270}{1996 - 1992} = -\dfrac{249{,}990}{4} = -62{,}498$ reported offenses per year.

(c) Between 1983 and 1992, reported offenses of violent crime in the US increased at an average rate of 74,909 offenses per year. After 1992, however, reported offenses *decreased*, at an average rate of 62,498 offenses per year between 1992 and 1996.

3. The total sales of household computers in the US, as measured by sales to retail consumer dealers, in millions of dollars, was 2,385 in 1984 and was 16,585 in 1997.[3] Find the average rate of change in sales between 1984 and 1997. Give units with your answer. Use your answer to estimate total sales in 1998.

ANSWER:

The average rate of change is $\dfrac{16{,}585 - 2{,}385}{1997 - 1984} = \dfrac{14{,}200}{13} = 1{,}092.3$ million dollars per year. To estimate the total sales in 1998, we add the estimated annual increase to the number of sales in 1997: $16{,}585 + 1{,}092.3 = 17{,}677.3$ million dollars.

4. The distance $d = f(t)$ in feet that a golf ball will fall in t seconds if dropped from a very high tower is given by the formula $f(t) = 16t^2$.

(a) Make and label a table of values of $f(t)$ giving distances fallen at 1-second intervals for 8 seconds.

(b) Sketch and label a graph of $f(t)$ for the time period $0 \leq t \leq 5$.

(c) Using the table in (a), determine

   (i) the change in height of the golf ball between times $t = 3$ and $t = 7$;

   (ii) the average rate of change in height of the golf ball between times $t = 3$ and $t = 7$.

(d) Using the graph in (b), determine

   (i) the change in height of the golf ball between times $t = 2$ and $t = 5$;

   (ii) the average rate of change in height of the golf ball between times $t = 2$ and $t = 5$.

ANSWER:

(a)

Table 1.3.3 *Distance fallen versus time falling*

t (sec)	0	1	2	3	4	5	6	7	8	9	10
$f(t)$ (feet)	0	16	64	144	256	400	576	784	1024	1296	1600

---

[3]*The World Almanac and Book of Facts 1998*, p. 650 (Mahwah, NJ: K-111 Reference Corporation).

(b)

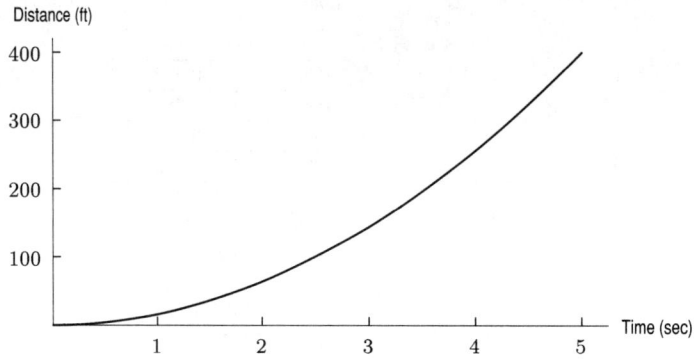

Distance (ft)

Figure 1.3.5

(c) (i) Change in height equals the change in the distance fallen:
$\Delta f = f(7) - f(3) = 784 - 144$ ft $= 640$ ft.

(ii) The average rate of change in height equals the average distance fallen per second: $\Delta f / \Delta t = (f(7) - f(3))/(7 - 3) = 640$ ft/4sec $= 160$ ft/sec. The values $f(3) = 144$ and $f(7) = 784$ have been read from the table of values of f.

(d) (i) Change in height equals the change in the distance fallen:
$\Delta f = f(5) - f(2) = 400 - 64$ ft $= 336$ ft.

(ii) The average rate of change in height equals the average distance fallen per second: $\Delta f / \Delta t = (f(5) - f(2))/(5 - 2) = 336$ ft/3sec $= 112$ ft/sec. The values $f(2) = 64$ and $f(5) = 400$ have been read from the graph of f.

5. On the axes below, sketch a smooth, continuous curve (i.e., no sharp corners, no breaks) which passes through the point $P(3, 4)$, and which clearly satisfies the following conditions:

- Concave up to the left of P
- Concave down to the right of P
- Increasing for $x > 0$
- Decreasing for $x < 0$
- Does *not* pass through the origin.

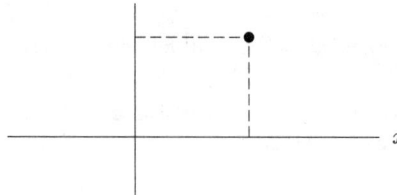

ANSWER:

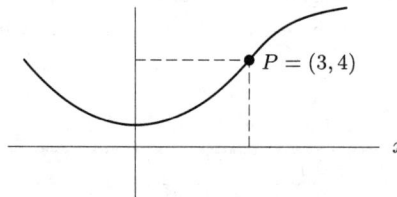

6. Sketch the graph of a function which is increasing and concave down.
ANSWER:
See Figure 1.3.6.

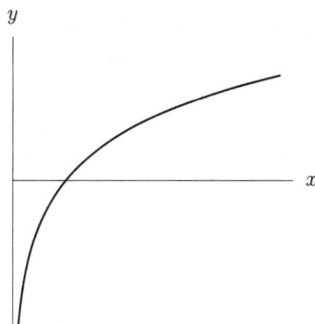

Figure 1.3.6

7. Values for $g(x)$ are given in Table 1.3.4. Do you think that $g(x)$ is concave up or concave down? How do you know?

Table 1.3.4

x	1	2	3	4	5	6
$g(x)$	100	90	81	73	66	60

ANSWER:

$g(x)$ is concave up. The decreasing rate is decreasing.

8. Sketch the graph of a continuous function f for which the following two properties hold.

(a) At $x < 0$, the function f increases at an increasing rate.

(b) At $x > 0$, the function f increases at a decreasing rate.

ANSWER:

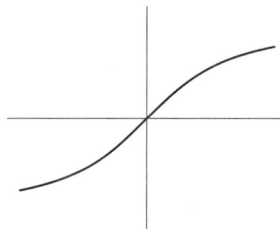

Figure 1.3.7

9. Identify the x-intervals on which the function graphed in Figure 1.3.8 is:

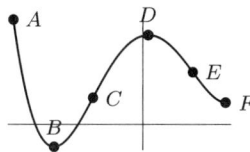

Figure 1.3.8

(a) increasing and concave downward.

(b) increasing and concave upward.

(c) decreasing and concave upward.

(d) decreasing and concave downward.

ANSWER:

(a) The function is increasing and concave downward on the x-interval between C and D.

(b) The function is increasing and concave upward on the x-interval between B and C.

(c) The function is decreasing and concave upward on the x-intervals between A and B and between E and F.

(d) The function is decreasing and concave downward on the x-intervals between D and E.

Problems and Solutions   for Section 1.4

1. A premium ice cream company finds that at a price of $4.00, demand for their ice cream cones is 4000. For each $0.25 reduction in price, the demand increases by 200. Find the price and the quantity sold that will maximize revenue.

   ANSWER:

   A table of values for the demand equation looks like Table 1.4.5.

Table 1.4.5

p	4.00	3.75	3.50	$\cdots$
p	4000	4200	4400	$\cdots$

A formula for this linear function is $q = 7200 - 800p$. We have revenue R equal to

$$R = p \cdot q = p \cdot (7200 - 800p) = 7200p - 800p^2.$$

A graph of this function shows that the maximum revenue occurs at a price of $4.50. See Figure 1.4.9.

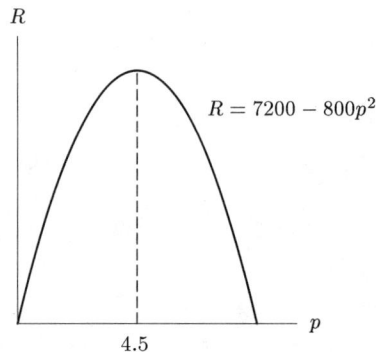

Figure 1.4.9

2. The graph in Figure 1.4.10 (a demand curve) shows the quantity of goods purchased by consumers at various prices.

   (a) Why is the graph decreasing as you move to the right?

   (b) If the price is $17 per item, how many items do the consumers purchase?

   (c) At what price do the consumers purchase 30 items?

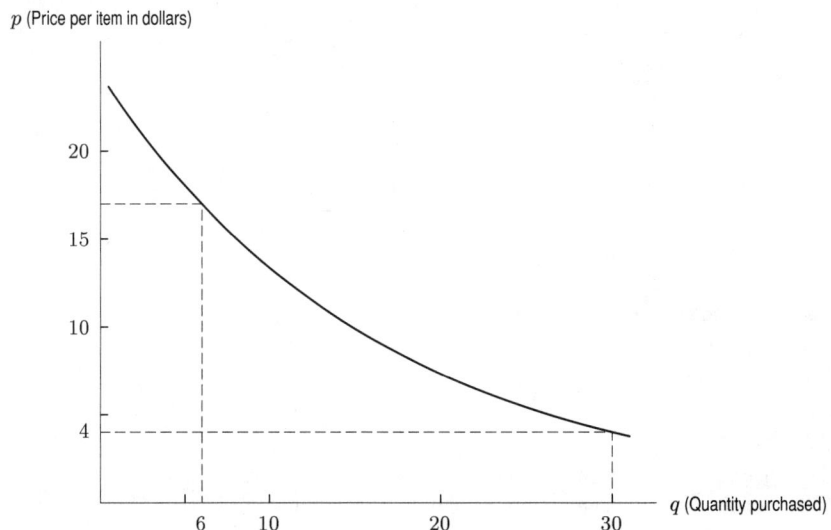

Figure 1.4.10

ANSWER:

(a) As prices go down, consumers buy more of the product.
(b) 6 items
(c) $4

3. Figure 1.4.11 gives both supply and demand curves for a certain economic good.

(a) Why is the supply curve increasing as you move to the right?
(b) If the price is $50 per item, the producers will supply _____ items.
(c) Why is the demand curve decreasing as you move to the right?
(d) If the price is $50 per item, the consumers will buy _____ items.
(e) If the price is $50 per item, would you expect market pressures to push the price higher or lower? Explain your answer.
(f) The equilibrium price will be _____.

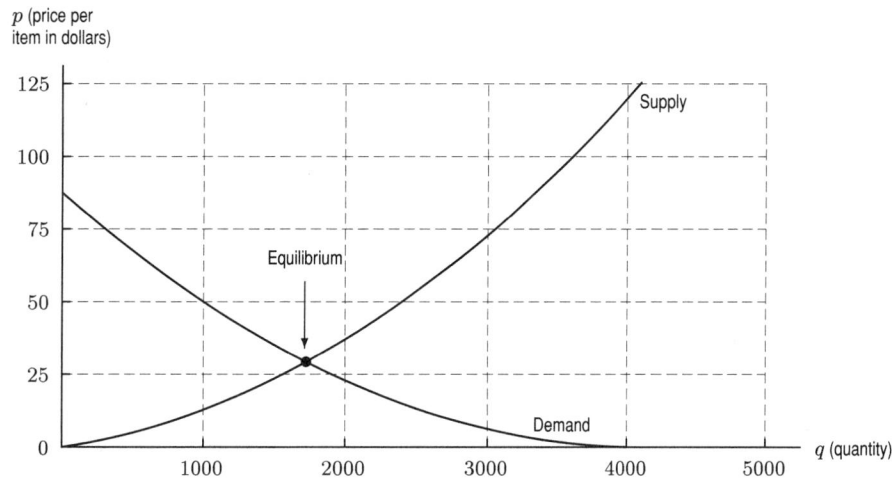

Figure 1.4.11

ANSWER:

(a) At higher prices, producers will make more products and hence supply more to the market.
(b) 2400
(c) At lower prices, consumers will buy more.
(d) 1000
(e) More will be supplied than demanded, oversupply will push prices down as sellers attempt to unload stocks.
(f) About $30.

4. Suppose that $S(q)$ is the price per unit (in dollars) of widgets which will induce producers to supply q thousand widgets to the market, and suppose that $D(q)$ is the price per unit at which consumers will buy q thousand items.

(a) Which is larger, $S(100)$ or $S(150)$, and why?
(b) Which is larger, $D(100)$ or $D(150)$, and why?
(c) If $D(100) = 10$ and $S(150) = 10$, what will you predict about the future selling price of widgets (currently at $10)?

ANSWER:

(a) $S(150)$ is larger, since a higher price will induce producers to supply more.
(b) $D(100)$ is larger, since consumers will buy more for a lower price.
(c) At $10 per unit, the supply is bigger than the demand, so the price will fall.

5. Suppose we buy quantities x_1 and x_2, respectively, of two goods. The following graph shows the budget constraint $p_1 x_1 + p_2 x_2 = k$, where p_1 and p_2 are the prices of the two goods and k is the available budget. On the graph, draw the lines that correspond to the following situations, and for each line, give the equation and the coordinates of both intercepts. Label each line clearly.

(a) The budget is doubled, but prices remain the same.

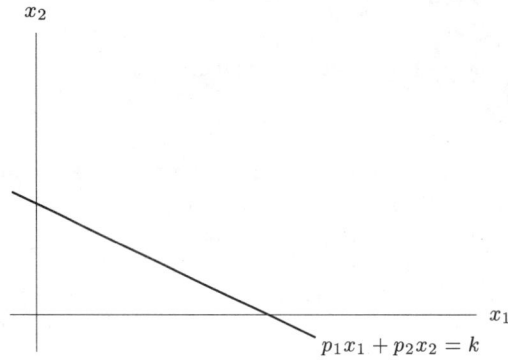

$$p_1 x_1 + p_2 x_2 = k$$

(b) The price of the first good is doubled, but everything else remains the same (the available budget is still k).

ANSWER:

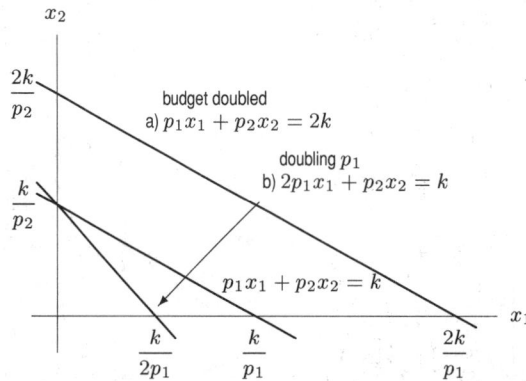

(a) $x_2 = -\dfrac{p_1}{p_2} x_1 + \dfrac{k}{p_2}$. If the prices remain the same, the slope of the line remains the same. If the budget is doubled, push the line up, keeping the slope the same, but doubling the x_1 and x_2 intercepts.

(b) $x_2 = -\dfrac{p_1}{p_2} x_1 + \dfrac{k}{p_2}$. When the price of the first good is doubled, we get $x_2 = -\dfrac{2p_1}{p_2} x_1 + \dfrac{k}{p_2}$. The slope of the line is double; y-intercept remains the same.

Problems and Solutions   for Section 1.5

1. A population is growing according to the function $P = 250(1.065)^t$, where P is the population at year t.

(a) What is the initial population?

(b) What is the annual growth rate?

(c) What is the population in year 10?

(d) How many years will it take for the population to reach 1000?

ANSWER:

(a) The initial population is 250 people.

(b) The annual growth rate is $100 \cdot 0.065 = 6.5\%$.

(c) We find the population in year 10 by substituting $t = 10$ in the formula, to get $P = 250(1.065)^{10} \approx 469.3$. So there are 469 people in year 10.

(d) We find the number of years it will take for the population to reach 1000 by substituting $P = 1000$ in the formula and solving for t:

$$1000 = 250(1.065)^t$$
$$4 = (1.065)^t$$
$$\ln 4 = t \ln 1.065$$
$$t = \frac{\ln 4}{\ln 1.065} \approx 22.01.$$

Since our solution for t is greater than 22, it will take 23 years for the population to reach 1000. (The population will actually be 999 people after 22 years.)

2. A town has 1000 people initially. In each of the cases below, find the formula for the population of the town, P, in terms of the number of years, t.

   (a) The town grows by 50 people a year.
   (b) The town grows at an annual rate of 8% a year.
   (c) The town shrinks by 75 people a year.
   (d) The town shrinks at an annual rate of 3% a year.

   ANSWER:

   (a) $P = 1000 + 50t$.
   (b) $P = 1000(1.08)^t$.
   (c) $P = 1000 - 75t$.
   (d) $P = 1000(0.97)^t$.

3. Tables of values of three functions are given in Table 1.5.6.

   (a) Which (if any) of these functions could be linear? For those functions which could be linear, find a possible formula.
   (b) Which (if any) of these functions could be exponential? For those functions which could be exponential, find a possible formula.

Table 1.5.6

x	$f(x)$	$g(x)$	$h(x)$
-2	12	16	37
-1	17	24	34
0	20	36	31
1	21	54	28
2	18	81	25

   ANSWER:

   (a) To find out which functions could be linear, we look at the differences of successive terms of $f(x)$, $g(x)$, and $h(x)$. Since the successive values given for x differ by equal amounts, all successive values of $f(x)$, $g(x)$ or $h(x)$ will have to have equal differences if it is to be linear. For $f(x)$, we have $17 - 12 = 5$, $20 - 17 = 3$, $21 - 20 = 1, \ldots$. Since the differences are not equal, $f(x)$ cannot be linear. Similarly, for $g(x)$, we have $24 - 16 = 8$, $36 - 24 = 12$, $54 - 36 = 18, \ldots$, and so $g(x)$ cannot be linear either. For $h(x)$, however, we have $34 - 37 = 31 - 34 = 28 - 31 = 25 - 28 = -3$, and so $h(x)$ could be a linear function. If $h(x)$ is a linear function, then it is of the form $b + mx$. We know that $b = 31$ from the table, and for each increase of x by 1, we know that $h(x)$ decreases by 3, so m must equal -3. Thus a possible formula for $h(x)$ is $h(x) = 31 - 3x$.

   (b) To find out which functions could be exponential, we look at the ratios of successive terms of $f(x)$, $g(x)$, and $h(x)$. Since the successive values given for x differ by equal amounts, all successive values of $f(x)$, $g(x)$ or $h(x)$ will have to have equal ratios if it is to be linear. For $f(x)$, we have $17/12 \approx 1.42$, $20/17 \approx 1.18$, $21/20 = 1.05, \ldots$. Since the ratios are not equal, $f(x)$ cannot be exponential. Similarly, for $h(x)$, we have $34/37 \approx 0.92$, $31/34 \approx 0.91$, $28/31 \approx 0.90, \ldots$, and so $h(x)$ cannot be exponential either. For $g(x)$, however, we have $24/16 = 36/24 = 54/36 = 81/54 = 1.5$, and so $g(x)$ could be an exponential function. If $g(x)$ is an exponential function, then it is of the form $P_0 a^x$. We know that $P_0 = 36$ from the table, and for each increase of x by 1, we know that $f(x)$ increases by a factor of 1.5, so a must equal 1.5. Thus a possible formula for $f(x)$ is $f(x) = 36(1.5)^x$.

4. A population of rabbits is growing. In 1996, there are 10,000,000 rabbits, and the increase is 20% per decade.

   (a) What will the rabbit population be in 2006?
   (b) What will the population be in 2096?
   (c) In 1997?
   (d) Find $P(t)$, the formula for the population t years after 1996.

   ANSWER:

   (a) In 2006, the population will be $10{,}000{,}000 \times 1.2 = 12{,}000{,}000$ rabbits.
   (b) In 2096, the population will be $10{,}000{,}000 \times (1.2)^{10} \approx 10{,}000{,}000 \times 6.1917364 = 61{,}917{,}364$ rabbits.
   (c) In 1997, the population will be $10{,}000{,}000 \times (1.2)^{\frac{1}{10}} \approx 10{,}183{,}994$ rabbits.
   (d) The formula for the population t years from 1996 is $P = 10{,}000{,}000 \times (1.2)^{\frac{t}{10}}$.

5. Table 1.5.7 defines three functions for $0 \leq x \leq 8$: $y_1 = f_1(x)$; $y_2 = f_2(x)$; and $y_3 = f_3(x)$. Identify which of the functions are linear, exponential, or neither. Write an equation for the functions which are exponential or linear.

Table 1.5.7

x	y_1	y_2	y_3
0	4.25	4.25	4.25
2	6.80	5.11	3.39
4	10.88	5.97	2.53
6	17.408	9.552	1.67
8	27.8528	15.2832	0.81

ANSWER:

For the y_1's: $\dfrac{6.8}{2.5} = 1.6$; $\dfrac{10.88}{6.8} = 1.6$; $\dfrac{17.408}{10.88} = 1.6$; $\dfrac{27.8528}{17.408} = 1.6$. Therefore, y_1 is an exponential function whose y-intercept is 4.25; $y_1 = f_1(x) = 4.25(1.6)^x$.

For the y_2's, there are no common ratios or common differences.

For the y_3's, the Δy's are: $3.39 - 4.25 = -0.86$; $2.53 - 3.39 = -0.86$; $1.67 - 2.53 = -0.86$; and $.81 - 1.67 = -0.86$.

Thus y_1 is a linear function with slope $\dfrac{-0.86}{2} = -0.43$ and y-intercept 4.25; $y_3 = f_3(x) = -0.86x + 4.25$.

6. You are the housing minister in the year 1996 for a country with 30 million people. You have been asked to predict the population 5 years and 10 years from 1996, as part of a 10 year master plan for housing. Census records show that the population was 22.684 million in 1986 and 26.087 million in 1991. Give your predictions, with justification in words.

ANSWER:

We have the following table.

Year	Population (in million)
1986	22.684
1991	26.087
1996	30

Since $\dfrac{30}{26.087} = \dfrac{26.087}{22.684} = 1.15$, we have the population $P = (22.684)(1.15)^{t/5}$ at t years after 1986. So in $1996 + 5 = 2001$,

$$P = 22.684(1.15)^{15/5} = 34.5 \text{ million,}$$

and at $1996 + 10 = 2006$,

$$P = 22.684(1.15)^4 = 39.67 \text{ million.}$$

7. In Table 1.5.8, we are given the population of a small country over a ten year period.

Table 1.5.8

Population by year

Year	Population
1985	100,004
1987	108,104
1989	116,860
1991	126,326
1993	136,559
1995	147,620

During this same period, each year, the farmers of this country have produced more than enough food to support its population. Table 1.5.9 gives the number of people that this country's agriculture were able to support during this same period:

Table 1.5.9 *Food production by year*

Year	Number of people farmers can feed
1985	105,000
1987	115,650
1989	125,253
1991	134,847
1993	145,506
1995	155,100

We are interested in determining how long the farmers will be able to produce enough food to feed this population.

(a) Knowing that populations tend to grow exponentially and assuming that food production is linear, find equations that model these two sets of data.

(b) After studying these two sets of data, what can be said about the food supply for this population during this period?

(c) Using the equations that model population and food supply, how long will there be enough food for this population?

ANSWER:

(a) Since we are assuming the population will grow exponentially, we consider ratios of population for consecutive periods:

$$\frac{108,160}{100,004} \approx 1.081$$

$$\frac{116,986}{108,160} \approx 1.081$$

Do more if needed, but this tells us that an exponential model for this population can be given by

$$N(t) = 100,004(1.082)^{\frac{t}{2}}$$

with $t = 0$ for 1985 and $t = 2$ for 1987, etc.

To find the linear model for food produced, we find an equation of a straight line from this data. Let F represent food produced. Then,

$$\frac{F - 115,650}{t - 2} = \frac{115,650 - 105,000}{2 - 0}$$

or

$$F - 115,650 = 5325(t - 2)$$

or

$$F(t) = 5325(t - 2) + 115,650$$
$$F(t) = 5325t + 105,000.$$

Again, $t = 0$ represents 1985; $t = 2$, 1987, etc.

(b) Since the number of people the farmers can feed is greater than the population on any one of the given years, one can expect a happy, healthy, growing population.

(c) Note that both N and F are increasing functions. Graphing both N and F, one can see that they intersect at about $t = 20$, or during 2005. After this year, the food supply will be inadequate for the population.

8. Complete the tables of values for the three exponential functions f, g, and h:

(a)

x	0	1	2	3	4
$f(x)$	10	20	?	?	?

(b)

x	0	5	10	15	20
$g(x)$	10	20	?	?	?

(c)

x	0	100	200	300	400
$h(x)$	10	20	?	?	?

ANSWER:

(a)

x	0	1	2	3	4
$f(x)$	10	20	40	80	160

(b)

x	0	5	10	15	20
$g(x)$	10	20	40	80	160

(c)

x	0	100	200	300	400
$h(x)$	10	20	40	80	160

9. Find formulas for the three exponential functions in Problem 8.
   ANSWER:

(a) $f(x) = 10 \cdot 2^x$
(b) $g(x) = 10 \cdot 2^{x/5} = 10(1.1487)^x$
(c) $h(x) = 10 \cdot 2^{x/100} = 10(1.006955)^x$

10. Complete the table of values for the function f, if f is

   (a) linear;   **(b)** exponential.

x	0	10	20	30	40
$f(x)$	50	100	?	?	?

   ANSWER:

(a)

x	0	10	20	30	40
$f(x)$	50	100	150	200	250

(b)

x	0	10	20	30	40
$f(x)$	50	100	200	400	800

11. Table 1.5.10 shows tables of values for two functions. Which function could be exponential? Find a possible formula for any function which could be exponential.

Table 1.5.10

t	$f(t)$	t	$g(t)$
-2	250	2	500
-1	300	3	750
0	360	4	1500
1	432	5	4500

ANSWER:
   We look at the ratios of successive terms of $f(t)$ and $g(t)$. Since the successive values given for t differ by equal amounts, any exponential function will have to have equal ratios of successive values of $f(t)$ or $g(t)$. For $f(t)$, we find that $300/250 = 360/300 = 432/360 = 1.2$, and so $f(t)$ could be exponential. For $g(t)$, however, we have $750/500 = 1.5$, $1500/750 = 2$, and $4500/1500 = 3$, so $g(t)$ cannot be an exponential function. If $f(t)$ is an exponential function, then it is of the form $P_0 a^t$. We know that $P_0 = 360$ from the table, and for each increase of t by 1, we know that $f(t)$ increases by a factor of 1.2, so a must equal 1.2. Thus a possible formula for $f(t)$ is $f(t) = 360(1.2)^t$.

12. A bar of soap starts out at 150 grams. In each of the following cases, write a formula for the quantity S grams of soap remaining after t days. The decrease is:

  (a) 10 grams per day
  (b) 10% per day
  (c) half the rate as in part (b)

  ANSWER:

  (a) This is a linear function with slope -10 grams per day and intercept 150 grams. The function is $S = 150 - 10t$.
  (b) Since the quantity is decreasing by a constant percent change, this is an exponential function with base 1 - 0.1 = 0.9. The function is $S = 150(0.9)^t$.
  (c) Since the quantity is decreasing by a constant percent change, this is also an exponential function. In this case, it is an exponential function with base 1 - 0.1/2 = 0.95. The function is $S = 150(0.95)^t$.

13. A photocopy machine can reduce copies to 90% or 70% of their original size. By copying an already reduced copy, further reductions can be made.

  (a) Write a formula for the size of the image, N, after the original image of size x has been reduced n times with the copy machine set on 90% reduction.
  (b) Write a formula for the size of the image, Q, after the original image of size x has been reduced q times with the copy machine set on 70% reduction.
  (c) Which will be larger: an image that has been reduced on the 90% setting 5 times or the same image after being reduced 7 times on the 70% setting?
  (d) If an image is reduced on the 90% setting and a copy of the same original image is reduced the same number of times on the 70% machine, will one image ever be less than 50% the size of the other? If so, how many copies on each of the settings will it take?

  ANSWER:

  (a) $N = x(.9)^n$
  (b) $Q = x(.7)^q$
  (c) If an image of size x is reduced 5 times on the 90% reduction setting, its new size will be

  $$x(.9)^5 = .59049x$$

  If the image of size x is reduced 7 times on the 70% reduction setting, its new size will be

  $$x(.7)^7 = .0823543x$$

  So, the first image will be larger.
  (d) We want to solve for n so that $(.5)(.9)^n \geq (.7)^n$.
  After one copy is made, the images sizes are $.81x$ and $.49x$ for the 90% and 70% setting respectively. After two copies, the image sizes are $.729x$ and $.343x$. Therefore, $n = 2$.

14. Give a possible formula for the function in the following figure:

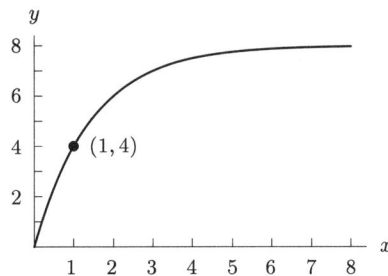

Figure 1.5.12

  ANSWER:
  The difference, D, between the horizontal asymptote and the graph appears to decrease exponentially, so we look for an equation of the form

  $$D = D_0 a^x$$

Where $D_0 = 8 = $ difference when $x = 0$. Since $D = 8 - y$, we have

$$8 - y = 8a^x \text{ or } y = 8 - 8a^x = 8(1 - a^x)$$

The point $(1, 4)$ is on the graph, so $4 = 8(1 - a^1)$, giving $a = 1/2$.
Therefore $y = 8(1 - (1/2)^x) = 8(1 - 2^{-x})$.

15. Give a possible formula for the function in the following figure:

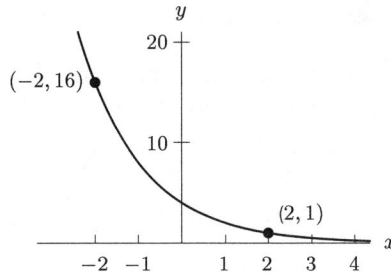

Figure 1.5.13

ANSWER:
We look for an equation of the form $y = y_0 a^x$ since the graph looks exponential. The points $(-2, 16)$ and $(2, 1)$ are on the graph, so

$$16 = y_0 a^{-2} \text{ and } 1 = y_0 a^2$$

Therefore $16/1 = y_0 a^{-2}/y_0 a^2 = 1/a^4$, giving $a = 1/2$, so $1 = y_0 a^2 = y_0(1/4)$, so $y_0 = 4$.

Hence, $y = 4(1/2)^x = 4(2^{-x})$.

16. Give a possible formula for the function in the following figure:

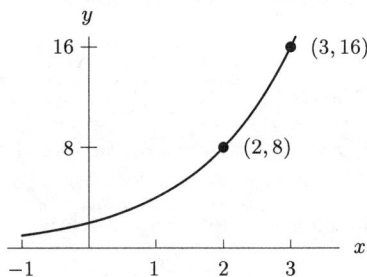

Figure 1.5.14

ANSWER:
We look for an equation of the form $y = y_0 a^x$ since the graph looks exponential. The points $(2, 8)$ and $(3, 16)$ are on the graph, so

$$8 = y_0 a^2 \text{ and } 16 = y_0 a^3$$

Therefore $8/16 = y_0 a^2/y_0 a^3 = a^{-1}$, giving $a = 2$, so $16 = y_0 a^3 = y_0(2)^3$, so $y_0 = 2$.

Hence, $y = 2(2^x)$.

17. These functions represent exponential growth or exponential decay.

$$P = 6(1.06)^t$$

$$Q = 4.2e^{0.04t}$$

$$S = 2e^{-0.2t}$$

$$R = 8(0.88)^t$$

Which functions represent growth and which represent decay?

ANSWER:

$P = 6(1.06)^t$ represents exponential growth because $1.06 > 1$. Since $e^{0.04t} = (e^{0.04})^t \approx (1.04)^t$, we have $Q = 4.2(1.04)^t$. This is exponential growth because $1.04 > 1$. $R = 8(0.88)^t$ represents exponential decay because $0.88 < 1$. Since $e^{-0.2t} = (e^{-0.2})^t \approx (0.82)^t$, we have $S = 2(0.82)^t$. This is exponential decay because $0.82 < 1$.

18. Joe and Sam each invested $20,000 in the stock market. Joe's investment increased in value by 5% per year for 10 years. Sam's investment decreased in value by 10% for 5 years and then increased by 10% for the next 5 years.

   (a) At the end of the 10 years, whose investment was worth more, Joe's or Sam's?
   (b) If Sam's initial investment was $30,000, but Joe's was still $20,000, would that change whose investment would be worth more at the end of the 10 years?

   ANSWER:

   (a) The value of Joe's investment after ten years is $20,000(1.05)^{10} = \$32,577.89$. The value of Sam's investment after five years is $20,000(0.90)^5 = \$11,809.80$ and the value after ten years is $11,809.80(1.10)^5 = \$19,019.80$. This means that after ten years, Joe's investment will be worth more than Sam's.
   (b) The value of Sam's investment after five years is $30,000(0.90)^5 = \$17,714.70$. After ten years, the value is $17,714.70(1.10)^5 = \$28,529.70$, therefore Joe's investment would still be worth more than Sam's at the end of ten years.

19. A bakery has 200 lbs of flour. If they use 5% of the available flour each day, how much do they have after 10 days? Write a formula for the amount of flour they have left after n days.

   ANSWER:
   Using $Q = Q_0(1 - r)^t$, we have

   $$Q = 200(1 - .05)^{10} = 200(0.95)^{10} = 119.75.$$

   The amount of flour left after 10 days is 119.75 lbs. The amount of flour left after n days is $200(0.95)^n$.

20. A substance has a half-life of 56 years.

   (a) Write a formula for the quantity, Q, of the substance left after t years, if the initial quantity is Q_0.
   (b) What percent of the original amount of the substance will remain after 20 years?
   (c) How many years will it be before less than 10% of the substance remains?

   ANSWER:

   (a) The formula is $Q = Q_0(1/2)^{(t/56)}$.
   (b) The percentage left after 20 years is

   $$\frac{Q_0(1/2)^{(20/56)}}{Q_0}$$

   The Q_0's cancel giving

   $$(1/2)^{(20/56)} \approx 0.781,$$

   so 78.1% is left.
   (c) The percent left after t years is

   $$(1/2)^{(t/56)}$$

   To find the number of years it takes for there to be less than 10% remaining, we solve

   $$(1/2)^{(t/56)} = 0.1$$
   $$t = 372,$$

   so after 372 years less than 10% of the substance will remain.

Problems and Solutions   for Section 1.6

1. (a) $7 \cdot 3^t = 5 \cdot 2^t$ $t = $ _____

(b) $8 \cdot (2.5^x) = a \cdot e^{kx}$ $a = $ _____ $k = $ _____

ANSWER:

(a)

$$7 \cdot 3^t = 5 \cdot 2^t$$

$$\frac{3^t}{2^t} = \frac{5}{7}$$

$$\left(\frac{3}{2}\right)^t = \frac{5}{7}$$

$$\ln\left[\left(\frac{3}{2}\right)^t\right] = \ln\left(\frac{5}{7}\right)$$

$$t \ln\left(\frac{3}{2}\right) = \ln\left(\frac{5}{7}\right)$$

$$t = \frac{\ln\left(\frac{5}{7}\right)}{\ln\left(\frac{3}{2}\right)}$$

$$t \simeq -0.829$$

(b) $8 \cdot (2.5^x) = a \cdot e^{kx}$ for any x implies $a = 8$ and $2.5^x = e^{kx}$. Therefore, $a = 8$ and $k = \ln 2.5$.

2. Find an equation for the line L shown below. Your answer will contain the positive constant b. Simplify your answer.

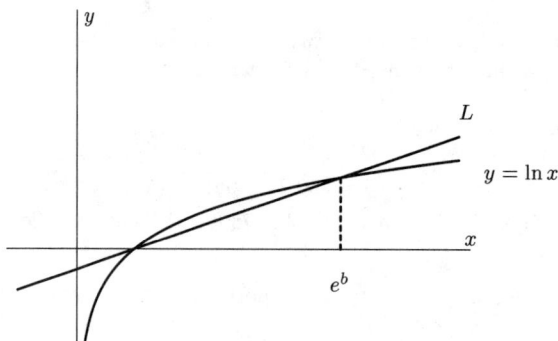

ANSWER:

The line L passes through the two points on the curve $y = \ln x$ specified by $y = 0$ and $x = e^b$, that is, $(1, 0)$ (since $1 = e^0$) and (e^b, b) respectively. The equation for this line is

$$\frac{y - 0}{x - 1} = \frac{b - 0}{e^b - 1},$$

so

$$y = \frac{b}{e^b - 1}(x - 1).$$

3. Find the equation of the line.

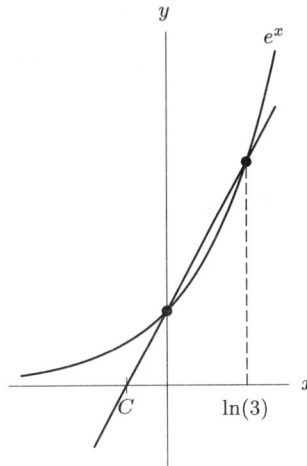

ANSWER:

The line passes through two points on the graph of $y = e^x$: $(\ln 3, e^{\ln 3}) = (\ln 3, 3)$ and $(0, e^0) = (0, 1)$. Therefore, the slope of the line is

$$m = \frac{3 - 1}{\ln 3 - 0},$$

so

$$m = \frac{2}{\ln 3}$$

and the equation of the line is

$$y = \frac{2}{\ln 3} x + 1.$$

4. Solve the following equation for t: $75 \cdot 14^t = 50 \cdot 12^t$.

   ANSWER:

$$75 \cdot 14^t = 50 \cdot 12^t$$
$$\frac{75}{50} = \left(\frac{12}{14}\right)^t$$
$$\frac{3}{2} = \left(\frac{6}{7}\right)^t$$
$$\ln\left(\frac{3}{2}\right) = t \ln\left(\frac{6}{7}\right)$$
$$t = \frac{\ln\left(\frac{3}{2}\right)}{\ln\left(\frac{6}{7}\right)} = \frac{\ln 3 - \ln 2}{\ln 6 - \ln 7} = -2.63.$$

5. Use logarithms to solve the equation $25(1.06)^x = 100$ for x.

   ANSWER:

$$25(1.06)^x = 100$$
$$1.06^x = 4$$
$$x = \frac{\ln 4}{\ln 1.06} \approx 23.78.$$

6. What interest rate, compounded annually, is equivalent to a 7% rate compounded continuously?

   ANSWER:

   Annual interest rates are of the form Ca^t for some a. A 7% continuously compounded interest rate is expressed in the form $Ce^{0.07t}$. To find the equivalent annual interest rate, we set these expressions equal and solve for a:

$$Ca^t = Ce^{0.07t}$$

$$a^t = e^{0.07t}$$

$$t \ln a = 0.07t \ln e = 0.07t$$

$$\ln a = \frac{0.07t}{t} = 0.07$$

$$a = e^{0.07} \approx 1.07251.$$

So the annual interest rate is 0.0725, or 7.25%.

7. The number of bacteria in milk grows at a rate of 10% per day once the milk has been bottled. When the milk is put in the bottles, it has an average bacteria count of 500 million per bottle.

   (a) Write an equation for $f(t)$, the number of bacteria t days after the milk is bottled.
   (b) Graph the number of bacteria against time. Label the axes and intercepts.
   (c) Suppose milk cannot be safely consumed if the bacteria count is greater than 3 billion per bottle. How many days will the milk be safe to drink once it has been bottled?

   ANSWER:

   (a) $f(t) = 500 \times 10^6 (1.1)^t$
   (b)

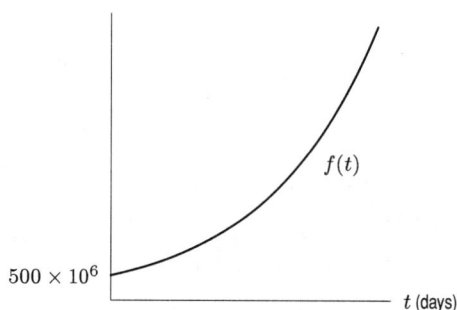

   (c) Find t making $f(t) = 3 \times 10^9$
   $$3 \times 10^9 = 500 \times 10^6 (1.1)^t$$

   $$\frac{3000}{500} = (1.1)^y \text{ so } 6 = (1.1)^t$$

   $$t = \frac{\ln 6}{\ln 1.1} \approx 18.8 \text{ days.}$$

   Thus milk will be safe for 18 days; during the 19th day it will turn bad, according to this model.

8. (a) Write the function $a(t) = 40(0.8)^t$ in the form Ae^{kt}.
   (b) Write the function $b(t) = -10e^{1.02t}$ in the form $P_0 a^t$.
   ANSWER:

   (a) $a(t) = 40(0.8)^t = 40e^{(\ln 0.8)t} = 40e^{-0.223t}$.
   (b) $b(t) = -10e^{1.02t} = -10(2.7732)^t$.

9. (a) Rewrite the function $g(t) = 15 \cdot 10^t$ in the form $A \cdot e^{kt}$.
   (b) Rewrite the function $h(t) = 100 \cdot e^{-0.3t}$ in the form $P_0 \cdot a^t$.
   ANSWER:

   (a) $g(t) = 15 \cdot 10^t = 15 \cdot e^{\ln 10^t} = 15 \cdot e^{2.3t}$.
   (b) $h(t) = 100 \cdot e^{-0.3t} = 100(0.74)^t$.

10. Simplify the expression as much as possible:

    (a) $6e^{\ln(a^3)}$
    (b) $3 \ln(e^a) + 5 \ln b^e$
    (c) $\ln(ab) + \ln(1/e)$

    ANSWER:

    (a) Using the identity $e^{\ln x} = x$, we have $6e^{\ln(a^3)} = 6a^3$.
    (b) Using the rules for ln, we have $3a + 5e \ln(b)$.
    (c) Using the rules for ln, we have

$$\ln(ab) + \ln(1/e) = \ln(a) + \ln(b) + \ln(1) - \ln(e)$$
$$= \ln(a) + \ln(b) + 0 - 1$$
$$= \ln(a) + \ln(b) - 1.$$

11. Solve for x using logs:

  (a) $6^x = 12$
  (b) $14 = 4^x$
  (c) $2e^{4x} = 8e^{6x}$
  (d) $3^{x+3} = e^{7x}$

  ANSWER:

  (a) Taking logs of both sides

$$\log 6^x = x \log 6 = \log 12$$

$$x = \frac{\log 12}{\log 6} \approx 1.39$$

  (b) Taking logs of both sides

$$\log 14 = \log 4^x$$

$$\log 14 = x \log 4$$

$$x = \frac{\log 14}{\log 4} \approx 1.90$$

  (c) Taking the natural logarithm of both sides

$$\ln(2e^{4x}) = \ln(8e^{6x})$$

$$\ln 2 + \ln(e^{4x}) = \ln 8 + \ln(e^{6x})$$

$$0.69 + 4x \approx 2.08 + 6x$$

$$2x \approx -1.39$$

$$x \approx -0.695$$

  (d) Using the rules for ln, we get

$$\ln(3^{x+3}) = \ln(e^{7x})$$

$$(x+3)\ln 3 = 7x$$

$$x \ln 3 + 3 \ln 3 = 7x$$

$$x(\ln 3 - 7) = -3 \ln 3$$

$$x = \frac{-3 \ln 3}{\ln 3 - 7} \approx 0.558$$

12. What is the doubling time of prices which are increasing by

(a) 7% a year

(b) 14% a year

ANSWER:

(a) Since the factor by which the prices have increased after time t is given by $(1.07)^t$, the time after which the prices have doubled solves

$$2 = (1.07)^t$$
$$\log 2 = \log(1.07^t) = t \log(1.07)$$
$$t = \frac{\log 2}{\log 1.07}$$
$$t \approx 10.24 \text{ y ears}$$

(b) Since the factor by which the prices have increased after time t is given by $(1.14)^t$, the time after which the prices have doubled solves

$$2 = (1.14)^t$$
$$\log 2 = \log(1.14^t) = t \log(1.14)$$
$$t = \frac{\log 2}{\log 1.14}$$
$$t \approx 5.29 \text{ years.}$$

13. If the size of a bacteria colony doubles in 8 hours, how long will it take for the number of bacteria to be 5 times the original amount?

ANSWER:

Given the doubling time of 8 hours, we can solve for the bacteria's growth rate;

$$2P_0 = P_0 e^{k8}$$

$$k = \frac{\ln 2}{8}$$

So the growth of the bacteria population is given by;

$$P = P_0 e^{\ln(2)t/8}$$

We want to find t such that

$$5P_0 = P_0 e^{\ln(2)t/8}$$

Therefore we cancel P_0 and apply ln . We get

$$t = \frac{8 \ln(5)}{\ln(2)} = 18.575$$

Problems and Solutions   for Section 1.7 ━━━━━━━━━━

1. Bank A offers 12% interest, compounded yearly, and bank B offers 11.8% interest, compounded continuously.

(a) You want to invest $1000 for 10 years. Which bank should you choose? Show your computations.

(b) What is the effective interest rate of bank B?

ANSWER:

(a) The balance at Bank A will be

$$B_A = 1000(1 + \frac{12}{100})^{10} = \$3105.85,$$

and at Bank B will be

$$B_B = 1000 e^{(0.118)(10)} = \$3254.37.$$

So you should choose Bank B.

(b) $e^{0.118} = 1.125$, so the effective interest rate of Bank B is 12.5%.

2. The population $P = f(t)$ of the United States in millions t years after 1790 during the years from 1790 to 1860 was given approximately by the exponential formula $f(t) = 3.9e^{0.0298t}$.

 (a) Make and label a table of values for the population of the US at 10 year intervals from 1790 to 1860.
 (b) Sketch and label a graph of population as a function of year from 1790 to 1860.
 (c) What was the annual percent growth rate of the US during this time period?
 (d) What was the approximate doubling time for the population?
 (e) What population would this model predict for the US in the year 2000? Comment on your answer.

   ANSWER:

 (a)

Table 1.7.11 *Population of the US: 1790-1860*

Year	1790	1800	1810	1820	1830	1840	1850	1860
Population (millions)	3.9	5.3	7.1	9.5	12.8	17.3	23.3	31.4

 (b)

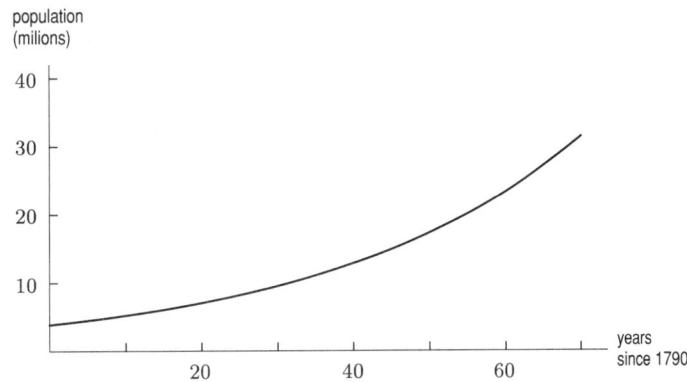

Figure 1.7.15

 (c) In one year the population increased by a factor of $e^{0.0298} = 1.03$, for an annual percent growth rate of 3%.
 (d) From the table, we see that the initial popuation of 3.9 million in 1790 had not yet doubled by 1810 but that it had more than doubled by 1820. The doubling time was between 20 and 30 years.
 (e) The prediction for the year 2000 would be $3.9e^{0.0298(210)} = 2036$ million $\approx$ 2 billion. This figure is much higher than the actual population. The exponential model that described so well the US population during the period 1790-1860 is not appropriate for the extended period up to the present time.

3. A standard cup of coffee contains about 100 mg of caffeine, and caffeine leaves the body at a rate of about 17% per hour.

 (a) Give a formula for the amount of caffeine in the body after t hours if this rate is an hourly rate (not continuous). How much caffeine is left in the body after 3 hours?
 (b) Give a formula for the amount of caffeine in the body after t hours if this rate is an hourly *continuous* rate. How much caffeine is left in the body after 3 hours?

   ANSWER:

 (a) The formula will be of the form $P = P_0 a^t$, with P in mg of caffeine and t in hours. We know that the initial value P_0 is 100. Every hour, 17% leaves the body, so 83% remains. Thus a must be 0.83. So our formula is $P = 100(0.83)^t$. To find the amount of caffeine left in the body after 3 hours, we substitute 3 fot t in the formula to get $P = 100(0.83)^3 \approx 57.18$. So there are 57.18 mg of caffeine left in the body after 3 hours.
 (b) The formula will be of the form $P = P_0 e^{-kt}$, with P in mg of caffeine and t in hours. We know that the initial value P_0 is 100. P is decaying at a continuous rate of 17%, so k must be 0.17. So our formula is $P = 100e^{-0.17t}$. To find the amount of caffeine left in the body after 3 hours, we substitute 3 fot t in the formula to get $P = 100e^{-(0.17)(3)} \approx 60.05$. So there arc 60.05 mg of caffeine left in the body after 3 hours.

4. A clean up of a polluted lake will remove 3% of the remaining contaminants every year, beginning in 1996.

 (a) In which year will the most be removed?   _____
 (b) The goal is to reduce the quantity of contaminants to $\frac{1}{10}$ its present level. When will this be achieved?   _____

   ANSWER:
 (a) We have $P = P_0(1 - r)^t$ and $r = 0.03$, therefore $P = P_0(0.97)^t$.

We may assume that most will be removed when there is only 1% of the initial quantity of contaminants left, so we need to solve for t:

$$0.01P_0 = P_0(0.97)^t$$
$$0.01 = (0.97)^t$$
$$t = \frac{\ln(0.01)}{\ln(0.97)}$$
$$t \simeq 151 \text{ years}$$

(b) Solve for t:

$$\frac{1}{10}P_0 = P_0(0.97)^t$$
$$0.1 = (0.97)^t$$
$$t = \frac{\ln(0.1)}{\ln(0.97)}$$
$$t \simeq 75.5 \text{ years}$$

5. A new species, introduced into an environment in which it has no natural predators, grows exponentially with continuous growth rate $k = 0.095$ per year.

   (a) If there were initially 45 individuals introduced, write the formula for the number of individuals after t years.
   (b) What is the approximate doubling time of this population?
   (c) About how long after introduction will the population reach 250 individuals?

   ANSWER:

   (a) $N(t) = 45e^{0.095t}$.
   (b) By the "70 rule", the doubling time is approximately $70/9.5 = 7.37$ years.
       More precisely, let $90 = 45e^{0.095t}$, or $2 = e^{0.095t}$, and solve for t:

   $$t = \frac{\ln 2}{0.095} = 7.296 \text{ years}.$$

   (c) Let $250 = 45e^{0.095t}$ and solve for t:

   $$t = \ln\left(\frac{250}{45}\right)/0.095 \approx 18 \text{ years}.$$

   So it will take about 18 years for the population to reach 250.

6. The population of Nicaragua was 3.6 million in 1990 and growing at 3.4% per year. Let P be the population in millions, and let t be the time in years since 1990.

   (a) Express P as a function in the form $P = P_0 a^t$.
   (b) Express P as an exponential function using base e.
   (c) How long does it take for the population of Nicaragua to increase by 50%?

   ANSWER:

   (a) We have $P_0 = 3.6$ million and $r = 0.034$. Then,

   $$P = P_0(1 + r)^t$$
   $$= 3.6(1 + 0.034)^t$$
   $$= 3.6(1.034)^t$$

   (b) We need to express 1.034 as e^k.

   $$e^k = 1.034$$
   $$k = \ln 1.034$$
   $$k \approx 0.0334$$
   $$P = 3.6(e^k)^t$$
   $$P \approx 3.6e^{0.0334t}$$

(c) Solve for t:

$$P_0 + \frac{50}{100}P_0 = P_0(1.034)^t$$

Cancelling P_0,

$$1.5 = (1.034)^t$$
$$t = \frac{\ln 1.5}{\ln 1.034}$$
$$t \simeq 12.127 \text{ years}$$

7. An exponentially decaying substance was weighed every hour and the results are given below:

Time	Weight (in grams)
9 am	10.000
10 am	8.958
11 am	8.025
12 noon	7.189
1 pm	6.440

(a) Determine a formula of the form

$$Q = Q_0 e^{-kt}$$

which would give the weight of the substance, Q, at time t in hours since 9 am.

(b) What is the approximate half-life of the substance?

ANSWER:

(a) $Q = 10e^{-kt}$ since $Q_0 = $ initial value $= 10$.
When $t = 1$, $Q = 8.958$, so $8.958 = 10e^{-k(1)}$ and
$0.8958 = e^{-k}$ so $k = -\ln 0.8958 = 0.11$
Thus $Q = 10e^{-0.11t}$.

(b) Half life when $Q = \frac{1}{2}Q_0$: $\frac{1}{2}Q_0 = Q_0 e^{-0.11t}$ so $\ln \frac{1}{2} = -0.11t$, so $t = 6.3$ hours.

8. In 1992, the Population Crisis Committee wrote:

Large cities in developing countries are growing much faster than cities in the industrialized world ever have. London, which in 1810 became the first industrial city to top 1 million, now has a population of 11 million. By contrast, Mexico City's population stood at only a million just 50 years ago and now is 20 million.

Assume that the instantaneous percentage growth rates of London and Mexico City were constant over the last two centuries.

(a) How many times greater is Mexico City's percentage growth rate than London's? Show your calculations and reasoning.

(b) When were the two cities the same size? Show your calculations and reasoning.

ANSWER:

(a) Letting α and β be the two growth rates for London and Mexico City, respectively, we approximate the population growth in millions by two exponentials, $e^{\alpha t}$ and $e^{\beta t}$, both of which are set to have population 1 million when $t = 0$. Since 182 and 50 are, respectively, the times that have passed since each city had 1 million people, we get

$$11 \approx 1 \cdot e^{\alpha \cdot 182} \quad \text{and} \quad 20 \approx 1 \cdot e^{\beta \cdot 50}$$

and

$$\beta = \frac{\ln 20}{50} \approx 0.0599, \alpha = \frac{\ln 11}{182} \approx 0.0132, \quad \text{and} \quad \frac{\beta}{\alpha} = \frac{182 \ln 20}{50 \ln 11} \approx 4.5.$$

(Note that these are growth rates, not percentages, but the ratio is the same as if we did it in terms of percentages.)

(b) We measure from 50 years ago (1942), when the population in London was $e^{0.0132(132)}$ and the population is Mexico City was 1 million. The functions describing population in the two cities are then:

$$\text{Mexico City : Population} = e^{0.0599t}$$
$$\text{London : Population} = e^{0.0132(132)}e^{0.0132t}$$

Setting these equal and solving, we get:

$$e^{0.0599t} = e^{0.0132(132)}e^{0.0132t}$$

$$0.0599t = 0.0132(132 + t)$$

$$t = \frac{0.0132 \cdot 132}{0.0599 - 0.0132} \approx 37.2$$

So the populations were equal 37.2 years after 1942, that is, in 1979.

For Problems 9– 10, decide whether each statement is true or false, and provide a short explanation or a counterexample.

9. The function described by the following table of values is exponential:

x	5.2	5.3	5.4	5.5	5.6
$f(x)$	27.8	29.2	30.6	32.0	33.4

ANSWER:
FALSE. The function is linear; for every increase of 0.1 in x, there is an increase of 1.4 in $f(x)$.

10. A quantity Q growing exponentially according to the formula $Q(t) = Q_0 5^t$ has a doubling time of $\frac{\ln 2}{\ln 5}$.
ANSWER:
TRUE. To calculate the doubling time, T, we use $2Q_0 = Q_0 5^T$ which gives $T = \frac{\ln 2}{\ln 5}$.

11. Which is worth more: $1200 invested at 10% annual interest or $1500 invested at 8% annual interest after

  (a) 5 years;
  (b) 25 years?
  (c) When would the two investments have equal value?
  (d) Make a table showing the values of the two investments every five years for forty years.

ANSWER:
After t years, the 10% investment would have a value of $f(t) = 1200(1.1)^t$ dollars. The 8% investment would have a value of $g(t) = 1500(1.08)^t$ dollars.

(a) $f(5) = \$1933$ and $g(5) = \$2204$. After 5 years the 8% investment will be worth more.
(b) $f(25) = \$13002$ and $g(25) = \$10273$. After 25 years the 10% investment will be worth more.
(c) We solve for the time t in the equation

$$1200(1.1)^t = 1500(1.08)^t$$

This can be done graphically (by finding the point where the graphs of f and g cross), numerically (by constructing a table as in part (d)), or analytically (by using logarithms.) Here is the analytic solution.

$$\ln(1200) + t\ln(1.1) = \ln(1500) + t\ln(1.08)$$

$$t(\ln(1.1) - \ln(1.08)) = \ln(1500) - \ln(1200)$$

$$0.01835t = 0.2231$$

$$t = \frac{0.2231}{0.01835} = 12.16.$$

After a little more than 12 years the two investments will have equal value.

(d)

Table 1.7.12 *Values of two investments at future times*

Time (years)	0	5	10	15	20	25	30	35	40
10% value (dollars)	1200	1933	3112	5013	8073	13002	20939	33723	54311
8% value (dollars)	1500	2204	3238	4758	6991	10273	15094	22178	32587

12. A cigarette contains about 0.4 mg of nicotine. The half-life of nicotine in the body is about 2 hours. How long does it take, after smoking a cigarette, for the level of nicotine in a smoker's body to be reduced to 0.08 mg?
ANSWER:
Let P (in mg) be the amount of nicotine in the body at time t (in hours). We can express this situation by the equation $P = 0.4e^{kt}$ for some constant k. To find k, we use the fact that the half-life of nicotine in the body is 2 hours; this means

that 2 hours after smoking a cigarette, there will be $0.4/2 = 0.2$ mg of nicotine in the body. We substitute 0.2 for P and 2 for t and solve the resulting equation for k:

$$0.2 = 0.4e^{k \cdot 2}$$
$$0.5 = e^{2k}$$
$$\ln 0.5 = 2k \ln e = 2k$$
$$k = \frac{\ln 0.5}{2} \approx -0.3466.$$

So the amount of nicotine in the body t hours after smoking a cigarette is $0.4e^{-0.3466t}$ mg. To find the amount of time it take for the level of nicotine to decrease to 0.08 mg, we set this expression equal to 0.08 and solve for t:

$$0.08 = 0.4e^{-0.3466t}$$
$$0.2 = e^{-0.3466t}$$
$$\ln 0.2 = -0.3466t$$
$$t = \frac{\ln 0.2}{-0.3466} \approx 4.644.$$

So it takes 4.644 hours for the level of nicotine to decrease to 0.08 mg.

Problems and Solutions   for Section 1.8

1. The graph of $f(x)$ is given below.

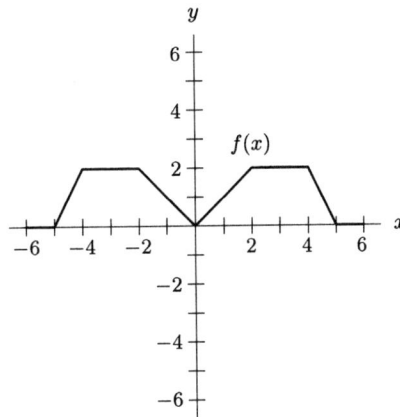

Sketch the graphs of:

(a) $2 + f(x)$ **(b)** $2f(x)$ **(c)** $1 - f(x)$ **(d)** $\dfrac{1}{f(x)}$

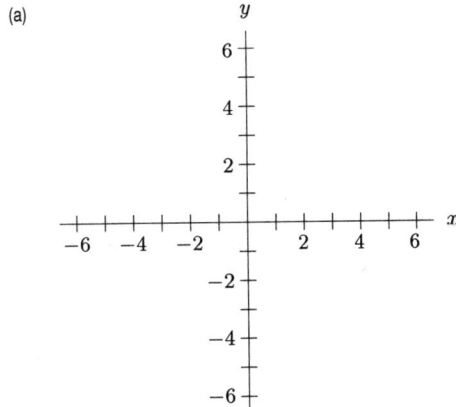

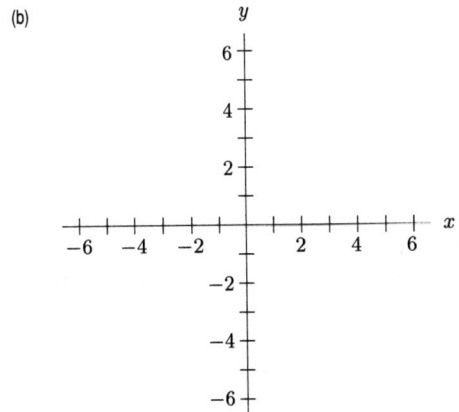

(c)

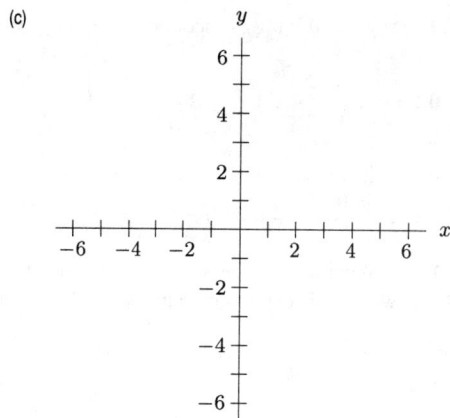

(d)

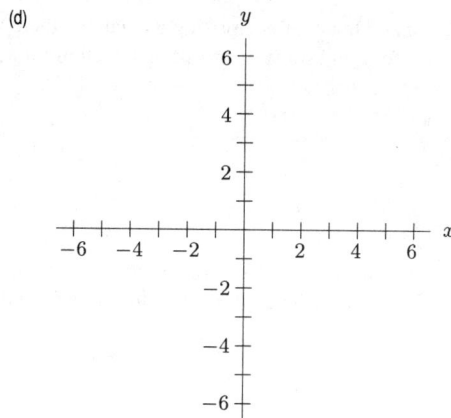

ANSWER:

(a)

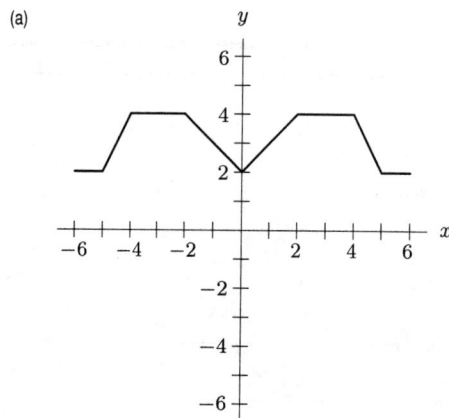

(b)

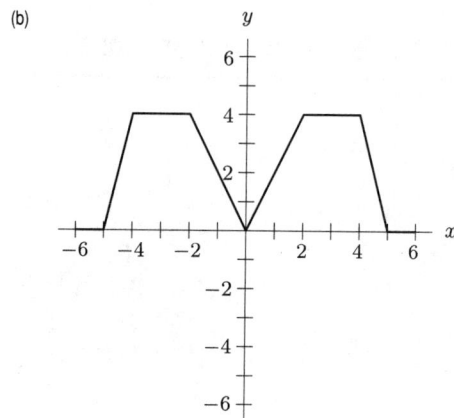

(c)

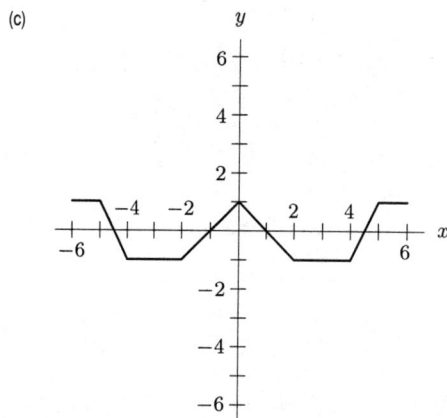

(d)

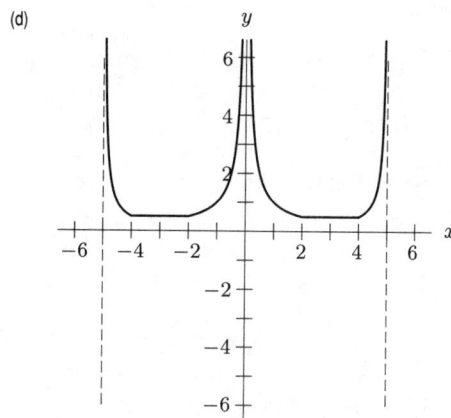

2. One of the graphs below shows the rate of flow, R, of blood from the heart in a man who bicycles for twenty minutes, starting at $t = 0$ minutes. The other graph shows the pressure, p, in the artery leading to a man's lungs as a function of the rate of flow of blood from the heart.

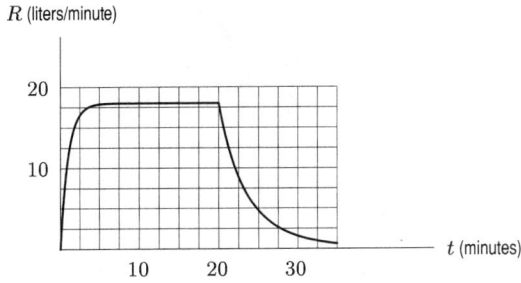

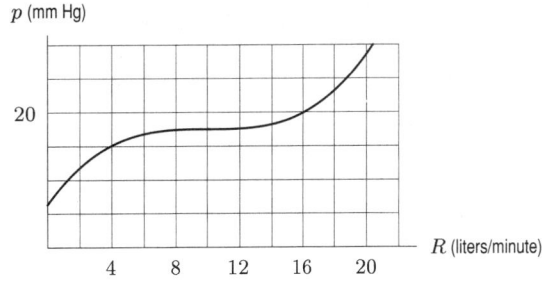

(a) Estimate $p(R(10))$ and $p(R(22))$.

(b) Explain what $p(R(10))$ represents in practical terms.

ANSWER:

(a) $p(R(10)) = p(18) = 23$ mm Hg

$p(R(22)) = p(10) = 17.5$ mm Hg

(b) $p(R(10))$ represents the pressure in the artery at $t = 10$.

3. Given the function $m(z) = z^2$, find and simplify $m(z + h) - m(z)$.

ANSWER:

$$m(z + h) - m(z) = (z + h)^2 - z^2$$
$$= z^2 + 2zh + h^2 - z^2$$
$$= 2zh + h^2$$

4. Let $f(x) = 2x + 1$ and $g(x) = x^2 + 3$. Find each of the following and simplify your answers. **(a)** $f(g(x))$ **(b)** $g(f(x))$ **(c)** $f(f(x))$

ANSWER:

(a) $f(g(x)) = f(x^2 + 3) = 2(x^2 + 3) + 1 = 2x^2 + 7$.

(b) $g(f(x)) = g(2x + 1) = (2x + 1)^2 + 3 = 4x^2 + 4x + 4$.

(c) $f(f(x)) = f(2x + 1) = 2(2x + 1) + 1 = 4x + 3$.

5. The graph of $y = f(x)$ is shown in Figure 1.8.16. Sketch graphs of each of the following. Label any intercepts or asymptotes that can be determined.

(a) $y = 3f(x) - 4$

(b) $y - 2 - 2f(x)$

(c) $y = f(x) + 3$

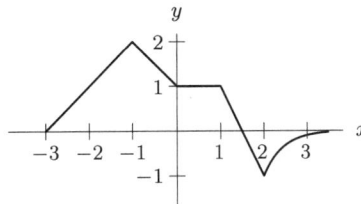

Figure 1.8.16

ANSWER:

Figure 1.8.17 shows the appropriate graphs. Note that asymptotes are shown as dashed lines and x- or y-intercepts are shown as filled circles.

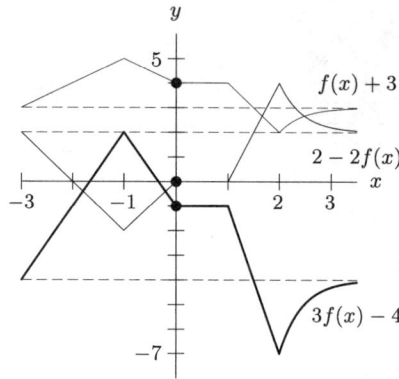

Figure 1.8.17

6. The graphs of $y = g(x)$ and $y = f(x)$ are given in Figure 1.8.18.

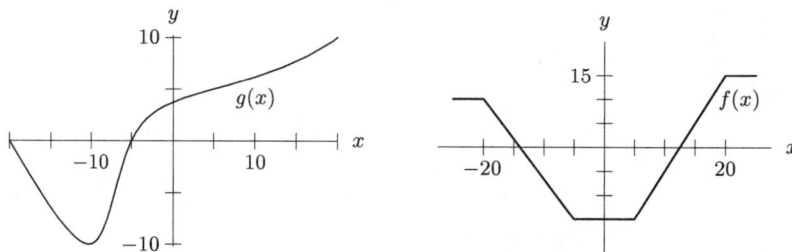

Figure 1.8.18

Estimate

(a) $f(g(5))$
(b) $g(f(5))$
(c) $f(g(-10))$
(d) $g(f(-5))$

ANSWER:

(a) $f(g(5)) \approx -15$
(b) $g(f(5)) \approx -7$
(c) $f(g(-10)) \approx -7$
(d) $g(f(-5)) \approx -7$

7. (a) Write an equation for the graph obtained by shifting the graph of $y = x^3$ vertically upward by 3 units, followed by vertically stretching the graph by a factor of 5.
   (b) Write the equation for a graph obtained by reflecting the graph for the function obtained in part (a) across the x-axis.
   ANSWER:

(a) After a vertical shift upward by 3 units the equation is $y = x^3 + 3$. After vertically stretching the new graph by a factor of 5, the resulting equation is $y = 5(x^3 + 3) = 5x^3 + 15$.
(b) $y = -(5x^3 + 15) = -5x^3 - 15$.

8. If the graph of $y = f(x)$ is shrunk vertically by a factor of 1/2, then shifted vertically by 4 units, then stretched vertically by a factor of 2, is the resulting graph the same as the original graph?
   ANSWER:

The final graph is not the same as the original graph. The equation for the graph after it is shrunk vertically by a factor of 1/2 is $y = .5f(x)$. After being shifted vertically by 4 units, the new equation is $y = .5f(x) + 4$. If the new graph is stretched vertically by a factor of 2, the resulting equation is $y = 2(.5f(x) + 4) = f(x) + 8$.

9. Given the function $q(x) = x^3$, find and simplify

(a) $q(2x + a) + q(x)$

(b) $q(x^2) + q(x + a)$

ANSWER:

(a) $q(2x + a) + q(x) = (2x + a)^3 + x^3 = 9x^3 + 12ax^2 + 6a^2x + a^3$.
(b) $q(x^2) + q(x + a) = (x^2)^3 + (x + a)^3 = x^6 + x^3 + 3ax^2 + 3xa^2 + a^3$.

10. The cost of shipping r kilograms of material is given by the function $C = f(r) = 200 + 4r$.

    (a) Find a formula for the inverse function.
    (b) Explain in practical terms what the inverse function tells you.

    ANSWER:

    (a) The function f tells us C in terms of r. To get its inverse, we want r in terms of C, which we find by solving for r:

$$C = 200 + 4r$$
$$C - 200 = 4r$$
$$r = (C - 200)/2$$
$$= f^{-1}(C).$$

    (b) The inverse function tells us the number of kilograms that can be shipped for a given cost.

11. For $g(x) = 2x^2 - 2x$ and $h(x) = 3x - 1$, find and simplify

    (a) $g(x) + 2h(x)$
    (b) $g(h(x))$
    (c) $h(g(x))$

    ANSWER:

    (a) $g(x) + 2h(x) = 2x^2 - 2x + 2(3x - 1) = 2x^2 - 2x + 6x - 2 = 2x^2 + 4x - 2$
    (b) $(g(h(x)) = 2(3x - 1)^2 - 2(3x - 1) = 2(9x^2 - 6x + 1) - 6x + 2 = 18x^2 - 18x + 4$
    (c) $h(g(x)) = 3(2x^2 - 2x) - 1 = 6x^2 - 6x - 1$

Problems and Solutions   for Section 1.9

1. Give rough sketches, for $x > 0$, of the graphs of $y = x^5$, $y = x$, $y = x^{1/3}$, $y = x^0$, and $y = x^{-2}$.
    ANSWER:

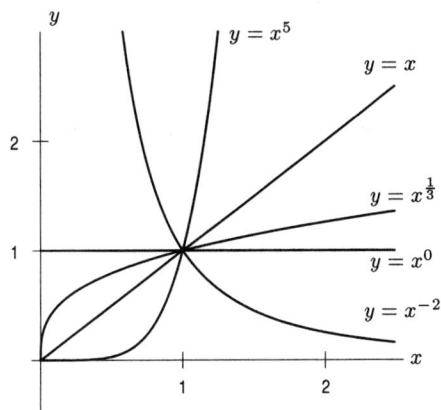

2. Simplify by hand the given expressions, and, where appropriate, leave your answer as a fraction. **(a)** $(3)^{-2}$ **(b)** $(4)^{-3/2}$
    ANSWER:

    (a) $(3)^{-2} = \dfrac{1}{3^2} = \dfrac{1}{9}$.
    (b) $(4)^{-3/2} = \dfrac{1}{4^{3/2}} = \dfrac{1}{\sqrt{4^3}} = \dfrac{1}{\sqrt{64}} = \dfrac{1}{8}$.

3. Write a function representing the following situation: the gravitational force, F, between two bodies is inversely proportional to the square of the distance, d, between them.

    ANSWER:
    $$F = k \left(\frac{1}{d^2} \right).$$

4. The number of species, S, on an island is proportional to the square root of the area, A, of the island. An island which has an area of 4 square miles contains 20 species.

    (a) Find a formula for S as a function of A.
    (b) If an island is 9 square miles in area, determine the number of species expected on the island.

    ANSWER:

    (a) Since S is proportional to the square root of A, we must have $S = k\sqrt{A}$ for some k. To find k, we substitute the known values of S and A to get $20 = k\sqrt{4}$. Solving this equation gives $k = 10$. Thus the formula is $S = 10\sqrt{A}$.
    (b) Use the formula from (a) and substitute 9 for A: $S = 10\sqrt{9} = 30$. So 30 species are expected on the island.

5. Poiseuille's law says that the rate of flow, F, of a gas through a cylindrical pipe is proportional to the fourth power of the raduis of the pipe, r. If the rate of flow is 400 cm^3/sec in a pipe of radius 3 cm for a certain gas, find an explicit formula for the rate of flow, F, as a function of the radius, r, and find the rate of flow through a pipe with a 5 cm radius.

    ANSWER:

    From Poiseuille's law, we know that the formula will be of the form $F = kr^4$ for some constant k. We substitute the known values $F = 400$ and $r = 3$ and solve for k:

    $$400 = k(3^4)$$
    $$400 = k \cdot 81$$
    $$k \approx 4.94.$$

    Thus the formula is $F = 4.94r^4$. To find the rate of flow through a pipe with a 5 cm radius, we substitute $r = 5$ and find F:

    $$F = 4.94(5^4)$$
    $$= 3087.5 \text{ cm}^3/\text{sec}.$$

6. A new music company wants to start selling compact discs. The profit π (in thousands of dollars) is $\pi(p) = 160p - 6.4p^2$, where p is the price of a compact disc (in dollars).

    (a) Sketch the graph of $\pi(p)$, showing the zeros of the function.
    (b) Find the coordinates of the maximum point on the graph.
    (c) Interpret the maximum point and its coordinates in practical economic terms.

    ANSWER:

    (a) A sketch of $\pi(p)$ is shown in Figure 1.9.19.

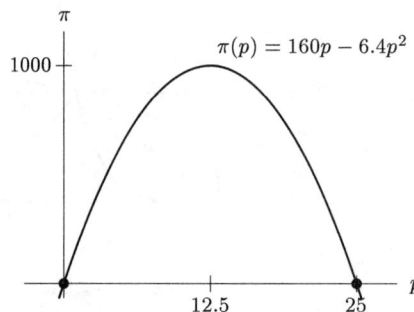

Figure 1.9.19

    (b) $\pi(p)$ is maximum at $(12.5, 1000)$.
    (c) The maximum profit is $1000, at which point the price of a compact disc is $12.5.

7. A ball is thrown into the air at time $t = 0$, and its height above ground (in feet) t seconds after it is thrown is given by
$f(t) = -16t^2 + 96t + 6$.

    (a) How long is it in the air?

    (b) How high does it go?

    (c) When does it reach its maximum height?

    ANSWER:

    We use a graphing calculator to get Figure 1.9.20. The answers can be read off the graph:

(a) About 6 seconds.

(b) 150 feet.

(c) 3 seconds.

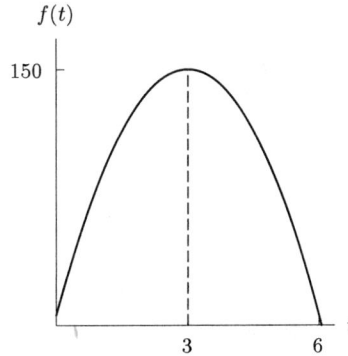

Figure 1.9.20

8. Use a graphing calculator to find all roots of the function $f(x) = x^3 - 23x^2 + 42x - 12$.

    ANSWER:

    The roots are approximately 0.35, 1.62 and 21.03.

Problems and Solutions   for Section 1.10

1. At high tide, the water level is 10 feet below a certain pier. At low tide the water level is 26 feet below the pier. Assuming sinusoidal behavior, sketch a graph of $y = f(t) =$ the water level, relative to the pier, at time t (in hours) if at $t = 0$ the water level is -18 feet and falling, until it reaches the first low tide at $t = 3$. Based on your sketch and the information provided above, give a formula for $f(t)$.

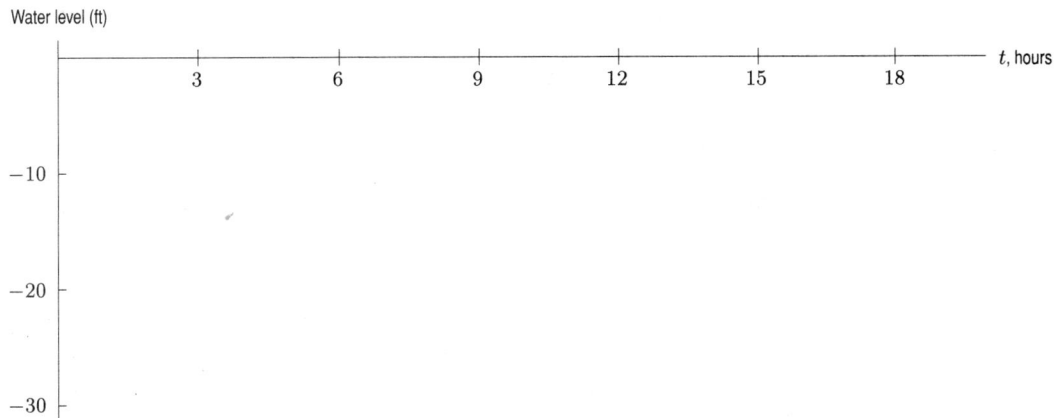

ANSWER:

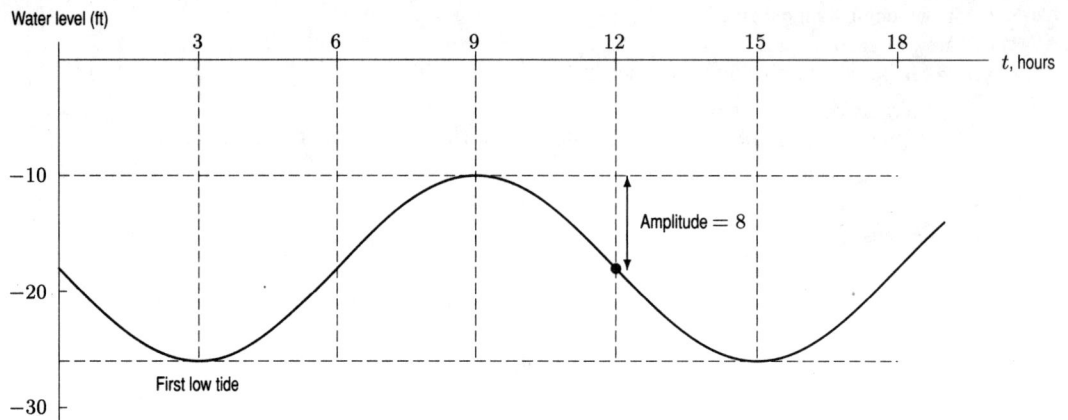

$$f(t) = A\sin(B(t+C)) + D$$

$$\text{period} = 12 \text{ hrs} \qquad \frac{2\pi}{B} = 12 \qquad B = \frac{\pi}{6}$$

$$f(t) = 8\sin\left(\frac{\pi}{6}(t+6)\right) - 18$$

$$= 8\sin\left(\frac{\pi}{6}t + \pi\right) - 18$$

$$= -8\sin\frac{\pi}{6}t - 18$$

2. In nature, the population of two animals, one of which preys on the other (such as foxes and rabbits) are observed to oscillate with time, and are found to be well approximated by trigonometric functions. The population of foxes is given by the graph below.

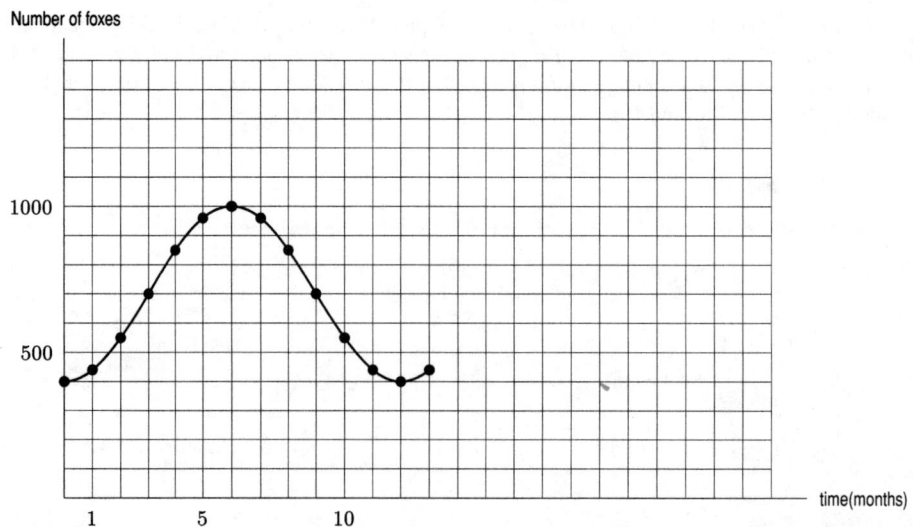

(a) Find the amplitude.

(b) Find the period.

(c) Give a formula for the function.

(d) Give an estimate for three times when the population is 500.

ANSWER:

(a) $\dfrac{1000 - 400}{2} = 300$

(b) Period=12 months

(c) Average value = 700 and graph looks like an upsidedown cosine, so $F = 700 - 300\cos(kt)$

   since $k(12) = 2\pi$, $k = \dfrac{\pi}{6}$ so $F = 700 - 300\cos\left(\dfrac{\pi t}{6}\right)$

(d) First time is between $t = 1$ and $t = 2$, so let's say about $t \approx 1.5$. From graph, next values are $t \approx 10.5$ and then $t \approx 13.5$ months

3. Find an equation which defines the function in Figure 1.10.21.

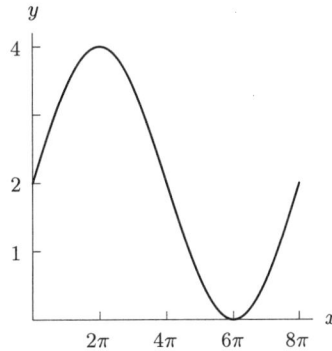

Figure 1.10.21

ANSWER:

This is a sine curve with amplitude 2, period 8π and vertical shift 2. Thus $y - 2 = 2\sin\left(\dfrac{x}{4}\right)$.

4. The graphs of $g(t)$ and $k(t)$ are shown in Figures 1.10.22 and 1.10.23. Find possible formulas for each.

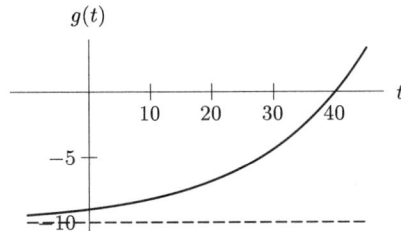

Figure 1.10.22

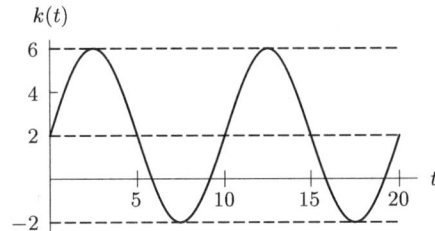

Figure 1.10.23

ANSWER:

Let $g(t) = e^{kt} - 10$. Then $e^{k(40)} - 10 = 0$ implies that $k = \ln 10/40$. We have

$$g(t) = e^{\frac{\ln 10}{40}t} - 10.$$

Let $k(t) = A\sin(Bt + C) + D$. Then $k(0) = 2$ implies that $D = 2$. The amplitude $A = 4$, and there is no phase shift, so $C = 0$. Since the period is 10, $2\pi/B = 10$, i.e. $B = 2\pi/10 = \pi/5$. We have

$$k(t) = 2 + 4\sin(\frac{\pi}{5}t).$$

5. The size of a bird population on an island can be described by a sinusoidal graph. The number of birds on the island decreased from a maximum of 20,000 in 1943 to a minimum of 12,000 in 1989, and then began increasing again.

   (a) Take 1943 as $t = 0$ and find a function that describes this behavior.
   (b) How many birds do you expect there to be on the island in the year 2000?
   (c) For what single period after 1989 will the population show a positive growth?

   ANSWER:

   (a) Since $1989 - 1943 = 46$, we have

$$f(t) = 16,000 + 4000 \cos\left(\frac{\pi}{46}t\right).$$

   (b) In the year 2000, $t = 2000 - 1943 = 57$, and we have

$$f(57) = 16,000 + 4000 \cos\left(\frac{57}{46}\pi\right) \approx 13,077.$$

   So there will be about 13,077 birds on the island.

   (c) The population will have a positive growth from 1989 to the year $1989 + 46 = 2035$.

6. What value of B would you use if $\sin Bt$ is to model a periodic function with period 1 year where t is measured in

   (a) months;   **(b)** days;   **(c)** years?

   ANSWER:

   (a) $B = 2\pi/12 = 0.5236$
   (b) $B = 2\pi/365 = 0.017214$
   (c) $B = 2\pi = 6.2832$

7. What value of B would you use if $\sin Bt$ is to model a periodic function where t is measured in hours and the period is

   (a) 1 day;   **(b)** 1/2 day;   **(c)** 5 days?

   ANSWER:

   (a) $B = 2\pi/24 = 0.2618$
   (b) $B = 2\pi/12 = 0.5236$
   (c) $B = 2\pi/120 = 0.05236$

8. Find possible formulas for the following sinusoidal function as a

   (a) transformation of $f(t) = \sin t$
   (b) transformation of $f(t) = \cos t$

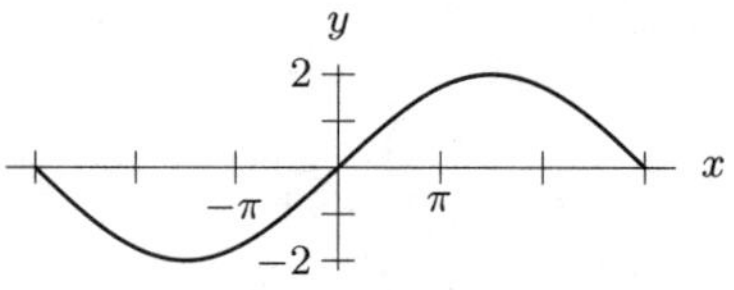

Figure 1.10.24

   ANSWER:

   (a) This function looks like a sine function with amplitude 2, so $f(t) = 2\sin(Bt)$. Since the function executes one full oscillation between $t = 0$ and $t = 6\pi$, when t changes by 6π, the quantity Bt changes by 2π. This means $B(6\pi) = 2\pi$, so $B = 1/3$. Therefore, $f(t) = 2\sin(t/3)$ has the graph shown.
   (b) The function also looks like a cosine function with amplitude 2 that has a period of 6π that been translated horizontally by 4.5π. Therefore, $f(t) = 2\cos(t/3 - 4.5\pi)$.

9. Temperatures in Town A oscillate daily between 30°F at 4am and 60°F at 4pm. Write the following formulas:

   (a) Temperature in Town A, in terms of time where time is measured in hours from 4am.
   (b) Temperature in Town A, in terms of time where time is measured in hours from midnight.
   (c) Temperature in Town B, where the temperatures are consistently 10°F colder than in Town A and measured from 4am.

   ANSWER:

   (a) We use a cosine of the form

$$H = A\cos(Bt) + C$$

and choose B so that the period is 24 hours, so $2\pi/B = 24$, giving $B = \pi/12$.

The temperature oscillates around an average value of 45°F, so $C = 45$. The amplitude of the oscillation is 15°F. To arrange that the temperature be at its lowest when $t = 0$, we take A negative,

so

$$H = -15\cos\left(\frac{\pi}{12}t\right) + 45.$$

(b) The formula is the answer from (a), shifted to the right by four hours,

$$H = -15\cos\left(\frac{\pi}{12}(t - 4)\right) + 45.$$

(c) The shape of the graph will be the same as (a), only translated down 10. So the formula is

$$H = -15\cos\left(\frac{\pi}{12}t\right) + 35.$$

10. Temperatures in a room oscillate between the low of $-10°$F (at 5am) and the high of 40°F (reached at 5pm).

(a) Find a possible formula for the temperature in the room in terms of time from 5am.

(b) Sketch a graph of temperature in terms of time.

ANSWER:

(a) We use a cosine of the form

$$H = A\cos(Bt) + C$$

and choose B so that the period is 24 hours, so $2\pi/B = 24$, giving $B = \pi/12$.

The temperature oscillates around an average value of 15°F, so $C = 15$. The amplitude of the oscillation is 25°F. To arrange that the temperature be at its lowest when $t = 0$, we take A negative,

so

$$H = -25\cos\left(\frac{\pi}{12}t\right) + 15.$$

(b)

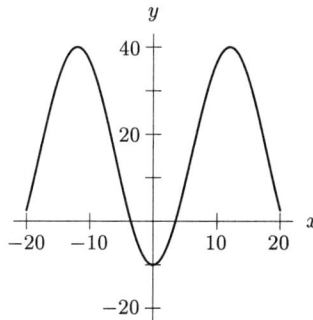

Figure 1.10.25

11. Consider the function $g(t) = 10 + \cos 2t$.

(a) What is its amplitude?

(b) What is its period?

(c) Sketch its graph.

(d) Write a short scenario that could be described by the behavior of $g(t)$.

ANSWER:

(a) The amplitude is 1.

(b) The period is $2\pi/2 = \pi$.

(c)

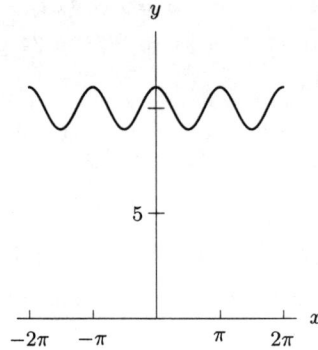

Figure 1.10.26

(d) Answers will vary.

12. Consider the functions $f(x) = 5 + \sin 3x$ and $g(x) = 3\sin x$. Describe how these functions are the same and how they are different. Include a description of amplitude, period and general shape of the graph.

ANSWER:

The amplitude of $g(x)$ is 3 times the amplitude of $f(x)$. The period of $g(x)$ is 2π while the period of $f(x)$ is $2\pi/3$, thus $f(x)$ is oscillating more quickly than $g(x)$. The y-intercept of $g(x)$ is zero whereas $f(x)$ has y-intercept 5.

Problems and Solutions to Review Problems for Chapter 1

1. The graph of the function $y = f(x)$ is shown in Figure 1.10.27.

(a) Estimate the domain of $f(x)$.
(b) Estimate the range of $f(x)$.
(c) Estimate $f(1)$.
(d) Is the graph increasing or decreasing at $x = 3$?
(e) Is the graph concave up or concave down at $x = 6$?

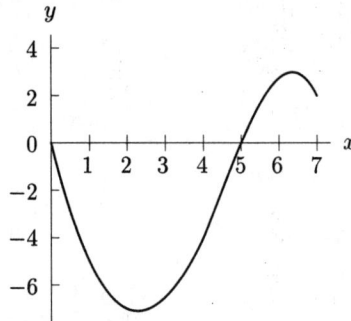

Figure 1.10.27

ANSWER:

(a) The domain of $f(x)$ is about $0 \le x \le 7$.
(b) The range of $f(x)$ is about $-7 \le y \le 3$.
(c) $f(1) \approx -5$.
(d) The graph is increasing at $x = 3$.
(e) The graph is concave down at $x = 6$.

2. Global pictures of six functions in the first quadrant are shown in Figure 1.10.28. Match the formulas below with the graphs.

(a) $y = 0.2x^5$ (b) $y = (1.2)^x$ (c) $y = x^3 + 25x^2 + 50x + 5$

(d) $y = \sqrt{x}$ (e) $y = \ln x$ (f) $y = 100x^2 + 10$

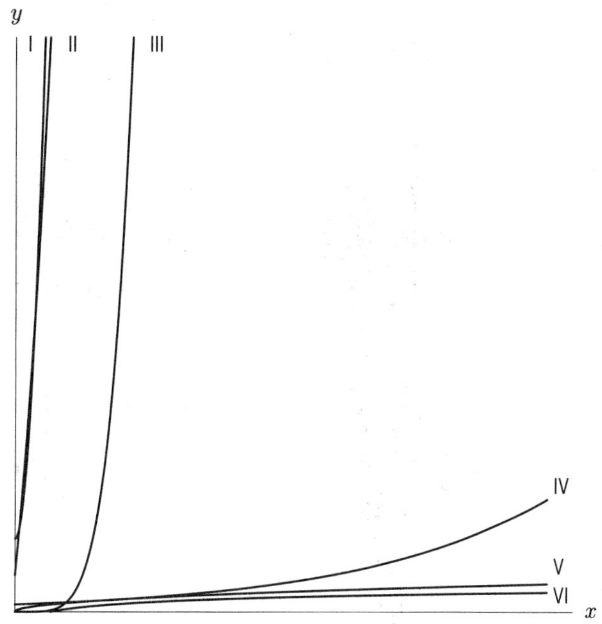

Figure 1.10.28

ANSWER:

(a) II   (b) I   (c) III

(d) V   (e) VI   (f) IV

3. Determine the end behavior of each of the following functions by answering the following questions:

(a) As $x \to \infty$, what does $f(x)$ approach?

(b) As $x \to -\infty$, what does $f(x)$ approach?

(a) $f(x) = -x^3$ (b) $f(x) = x^{-4}$

(c) $f(x) = 370 - 5x^2 - 80x^3 + 10x^4$ (d) $f(x) = e^x$

ANSWER:

(a) (i) As $x \to \infty$, $f(x) \to -\infty$.

    (ii) As $x \to -\infty$, $f(x) \to \infty$.

(b) (i) As $x \to \infty$, $f(x) \to 0$.

    (ii) As $x \to -\infty$, $f(x) \to 0$.

(c) (i) As $x \to \infty$, $f(x) \to \infty$.

    (ii) As $x \to -\infty$, $f(x) \to \infty$.

(d) (i) As $x \to \infty$, $f(x) \to \infty$.

    (ii) As $x \to -\infty$, $f(x) \to 0$.

4. Tables of values for three different functions are given in Table 1.10.13.

(a) Which of these functions could be linear? Find a formula for this function.

(b) Which of these functions could be exponential? Find a formula for this function.

(c) Is the third function increasing or decreasing? Is it concave up or concave down?

Table 1.10.13

t	$f(t)$	$g(t)$	$h(t)$
-1	15	22.0	1000
0	9	24.1	600
1	5	26.2	360
2	4	28.3	216

ANSWER:

(a) Since the successive values given in the table for t each differ by the same amount, any function which has successive values that differ by the same amount could be linear. We find that this is the case for $g(t)$ ($22.0 - 24.1 = 24.1 - 26.2 = 26.2 - 28.3 = -2.1$) but not for $f(t)$ or $h(t)$. So only $g(t)$ could be linear. If $g(t)$ is linear, then it is of the form $b + tx$. We know that $b = 24.1$ from the table, and since the successive values of t given in the table increase by 1, we know that $m = -2.1$ (from our previous calculations). So a possible formula for $g(t)$ is $g(t) = 24.1 + 2.1t$.

(b) Again, since all successive values of t differ by the same amount, any function which has successive values whose ratios are all the same could be exponential. This is the case for $h(t)$ ($600/1000 = 360/600 = 216/360 = 0.6$), but not for $f(t)$ or $g(t)$. So only $h(t)$ could be exponential. If $h(t)$ is exponential, then it is of the form $P_0 a^t$. We know that $P_0 = 600$ from the table, and since the successive values of t given in the table increase by 1, we know that $a = 0.6$ (from our previous calculations). So a possible formula for $h(t)$ is $h(t) = 600(0.6)^t$.

(c) The third function is decreasing, and since it decreases by smaller and smaller amounts as t increases, it is concave up.

5. Write a formula for population, P, as a function of time, t, in each of the following cases.

(a) The population starts at 5000 people and grows by 50 people each year.
(b) The population starts at 5000 people and grows by 5% each year.

ANSWER:

(a) $P = 5000 + 50t$.
(b) $P = 5000(1.05)^t$.

6. Give expressions for $f(x), g(x), h(x)$ which agree with the following table of values.

x	$f(x)$	$g(x)$	$h(x)$
0	-7	0	—
1	-4	2	5
2	-1	8	2.50
3	2	18	$1.66\ldots$
4	5	32	1.25
5	8	50	1

ANSWER:

All the values of $f(x)$ jump by 3 for a change in x of 1, so $f(x)$ is linear with slope 3. Since $f(0) = -7$, we have $f(x) = 3x - 7$.

The differences between successive values of $g(x)$ are as follows: 2, 6, 10, 14, 18., so $g(x)$ is not linear. Notice that the values in the g column, 0, 2, 8, 18 ... are exactly twice the values of the well known function, x^2. So $g(x) = 2x^2$.

$h(x)$ is a decreasing, concave up function that is infinite when $x = 0$ and equal to 1 when $x = 5$. $h(x) = 5/x$ fits the requirements.

7. One of the following tables of data is linear and one is exponential. Say which is which and give an equation that best fits each table.

(a)

x	0	0.50	1.00	1.50	2.00
y	3.12	2.62	2.20	1.85	1.55

(b)

x	0	0.50	1.00	1.50	2.00
y	2.71	3.94	5.17	6.40	7.63

ANSWER:

(a) This table is exponential and we find that the ratios of successive y-values are all 0.84 (when rounded to two decimals). An appropriate equation is therefore

$$y = 3.12(0.84)^{2x} = 3.12(0.7056)^x,$$

since $y(0) = 3.12$. (Check: When $x = 2$, $y(2) = 3.12(0.7056)^2 \approx 1.5534$.) This could equally well be written $y = 3.12e^{-0.3487x}$. Actually, there is a range of possible answers: $y = 3.12a^x$ for any a between 0.7046 and 0.7059 will give the values shown in the table, when rounded to two decimals.

(b) The second table is linear. Pick a line of the form $y = mx + b$. Since $y(0) = 2.71$, $b = 2.71$, and $y = mx + 2.71$. Using the first two points gives

$$m = \frac{3.94 - 2.71}{0.50 - 0} = 2.46,$$

so $y = 2.46x + 2.71$.

8. Match the following graphs with the formulas.

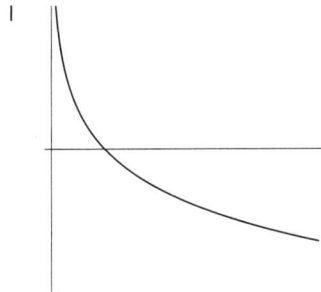

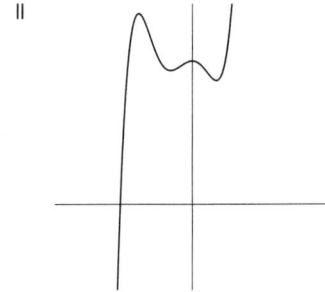

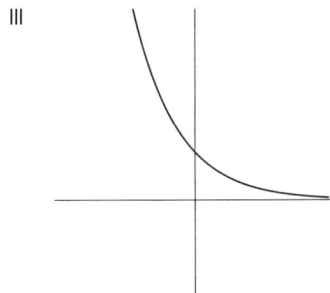

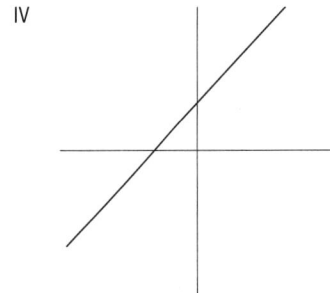

I

II

III

IV

(a) $\ln(e^x) + 1$
(b) $-2\ln x$
(c) e^{-x}
(d) $x^5 + 2x^4 - x^3 - 2x^2 + 5$

ANSWER:

I.   This curve has the appearance of an upside-down ln curve and crosses the x-axis at a positive x-value. Thus (b) is the corresponding equation.

II.   This curve has four "wiggles" in it and thus looks like it corresponds to a degree 5 polynomial. Hence, (d) is the correct equation.

III.   This curve is always positive, decreasing, and concave up. we conclude that (c) is the corresponding equation.

IV.   This is a linear function. Equation (a), $\ln(e^x) + 1$ is actually a linear function in disguise, since $\ln(e^x) = x$. Thus, (a) is the correct equation.

9. One of the functions below is a quadratic, one is a cubic, and one is a trigonometric function. Which is which? Why? [Note: You don't have to find formulas for these functions.]

x	$f(x)$
0.2	−0.42
0.4	−0.65
0.6	0.96
0.8	−0.15
1.2	0.84

x	$g(x)$
1.3	0.41
1.7	0.81
2.5	0.65
3.0	−0.10
3.5	−1.35

x	$h(x)$
0.5	−1.13
1.2	0.13
1.8	0.03
2.0	0.00
2.2	0.05

ANSWER:

$f(x)$ changes direction three times, so it cannot be cubic or quadratic, and thus it is periodic. $h(x)$ changes direction twice, so it cannot be a quadratic, and thus it must be cubic. This leaves only $g(x)$ (which changes direction only once), so it must be a quadratic.

10. Give a possible function for each curve.

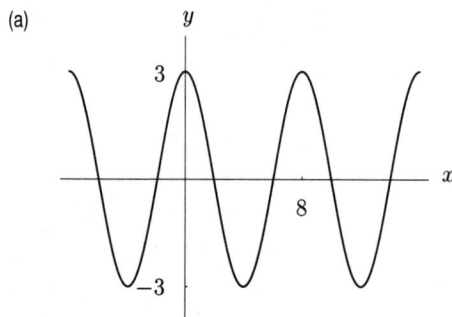

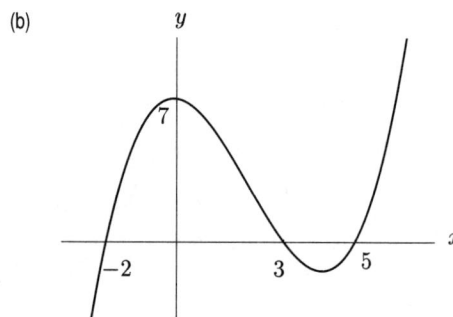

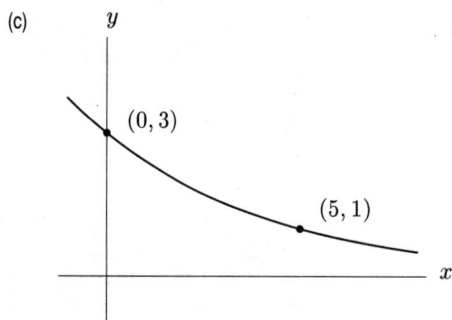

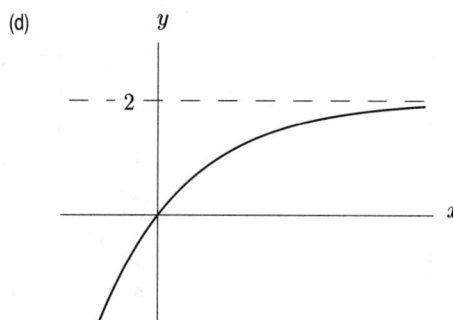

ANSWER:

(a) This graph is periodic with amplitude 3 and period 8 and has a maximum at $x = 0$. So a reasonable solution is $y = 3\cos\left(\frac{\pi}{4}x\right)$.

(b) This curve appears to be a cubic polynomial with roots at $x = -2, 3$ and 5. Thus $y = k(x + 2)(x - 3)(x - 5)$ is a first guess. Since $y(0) = 7$,

$$7 = k(2)(-3)(-5)$$
$$k = \frac{7}{30}$$

So, $y = \frac{7}{30}(x + 2)(x - 3)(x - 5)$ is a possible answer.

(c) This appears to be an exponential decay curve of the form $y = Ak^{-x}$. Since $y(0) = 3$, $y = 3k^{-x}$. Since $y(5) = 1$, we have

$$1 = 3k^{-5}$$
$$k = \left(\frac{1}{3}\right)^{-\frac{1}{5}}$$

So, $y = 3\left(\frac{1}{3}\right)^{\frac{1}{5}x} = 3^{(1 - \frac{x}{5})}$ is a possible answer.

(d) This graph appears to be of the form $y = a(1 - e^{-kx})$. As $x \to \infty$, $y \to a$, and the graph approaches 2, so $y = 2(1 - e^{-kx})$. Any positive k will work, since no scale is indicated for the x-axis.

Problems and Solutions on Fitting Formulas to Data

1. Here are some data from a recent Scientific American article on Old World monkeys.

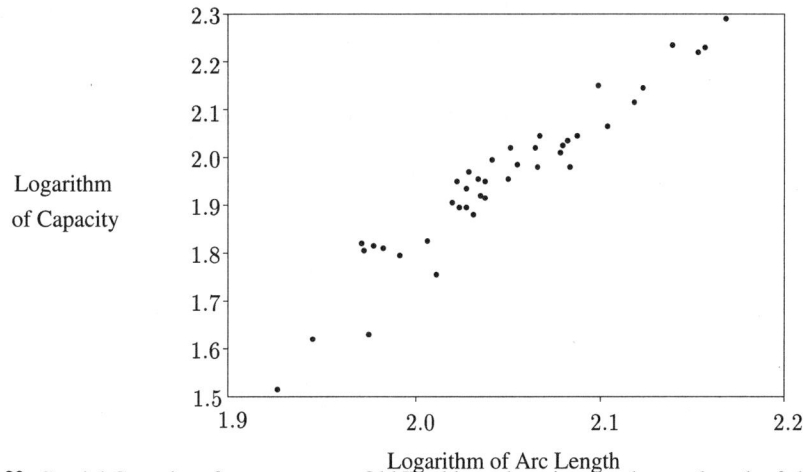

Figure 1.10.29: Cranial Capacity of contemporary Old World monkeys is related to arc length of skull as shown.

(a) From the data presented give an approximate formula for

$$C = \text{cranial capacity (in cm}^3)$$

as a function of

$$A = \text{arc length of skull (in cm)}.$$

[Hint: Fit a line through the data points. Logarithms are to base 10.]

(b) What type of function is $C = f(A)$ (logarithmic, exponential, trigonometric, power function,...)?

ANSWER:

(a) A line fit through the data points goes through $(2.18, 2.3)$ and $(1.9, 1.66)$, and so has slope $\frac{\Delta y}{\Delta x} = \frac{0.64}{0.28} = 2.3$. The equation is

$$y = 2.3x + b.$$

Solving for b yields

$$2.3 = (2.3)2.18 + b$$
$$b \approx -2.7.$$

So $y = 2.3x - 2.7$. But $x = \log(A)$ and $y = \log(C)$, so

$$\log C = 2.3 \log A + 2.7.$$

Exponentiating both sides yields

$$C = 10^{2.3 \log A + 2.7} \approx A^{2.3} (501).$$

(b) This is a power function.

2. The height (in inches) and weight (in pounds) of 8 students is given in Table 1.10.14.

(a) Plot these data, with height on the horizontal axis and weight on the vertical axis. Does a line fit the data reasonably well?

(b) Find the regression line for this data. Graph it with the data.

(c) Interpret the slope of the regression line in terms of height and weight.

(d) Use the regression line to estimate the weight of a person who is 5 feet 7 inches tall.

Table 1.10.14

Height (inches)	64	68	62	70	69	65	73	71
Weight (lbs)	110	150	115	185	160	125	200	170

ANSWER:

(a) See Figure 1.10.30. It appears that a line will fit the data reasonably well.

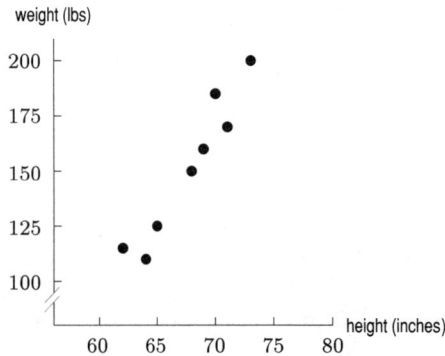

Figure 1.10.30

(b) A calculator or computer gives the regression line Weight $= -419 + 8.43$(Height).

(c) The slope is $\dfrac{8.43}{1} = \dfrac{\text{change in weight}}{\text{change in height}}$. We expect the weight to increase by 8.43 lbs for a one inch increase in height.

(d) To find the weight of a person who is 5 ft 7 inches tall, or 67 inches tall, we substitute 67 into the formula for the regression line found in part (b): Weight $= -419 + 8.43(67) = 145.81$ lbs.

Problems and Solutions on Compound Interest and the Number e

1. You have \$500 invested in a bank account earning 8.2% compounded annually.

(a) Write an equation for the money M in your account after t years.

(b) How long will it take to triple your money?

(c) Suppose the interest were compounded monthly instead, that is you earned $\frac{8.2}{12}$% interest each month. What interest would you then earn for 1 year?

ANSWER:

(a) $M = 500(1.082)^t$.

(b) To triple your money, set $M = 1500$, so

$$\frac{1500}{500} = (1.082)^t$$
$$3 = e^{(\ln 1.082)t}$$
$$\ln 3 = t \ln 1.082$$
$$t = \frac{\ln 3}{\ln 1.082}$$
$$\approx 13.9 \text{ years}$$

(c) If interest is compounded monthly, then we get $M = 500\left(1 + \frac{0.082}{12}\right)^{12t}$, where t is still measured in years. So $M \approx 500(1.0068333)^{12t}$. After $t = 1$ year,

$$M \approx 500(1.0068333)^{12}$$
$$\approx 500(1.08516)$$
$$\approx 542.58$$

so the interest earned is $42.58.

2. A bank pays 6% annual interest. If $1000 is deposited in an account in this bank, find the amount in the account after 5 years if interest is compounded

(a) annually; (b) monthly; (c) weekly; (d) continuously.

ANSWER:

Let B be the amount in the account after 5 years. We find it in each of the four cases by using the formula for the type of interest that is being compounded:

(a) $B = 1000(1.06)^5 = \$1338.23$.

(b) $B = 1000(1 + \dfrac{0.06}{12})^{12(5)} = \$1348.85$.

(c) $B = 1000(1 + \dfrac{0.06}{52})^{52(5)} = \$1349.63$.

(d) $B = 1000e^{(0.06)(5)} = \$1349.86$.

Problems and Solutions on Limits to Infinity and End Behavior

1. A spherical cell takes in nutrients through its cell wall at a rate proportional to the area of the cell wall. The rate at which the cell uses nutrients is proportional to its volume.

(a) Write an expression for the rate at which nutrients enter the cell as a function of its radius, r.

(b) Write an expression for the rate at which the cell uses nutrients as a function of its radius, r.

(c) Sketch a possible graph showing the rate at which nutrients enter the cell against the radius r (put r along the horizontal axis). On the same axes, sketch a possible graph for the rate at which the cell uses nutrients.

(d) Show algebraically why there must be a radius r_0 (other than $r_0 = 0$) at which the rate at which nutrients are used equals the rate at which nutrients enter the cell. Mark r_0 on your graph.

(e) What happens to the cell when $r > r_0$? When $r < r_0$? What does this tell you about the radius of the cell in the long run?

ANSWER:

(a) Rate at which nutrients enter cell $= k4\pi r^2 = Ar^2$ $(A > 0)$ k, c are constants of proportionality.

(b) Rate at which nutrients are used $= c\frac{4}{3}\pi r^3 = Br^3$ $(B > 0)$.

(c)

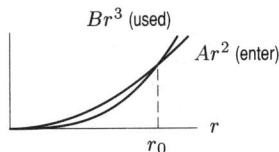

(d) $Ar^2 = Br^3$ for $r = 0$ and $r = \dfrac{A}{B}$ so $r_0 = \dfrac{A}{B}$

(e) When $r > r_0$, rate used > rate enter so cell shrinks.
When $r < r_0$, rate used < rate enter so cell grows.
In long run, cell's radius $\to r_0$.

2. Determine the end behavior of the following functions by answering the questions.

(a) $f(x) = -17x^4$

(i) As $x \to \infty$, what does $f(x)$ approach?

(ii) As $x \to -\infty$, what does $f(x)$ approach?

(b) $g(x) = 5x^7$

(i) As $x \to \infty$, what does $g(x)$ approach?

(ii) As $x \to -\infty$, what does $g(x)$ approach?

ANSWER:

(a) (i) As $x \to \infty$, $f(x) \to -\infty$.

(ii) As $x \to -\infty$, $f(x) \to -\infty$.

(b) (i) As $x \to \infty$, $g(x) \to \infty$.

(ii) As $x \to -\infty$, $g(x) \to -\infty$.

3. (a) Sketch global pictures of the three functions $f(x) = -100x^3$, $g(x) = 28x^4$ and $h(x) = 2x^5$ on the same set of axes. Label each function on both ends so that it is obvious which is which.

(b) Which function has the largest values as $x \to \infty$?

ANSWER:

(a) See Figure 1.10.31.

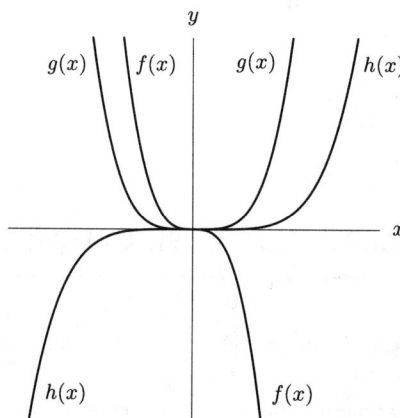

Figure 1.10.31

(b) $h(x)$ approaches ∞ the quickest, in other words, has the largest values as $x \to \infty$.

4. (a) Use your calculator to find all the solutions to the equation

$$2^x = x^2.$$

Give your answers to one decimal place. Sketch the graphs drawn by your calculator as part of the explanation for your answer.

(b) For what values of x is $2^x > x^2$?

ANSWER:

(a)

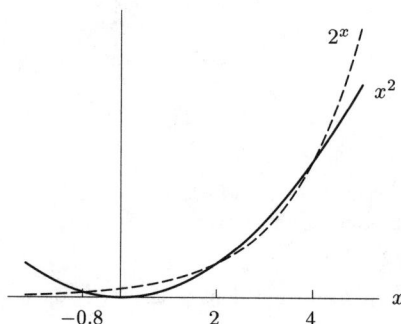

Solutions : $x = -0.8$ (by zooming), $x = 2, 4$.

(b) $2^x > x^2$ for $-0.8 < x < 2$ or $x > 4$.

5. Sketch by hand a global picture of $y = x^4$ and $y = 3^x$ on the same axes.

  ANSWER:

  See Figure 1.10.32.

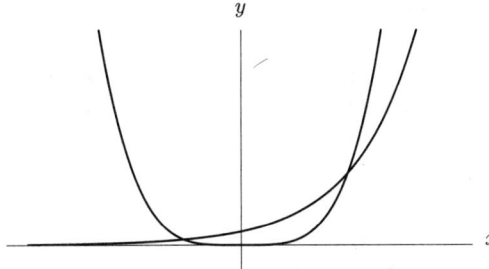

Figure 1.10.32

6. Determine the end behavior of the following functions by answering the questions.

  (a) $f(x) = 5 + 24x + 78x^3 - 17x^4$

   (i) As $x \to \infty$, what does $f(x)$ approach?

   (ii) As $x \to -\infty$, what does $f(x)$ approach?

  (b) $g(x) = 5x^5 - 30x^4 + 2x^2 - 8x + 100$

   (i) As $x \to \infty$, what does $g(x)$ approach?

   (ii) As $x \to -\infty$, what does $g(x)$ approach?

  ANSWER:

  (a) (i) As $x \to \infty$, $f(x) \to -\infty$.

    (ii) As $x \to -\infty$, $f(x) \to -\infty$.

  (b) (i) As $x \to \infty$, $g(x) \to \infty$.

    (ii) As $x \to -\infty$, $g(x) \to -\infty$.

Chapter 2 Exam Questions

Problems and Solutions for Section 2.1

1. Recently Esther swam a lap in an Olympic swimming pool (the length of the pool is 50 meters, and the length of the lap is 100 meters); her times for various positions s (in meters from her starting point) during the lap are given in Table 2.1.15.

Table 2.1.15

t(sec)	0	6.4	13.2	20.4	27.6	34.8	41.6	48.4	55.6	62.8	69.6
s(m)	0	10	20	30	40	50	40	30	20	10	0

(a) Sketch a graph of Esther's position as a function of time. Label and scale your axes.

(b) Complete Table 2.1.16, showing Esther's approximate velocities. (The listed times are midpoints of the time intervals from Table 2.1.15.)

Table 2.1.16

t(sec)	3.2	9.8	16.8	24.0	31.2	38.2	45.0	52.0	59.2	66.2
v(m/sec)										

(c) Sketch an approximate graph of Esther's velocity as a function of time. Label and scale your axes.

ANSWER:

(a)

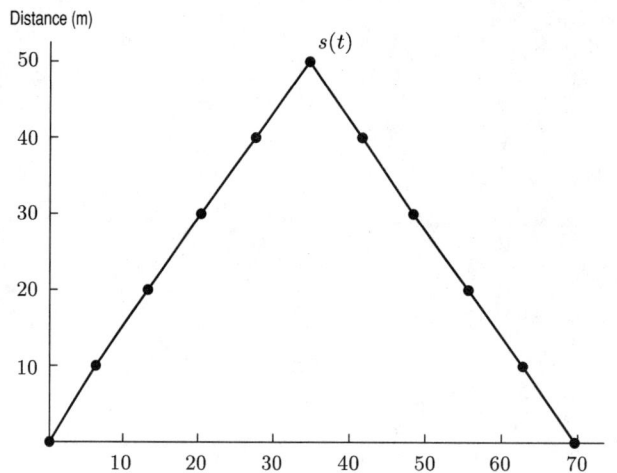

Figure 2.1.33

(b) Calculating the average velocity in each interval, we have Table 2.1.17.

Table 2.1.17

t(sec)	3.2	9.8	16.8	24.0	31.2	38.2	45.0	52.0	59.2	66.2
v(m/sec)	1.563	1.471	1.389	1.389	1.389	1.471	1.471	1.389	1.389	1.389

(c) Use Table 2.1.17, we have Figure 2.1.34.

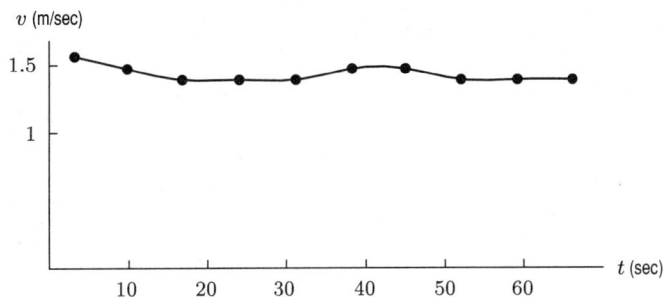

Figure 2.1.34

2. Let $f(t) = t^2 + t$.

   (a) What is the change in $f(t)$ between $t = 2$ and $t = 5$?
   (b) What is the average rate of change in $f(t)$ between $t = 2$ and $t = 5$?
   (c) What is an approximate value for the rate of change of f at $t = 2$?
   (d) How could you improve your estimate in part (c)?

   ANSWER:

   (a) $f(5) - f(2) = 30 - 6 = 24$.
   (b) $\dfrac{24}{5 - 2} = 8$.
   (c) $\dfrac{f(3) - f(2)}{1} = 6$.
   (d) Choose a smaller h in $\dfrac{f(2 + h) - f(2)}{h}$.

3. An amount of $500 was invested in 1970 and increased as shown in Table 2.1.18. (Amounts are given for the beginning of the year.)

   Table 2.1.18

Year	1970	1975	1980	1985	1990	1995
Capital	500	966	1856	3578	6876	13,233

   (a) Find the average rate at which the capital increased:
      (i) During the seventies.
      (ii) During the nineties (using the information up to 1995).
   (b) Compare the growth rate over both these intervals as percentages of the amounts at the beginning of each of the two periods and comment on this.
   (c) Find an approximation to the instantaneous growth rate at the beginning of 1990.
   (d) What additional information would you need in order to make a better approximation in part (c)?

   ANSWER:

   (a) (i) The average rate is $\dfrac{1856 - 500}{10} = \$135.60$ per year during the seventies.

      (ii) For the first half of the nineties, the average rate is $\dfrac{13{,}233 - 6876}{5} = \$1271.40/\text{year}$.

   (b) The percentages are $\dfrac{135.6}{500} = 0.2712$, i.e. 27%, and $\dfrac{1271.4}{6876} = 0.1849$, i.e. 18.5%.
      Since the average rate is taken over 10 years for the seventies compared with 5 years for the nineties, the percentage for the seventies is bigger than that for the nineties.

   (c) Take the average:
   $$\frac{13{,}233 - 6876}{5} = \$1271.4.$$

   At the begining of 1990, the instantaneous rate is approximately $1271.40 per year.
   (d) To make a better approximation, we need more data around 1990 or the formula for the balance.

4. (a) Explain how the average rate of change of a function f can be used to find the instantaneous rate of change of f at a point x_0.
   (b) Give a geometric interpretation of the instantaneous rate of change.
   ANSWER:

   (a) By taking points $x_1, x_2, \ldots$ closer and closer to x_0 and calculating the average rate of change of f over the interval $[x_0, x_n]$, we get a sequence which approaches the instantaneous rate of change of f at x_0.
   (b) The instantaneous rate of change at a given point is the slope of a tangent to the curve at that point.

5. (a) If $x(V) = V^{1/3}$ is the length of the side of a cube in terms of its volume, V, then calculate the average rate of change of x with respect to V over the intervals $0 < V < 1$ and $1 < V < 2$.
   (b) What might we conclude about this rate as the volume V increases? Is it increasing? Decreasing?
   ANSWER:

   (a) Average rate of change of $x = \dfrac{1^{1/3} - 0^{1/3}}{1 - 0} = 1$ for $0 < V < 1$.

   average rate of change of $x = \dfrac{2^{1/3} - 1^{1/3}}{2 - 1} \approx 0.26$ for $1 < V < 2$.
   (b) We conclude that as V increases, the rate of change of x decreases.

6. The graph in Figure 2.1.35 is the graph of $N = C(t)$, the cumulative number of customers served in a certain store during business hours one day, as a function of the hour of the day.

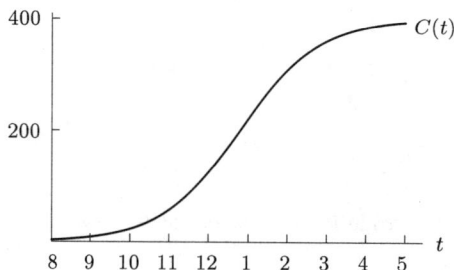

Figure 2.1.35

   (a) About when was the store the busiest?
   (b) What does $C'(t)$ mean in practical terms?
   (c) Estimate $C'(11)$.

   ANSWER:

   (a) The store was the busiest when $C(t)$ was increasing the fastest, which was at about 1 pm.
   (b) $C'(t)$ is the rate of increase, at time t, of the cumulative number of customers served in the store. This can perhaps more easily be understood as the instantaneous rate of service of the store, at time t, in number of customers per hour.
   (c) $C'(11) = $ the slope of a tangent line to $C(t)$ at $t = 11$, which appears to be about $\dfrac{200}{4} = 50$ customers per hour.

7. If the graph of $y = f(x)$ is shown below, arrange in ascending order (i.e., smallest first, largest last):

$$f'(A) \qquad f'(B) \qquad f'(C) \qquad \text{slope } AB \qquad \text{the number 1} \qquad \text{the number 0}$$

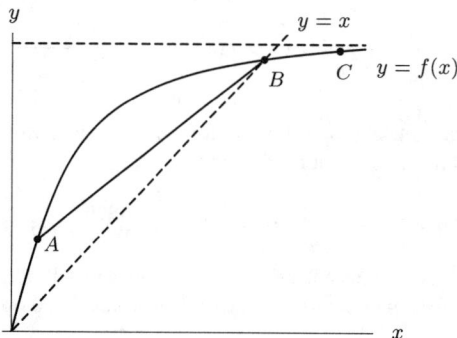

ANSWER:

By eye, we can see that $f'(C) < f'(B) < f'(A)$. We can also see that $f'(B) <$ slope $AB < f'(A)$, so we have $f'(C) < f'(B) <$ slope $AB < f'(A)$. Finally, we note that all the slopes on this graph are positive, and that $f'(A)$ is the only slope that is greater than the slope of $y = x$, namely 1. So we have $0 < f'(C) < f'(B) <$ slope $AB < 1 < f'(A)$.

8. **(a)** Estimate $f'(0)$ when $f(x) = 2^{-x}$.
(b) Will your estimate be larger or smaller than $f'(0)$? Explain.

ANSWER:

(a) To estimate $f'(0)$, find the average slope over intervals that get smaller and smaller but still contain $x = 0$:

Interval Size	Average Slope
0.1	$\frac{f(0.1)-f(0)}{0.1} \approx -0.670$
0.01	$\frac{f(0.01)-f(0)}{0.01} \approx -0.691$
0.001	$\frac{f(0.001)-f(0)}{0.001} \approx -0.693$

$f'(0)$ appears to be about -0.693.

(b) The average slopes in the chart above seem to approach a limiting value (which turns out to be $\ln\frac{1}{2} \approx -0.69315$) from above; this indicates that our estimate of $f'(0)$ is probably an overestimate.

9. Given the following data about a function f,

x	3.0	3.2	3.4	3.6	3.8
$f(x)$	8.2	9.5	10.5	11.0	13.2

(a) Estimate $f'(3.2)$ and $f'(3.5)$.
(b) Give the average rate of change of f between $x = 3.0$ and $x = 3.8$.
(c) Give the equation of the tangent line at $x = 3.2$.

ANSWER:

(a) Estimate the slope at 3.2 by finding the average slope over the interval $[3.2, 3.4]$:

$$\text{Slope} = \frac{f(3.4) - f(3.2)}{3.4 - 3.2} = \frac{10.5 - 9.5}{0.2} = 5$$

To estimate the slope at 3.5, we have to look at the average slope over $[3.4, 3.6]$, which is $\frac{11.0 - 10.5}{0.2} = 2.5$.

(b) The average rate of change is $\frac{13.2-8.2}{3.8-3.0} = 6.25$.

(c) At $x = 3.2$, $f(x) = 9.5$ and the slope ≈ 5 by part (a). So

$$y - 9.5 = 5(x - 3.2)$$
$$y = 9.5 + 5x - 16$$
$$= 5x - 6.5$$

10. A certain function f is decreasing and concave down. In addition, $f'(3) = -2$ and $f(3) = 5$.

(a) Sketch the graph of f.
(b) Estimate $f(2)$, namely give two values you're sure $f(2)$ is between.
(c) Estimate the zeros of f. (First say how many there are and why.)

ANSWER:

(a)

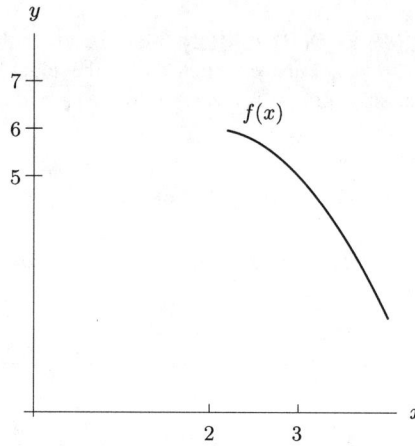

Figure 2.1.36

(b) Since f is concave down, it lies below any of its tangent lines (except at the point of contact). The tangent line at $(3, 5)$ has the equation $y = -2x + 11$. Hence $f(2) < -2(2) + 11 = 7$. Since f is decreasing, $f(2) > f(3) = 5$. Thus we know that $5 < f(2) < 7$. See Figure 2.1.37.

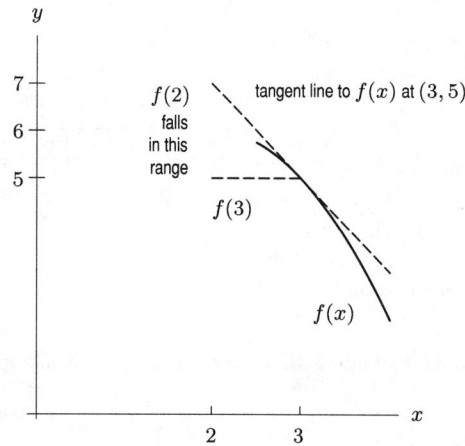

Figure 2.1.37

(c) Since f is decreasing, it can cross the x-axis only once, so there is only one zero. Since the curve lies everywhere below the tangent line $y = -2x + 11$ (considered in Part (b)), and this tangent line crosses the x-axis at $x = 5.5$, we know that $f(x)$ must cross the x-axis somewhere between $x = 3$ and $x = 5.5$. See Figure 2.1.38. The root can thus lie anywhere in the interval from 3 to 5.5.

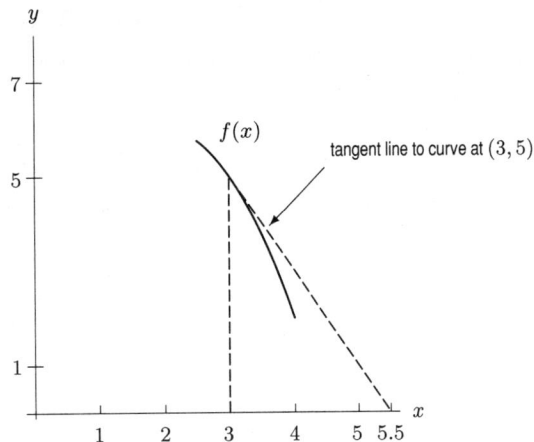

Figure 2.1.38

11. The growth graph in Figure 2.1.39 shows the height in inches of a bean plant during 30 days. How fast was the plant growing on the 15th day? Be sure to give the units of measurement as part of your answer.

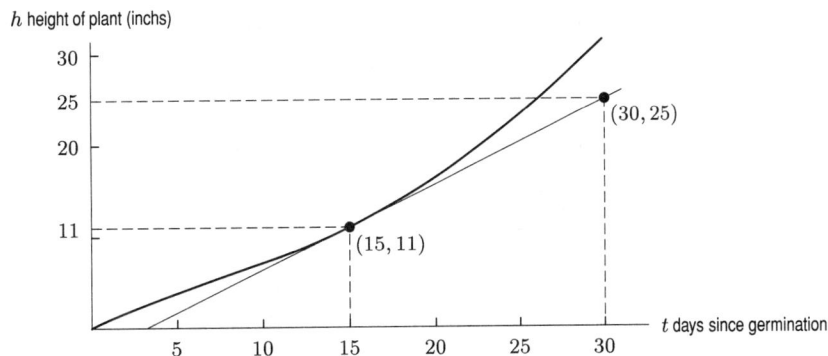

Figure 2.1.39

ANSWER:

Growth rate $= \frac{\Delta h}{\Delta t} = \frac{25-11}{30-15}\frac{\text{in}}{\text{days}} = 0.93\text{in/day}$.

12. (a) The graph of $h(x)$ is given in Figure 2.1.40. Indicate on the graph:

   (i) A line segment whose length equals the change Δh in $h(x)$ between $x = 20$ and $x = 40$.

   (ii) A line segment whose slope equals the average rate of change $\dfrac{\Delta h}{\Delta x}$ of $h(x)$ between $x = 20$ and $x = 40$.

   (iii) A line whose slope equals the derivative $h'(10)$.

   (iv) A point on the graph where $h' = 0$.

(b) $h'(30) \approx$ _____.

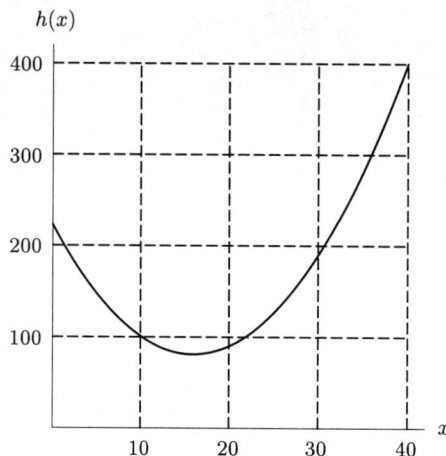

Figure 2.1.40

ANSWER:

(a)

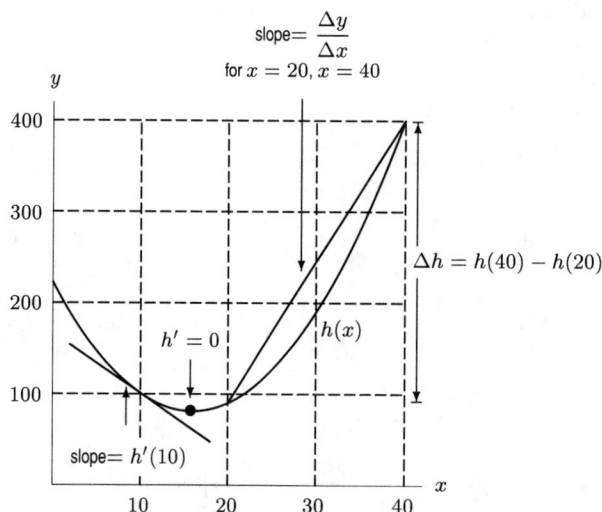

Figure 2.1.41

(b) $h'(30) \approx \dfrac{300}{20} = 15$.

13.

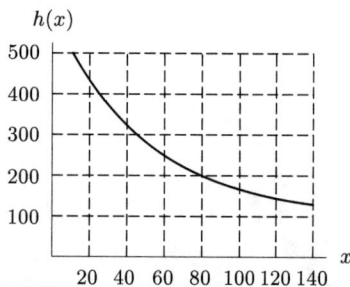

$h'(80) \approx$ _____

ANSWER:

We find $h'(80)$ by approximating the slope of the tangent line to the graph at $x = 80$.

$$h'(80) \approx \frac{100 - 260}{120 - 40}$$

$$h'(80) \approx -2$$

14. What does the expression $\dfrac{f(2) - f(1.99)}{0.01}$ represent if f is a function?

ANSWER:

$\dfrac{f(2) - f(1.99)}{0.01}$ represents the slope of the line through $(2, f(2))$ and $(1.99, f(1.99))$, which is an estimate of $f'(2)$ or $f'(1.99)$.

15. Using a difference quotient, compute $f'(1) \approx$ _____ for $f(x) = \sin(3x)$.

ANSWER:

$$f'(1) \approx \frac{\sin\left(3\left(1 + 0.01\right)\right) - \sin 3}{0.01}$$

$$f'(1) \approx -2.97$$

16. The height of an object in feet above the ground is given in Table 2.1.19.

Table 2.1.19

t (sec)	0	1	2	3	4	5	6
y (feet)	10	45	70	85	90	85	70

(a) Compute the average velocity over the interval $0 \le t \le 3$.

(b) Compute the average velocity over the interval $2 \le t \le 4$.

(c) If the height of the object is doubled, how do the answers to (a) and (b) change?

ANSWER:

(a) During the interval of $(3-0) = 3$ seconds, the object moves $(85 - 10) = 75$ feet. The average velocity is $75/3 = 25$ ft/sec.

(b) During the interval of $(4-2) = 2$ seconds, the object moves $(90 - 70) = 20$ feet. The average velocity is $20/2 = 10$ ft/sec.

(c) If the height is doubled, the average velocity also doubles. The answer to (a) would be 50 ft/sec., and the answer to (b) would be 20 ft/sec.

17. (a) Sketch a graph of height in terms of time for the object in the previous problem over the interval $0 \le t \le 6$.

(b) Indicate the average velocity for $0 \le t \le 3$ on the graph from (a).

ANSWER:

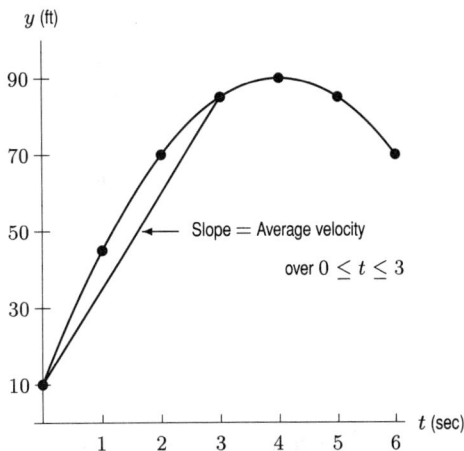

Figure 2.1.42

18. The graph of $p(t)$ in Figure 2.1.43 gives the position of a particle at time t. List the following quantities in order, smallest to largest:

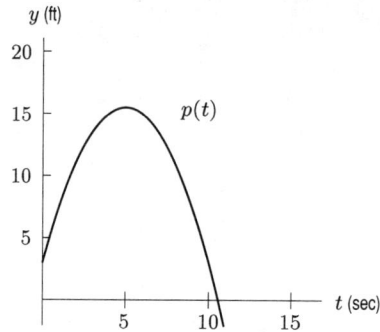

Figure 2.1.43

A, average velocity on $1 \leq t \leq 3$.

   B, average velocity on $8 \leq t \leq 10$.

   C, instantaneous velocity at $t = 1$.

   D, instantaneous velocity at $t = 3$.

   E, instantaneous velocity at $t = 10$.
   ANSWER:

   Since $p(t)$ is concave down on $1 \leq t \leq 3$, the average velocity between the two times should be less than the instantaneous velocity at $t = 1$, but greater than the instantaneous velocity at $t = 3$, so **D** < **A** < **C**. For analogous reasons, **E** < **B**. Finally, note that $p(t)$ is decreasing over the interval $8 \leq t \leq 10$, but increasing at $t = 0$, so **D** > 0. Therefore, **E** < **B** < **D** < **A** < **C**.

19. The following graph describes the position of a car at time t. Write a short story of the trip that corresponds to the graph. Be sure to discuss the average velocity.

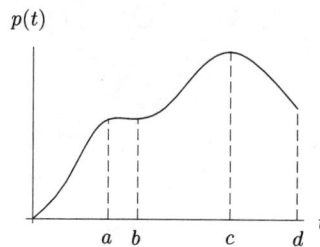

Figure 2.1.44

   ANSWER:

   Answers will vary. Average velocity on $0 \leq t \leq a$ is positive, on $a \leq t \leq b$ is approximately zero, positive on $b \leq t \leq c$ and negative on $c \leq t \leq d$. Average velocity on $b \leq t \leq c$ is smaller than on $0 \leq t \leq a$.

Problems and Solutions   for Section 2.2 ————————————————

1. A certain bacterial colony was observed for several hours and the following conditions were reported:

   • There were 1000 bacteria after 5 hours.

- The growth rate was never negative and never exceeded 100 per hour.
- The growth rate was decreasing for the first 5 hours.
- At 7 hours, the rate of growth was zero.

(a) Sketch a possible graph of the number $N(t)$ of bacteria as a function of time t.

(b) Which of the following are possible? Explain your answer in each case.

    (i) $N(7) = 1250$

    (ii) $N(0) = 450$

    (iii) $N'(0) = 0$

    (iv) $N'(7) = 0$

    ANSWER:

(a)

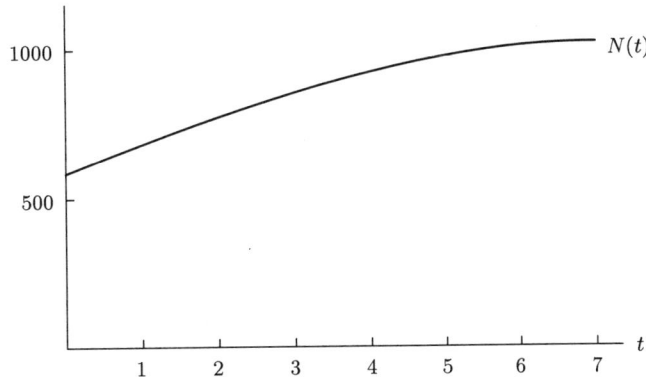

Figure 2.2.45

(b)   (i) Impossible, since in 2 hours N can not increase by 250.

    (ii) $N(0) = 450$ is impossible. If it were possible, we assume that the growth rate reached its maximum during the observation, that is the growth rate were 100 per hour, then the net number of the bacteria would have increased 500 in 5 hours. Add this 500 to $N(0) = 450$m and we would have only 950 bacteria after 5 hours; but this is not true according to the information given in the problem.

    (iii) $N'(0) = 0$ is impossible. Since the growth rate was decreasing for the first 5 hours, if $N'(0)$ were 0, then the growth rate would have been negative.

    (iv) $N'(7) = 0$ is given.

2. Which of the functions below could be the derivative of which of the others? (Hint: try all combinations.)

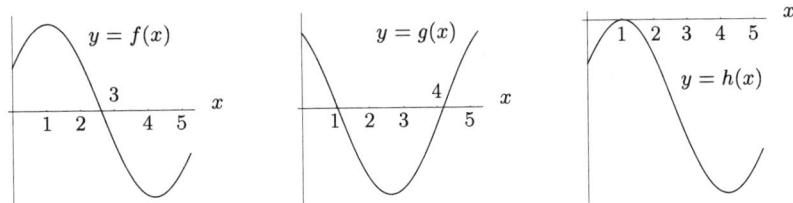

    ANSWER:

    $g(x)$ could be the derivative of $h(x)$ or $f(x)$

3. Below is the graph of a function f. Sketch the graph of its derivative f' on the same axes.

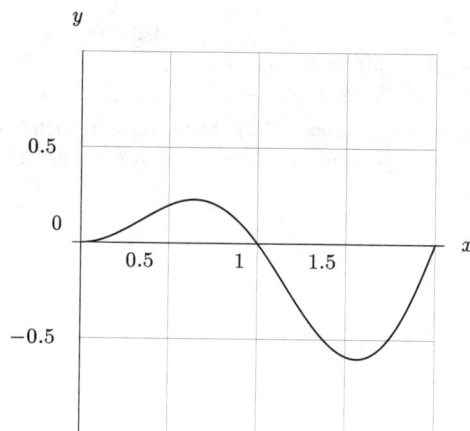

ANSWER:

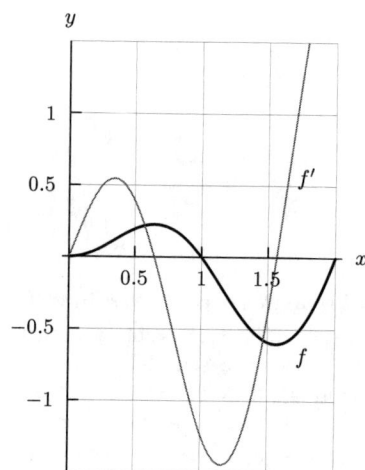

4. Sketch the graph of the derivative, $y = f'(x)$, for each of the functions $y = f(x)$ whose graphs are given below.

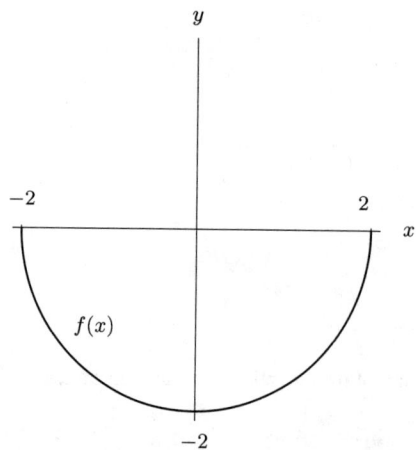

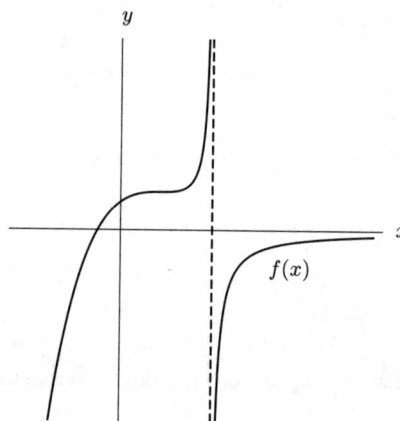

ANSWER:

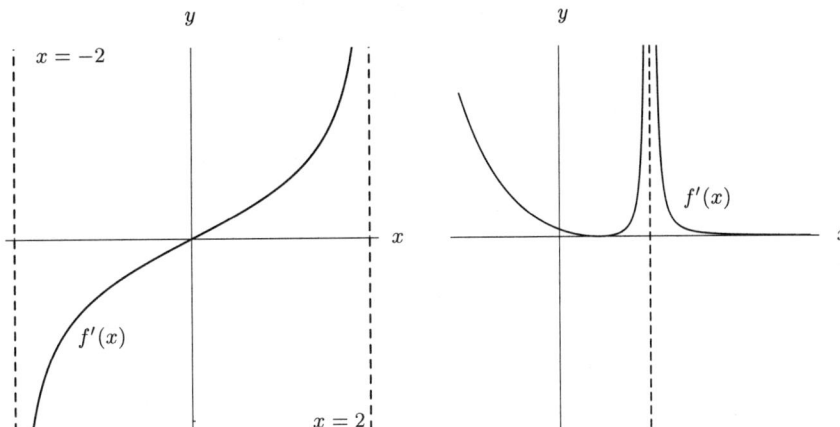

5. Indicate a scale on the axes for the graph of the derivative and sketch the graph of $f'(x)$.

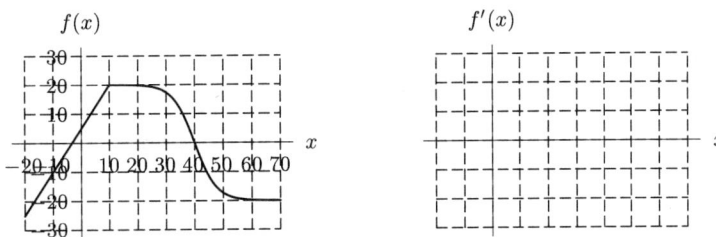

ANSWER:

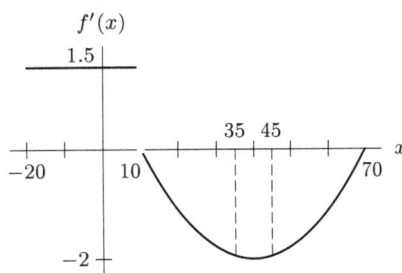

Figure 2.2.46

6. Sketch the graph of f' if the graph in Figure 2.2.47 represents f.

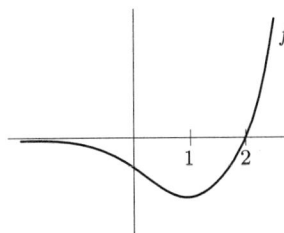

Figure 2.2.47

ANSWER:

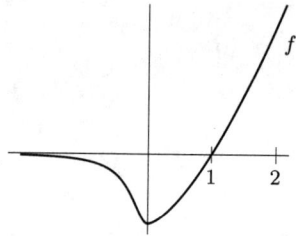

Figure 2.2.48

7. Sketch the graph of a function f given the information in Figure 2.2.49. Mark x_1 and x_2 on your graph of f.

$f' > 0$	$f' = 0$	$f' < 0$	f' is undefined	$f' < 0$

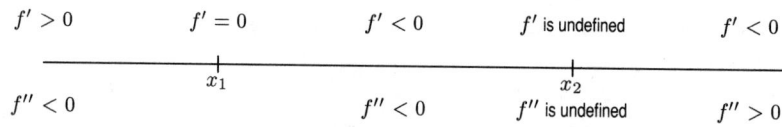

$f'' < 0$		$f'' < 0$	f'' is undefined	$f'' > 0$

Figure 2.2.49

ANSWER:

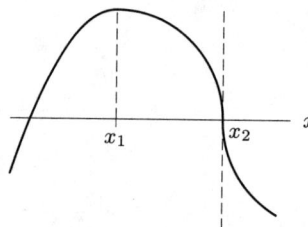

Figure 2.2.50

8. Sketch the graph of f' if the graph in Figure 2.2.51 represents the function f.

Figure 2.2.51

ANSWER:

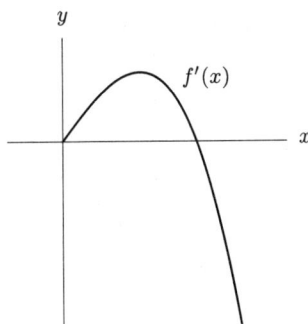

Figure 2.2.52

9. Draw the graph of a continuous function $y = g(x)$ that satisfies the following three conditions:

- $g'(x) = 0$ for $x < 0$
- $g'(x) > 0$ for $0 < x < 2$
- $g'(x) < 0$ for $x > 2$

ANSWER:

From the given information, we know that g is constant for $x < 0$, is increasing between $x = 0$ and $x = 2$, and is decreasing for $x > 2$. Figure 2.2.53 shows a possible graph—answers may vary.

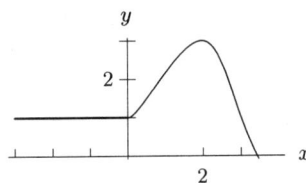

Figure 2.2.53

Problems and Solutions for Section 2.3

1. Let $g(v)$ be the fuel efficiency of a car moving at v miles per hour, with efficiency measured in miles per gallon.

 (a) Give the meaning, in plain English, of the equation $g(55) = 27$.
 (b) Give the meaning, in plain English, of the equation $g'(55) = -0.54$.
 (c) Give the units for $g'(v)$.
 (d) Why is $g'(55)$ negative?

 ANSWER:

 (a) $g(55) = 27$ means that, if the car travels at 55 miles per hour, the fuel efficiency is 27 miles per gallon, i.e. the car consumes 1 gallon of gasoline for every 27 miles.
 (b) $g'(55) = -0.54$ means that increasing the speed by 1 mile per hour will reduce the fuel efficiency by 0.54 mile per gallon, i.e., if the car travels at 56 miles per hour, it will consume 1 gallon of gasoline for every 26.46 miles.
 (c) The units for $g'(v)$ are $\dfrac{\text{miles/gallon}}{\text{miles/hour}}$.
 (d) The most fuel efficient speed of the car is less than 55 mph, so increasing the speed will reduce the fuel efficiency, which implies that $g'(55)$ is negative.

2. Suppose $g(t)$ is the height in inches of a person who is t years old.

 (a) Give a reasonable approximation for the following:
   (i) g(0)
   (ii) g(30)

(b) Give the meaning, in plain English, of the equation $g'(10) = 2$.

(c) What is $g'(40)$?

  ANSWER:

(a) (i) $g(0) = 20$ inches.

  (ii) $g(30) = 70$ inches.

(b) $g'(10) = 2$ means that at age 10 the person's height increases by 2 inches per year.

(c) $g'(40) = 0$, since at age 40 a person is not growing in height.

3. Let $f(T)$ be the time, in minutes, that it takes for an oven to heat up to $T°F$. What is/are the

  units of $f'(T)$ _____

  sign of $f'(T)$ _____

  meaning of $f(300) = 10$

  meaning of $f'(300) = 0.1$.

  ANSWER:

  $f'(T)$ is measured in minutes per $°F$. The sign of $f'(T)$ is positive because it takes longer for an oven to heat up to a higher temperature. $f(300) = 10$ means that it takes 10 minutes for an oven to heat up to $300°F$. $f'(300) = 0.1$ means that near $300°F$, it takes about 0.1 minute for the temperature to increase by one more degree F.

4. Suppose that $f(T)$ is the cost to heat my house, in dollars per day, when the outside temperature is T degrees Fahrenheit.

(a) What does $f'(23) = -0.17$ mean?

(b) If $f(23) = 7.54$ and $f'(23) = -0.17$, approximately what is the cost to heat my house when the outside temperature is $20°F$?

  ANSWER:

(a) $f'(23) = -0.17$ means that when the temperature outside is 23 degrees, the cost of heating the house will decrease by a rate of approximately 17 cents per day for each degree above 23. Since we know nothing about how $f(T)$ behaves at temperatures other than $T = 23$, it is impossible to know over which range of temperatures this approximation is valid. It seems reasonable to assume, however, that $f(T)$ will be relatively smooth over a range of a few degrees.

(b) If the temperature goes down by $3°$ (i.e., to $20°$), then the cost will increase by about $(-3)(-0.17) = 0.51$, resulting in a cost of $\$7.54 + \$0.51 = \$8.05$.

5. To study traffic flow along a major road, the city installs a device at the edge of the road at 4:00 a.m. The device counts the cars driving past, and records the total periodically. The resulting data is plotted on a graph, with time (in hours) on the horizontal axis and the number of cars on the vertical axis. The graph is shown below; it is the graph of the function

$$C(t) = \text{Total number of cars that have passed by after } t \text{ hours.}$$

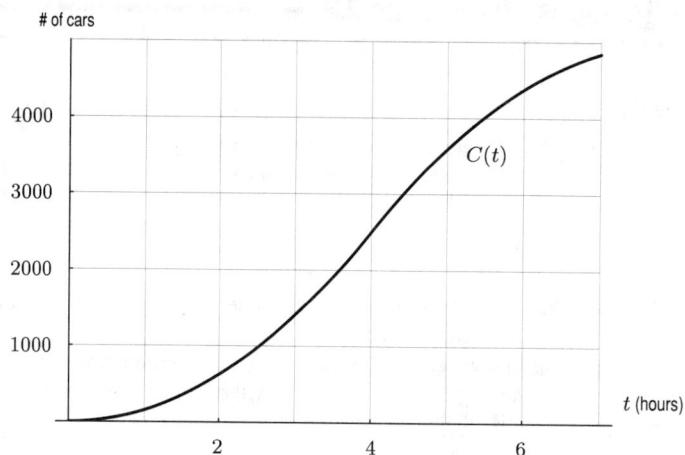

Figure 2.3.54: Traffic Along Speedway

(a) When is the traffic flow greatest?

(b) From the graph, estimate $C'(3)$.

(c) What is the meaning of $C'(3)$? What are its units? What does the value of $C'(3)$ you obtained in (b) mean in practical terms?

ANSWER:

(a) Traffic flow is greatest when the slope of $C(t)$ is greatest, which occurs at about $t = 4$. Since t is in hours past 4:00 a.m., the flow is greatest at about 8:00 a.m.

(b) $C'(3) \approx 1000$

(c) $C'(3)$ tells us how many cars per hour are flowing past at 7:00 a.m. The value we obtained above for $C'(3)$ tells us that traffic flow at that time is about 1000 cars per hour.

6. Every day the Office of Undergraduate Admissions receives inquiries from eager high school students (e.g. "Please, please send me an application", etc.) They keep a running account of the number of inquiries received each day, along with the total number received until that point. To the right is a table of *weekly* figures from about the end of August to about the end of October of a recent year.

Week of	Inquiries That Week	Total for Year
8/28–9/01	1085	11,928
9/04–9/08	1193	13,121
9/11–9/15	1312	14,433
9/18–9/22	1443	15,876
9/25–9/29	1588	17,464
10/02–10/06	1746	19,210
10/09–10/13	1921	21,131
10/16–10/20	2113	23,244
10/23–10/27	2325	25,569

(a) One of these columns can be interpreted as a rate of change. Which one? Of what? Explain.

(b) Based on the table write a formula that gives approximately the total number of inquiries received by a given week. Explain.

(c) Using your answer in part (b), roughly how many inquiries will the admissions office receive over the entire year?

(d) The actual number of inquiries that year was about 34,000. Discuss this, using your knowledge of how people apply to college.

ANSWER:

(a) The second column – Inquiries That Week – is the weekly rate of change of the total for the year since, for example, $13,121 - 11,928 = 1193$; we see that 1193 is the difference between the total number of inquiries as of 9/04 and 9/11.

(b) We have that the ratio of consecutive entries in the second column (total applicants for the year) is always about 1.1. So if T is the total number of applicants then we can try the exponential model

$$T(t) = (11,928)(1.1)^t$$

with $t = 0$ corresponding to the week of 8/28 to 9/01.

(c) There are about 18 weeks from the start $(t = 0)$ until the end of the year. Putting $t = 18$ into the formula for T above gives $T = 66,319$ for the total number of applications for the year.

(d) Since most students send for applications in October and November to apply by the first of January, requests should fall off in November, not continue to rise as our formula suggests. So the true figure (34,000) should be much less than the calculated figure (66,319).

7. The cost of extracting T tons of ore from a copper mine is $f(T)$ dollars.

(a) $f(2000) = 30,000$. What does this equation mean in terms of copper and money?

(b) What are the units of measurement of $f'(2000)$?

(c) Would you expect $f'(2000)$ to be positive or negative? Justify your answer.

(d) $f'(2000) = 10$. What does this equation mean in terms of copper and money?

ANSWER:

(a) It costs \$30,000 to extract 2000 tons of ore.

(b) $\dfrac{\Delta f}{\Delta T} = \dfrac{\text{dollars}}{\text{ton}}$

(c) Positive. It costs more to extract more ore.

(d) At a production level of 2000 tons of ore, it costs about \$10 to extract an extra additional ton.

8. Table 2.3.20 shows the number of bags of oranges sold in one month, $f(p)$, against the price per bag, p (in cents).

(a) Find an approximation for $f'(900)$.

(b) Describe what is meant by the equation $f'(800) = -80$.

Table 2.3.20

Price p (in cents)	750	800	850	900	950
Number of bags, $f(p)$	50,000	48,000	44,000	37,000	29,000

ANSWER:

(a) $f'(900) \approx \dfrac{29,000 - 37,000}{950 - 900} = -160$ bags/cent.

(b) $f'(800) = -80$ means that if the price is 800 cents per bag, increasing the price by 1 cent will reduce the sale by 80 bags per month.

9. This table gives the wind chill factor (°F) as a function of wind speed (miles/hour) when air temperature is $20°$ F.

Wind speed (mph)	5	10	15	20	25
Wind chill factor (°F)	16	3	−5	−10	−15

(a) Give an approximation (including units) of the derivative of wind chill with respect to wind speed when the air temperature is $20°$ F and the wind speed is 10 miles/hour.

(b) Explain in practical every day language the meaning of the number you computed in part(a).

ANSWER:

(a) From the table we can get an approximation:

$$\text{Rate of change } = \frac{-5 - 3}{15 - 10} = -1.8°\text{F/mph.}$$

The derivative of windchill at 10 mph will then be approximately $-1.8°$ F/mph.

(b) When the wind speed is 10 miles/hour, an additional mile/hour will decrease the wind chill by $1.8°$F.

10. Let $t(h)$ be the temperature in degrees Celsius at a height h (in meters) above the surface of the earth. What do each of the following quantities mean to a sky diver? Give units for the quantities.

(a) $t(1000)$ (b) $t'(20)$

(c) $t(h) + 20$ (d) h such that $t'(h) = 20$

ANSWER:

(a) The temperature in degrees Celsius at a height of 1000 meters.

(b) The rate of change of temperature with respect to height at 20 meters above the surface of the earth, in units of degrees per meter.

(c) The temperature at height h plus a temperature of $20°$C.

(d) The height, in meters, at which the rate of change of temperature with respect to height is 20 degrees per meter.

11. Let $L(r)$ be the amount of lumber, in board-feet, produced from a tree of radius r (measured in inches). Interpret the following in practical terms, giving units.

(a) $L(6)$ (b) $L'(24)$

(c) r such that $L(r) = 100$ (d) r such that $L'(r) = 10$

ANSWER:

(a) The number of board-feet of lumber obtained from a tree of radius 6 inches.

(b) The rate of change in the amount of lumber, with respect to radius when radius is 24 inches, in board-feet per inch.

(c) The radius (in inches) of a tree that produces 100 board-feet of lumber.

(d) The radius (in inches) at which the rate of change in board-feet per inch is 10.

12. The noise level, N, in decibels, of a rock concert is given by $N = f(d)$, where d is the distance in meters from the concert speakers. What do the following quantities mean to someone who lives in the neighborhood near the concert? Give units for the quantities.

(a) $f'(100)$

(b) $f'(1000)$
(c) d such that $f(d) = 100$

ANSWER:

(a) The rate of change, in decibels per meter, of noise 100 meters away from the speakers.
(b) The rate of change, in decibels per meter, of noise 1000 meters away from the speakers.
(c) The distance, in meters, away from the speakers at which the noise is 100 decibels.

13. In the scenario from Exercise 12,

   (a) Explain in words what it means if $f'(100) > f'(1000)$.
   (b) What situation might explain the expression $f(d) + 50$ describe?

   ANSWER:

   (a) The rate of change in sound, in decibels, at 100 meters from the speakers is greater than the rate of change at 1000 meters from the speakers.
   (b) The noise level, in decibels, d meters from the speakers plus some additional noise of 50 decibels.

14. The population of a certain town is given by the function $P(t)$ where t is the number of years since the town was incorporated.

   (a) What does it mean to say $P'(175) = -50$?
   (b) What does it mean to say $P'(185) = 100$?
   (c) If $P'(t)$ is constant for $185 < t$, what will $P(200)$ be if $P(185) = 25,500$?

   ANSWER:

   (a) 175 years after incorporation, the population is decreasing at a rate of 50 people per year.
   (b) 185 years after incorporation, the population is increasing at a rate of 100 people per year.
   (c) From (b) we know $P'(t) = 100$ for $185 < t$. This means the population will increase by 100 people per year.
   $P(200) = (200 - 185)(100) + 25,500 = 1,500 + 25,500 = 27,000$ people.

Problems and Solutions   for Section 2.4

1. **(a)** The graph of $f(x)$ is given in Figure 2.4.55. Sketch a graph of $f'(x)$ on the axes provided in Figure 2.4.56.

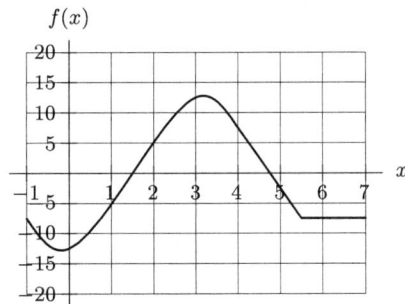

Figure 2.4.55

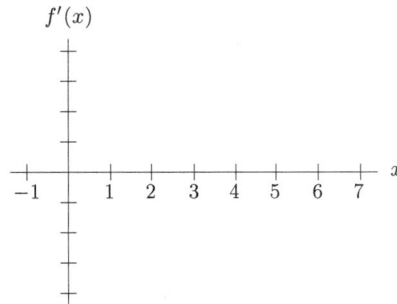

Figure 2.4.56

   (b) Using the graphs in part (a), complete the tables of values below.

Table 2.4.21

x	0	1	2	3	4	5	6	7
$f(x)$								

Table 2.4.22

x	1	2	3	4
$f'(x)$				

Chapter Two /SOLUTIONS

(c) Determine if the following are positive, negative or zero, and explain how you know.

(a) $f''(0.5)$ (b) $f''(2.5)$ (c) $f''(4.5)$

ANSWER:

(a) See Figure 2.4.57.

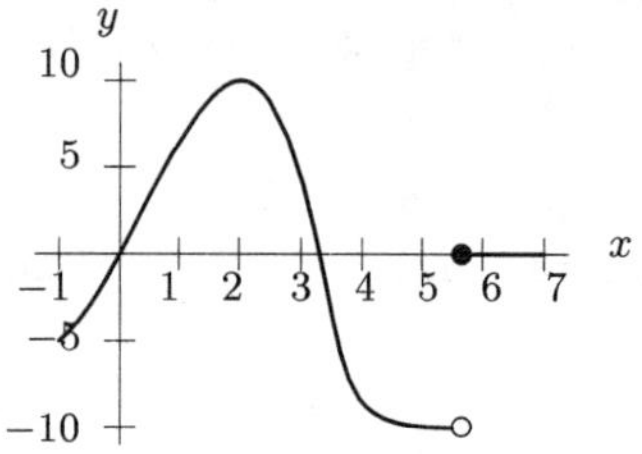

Figure 2.4.57

(b) See Tables 2.4.23 and 2.4.24.

Table 2.4.23

x	0	1	2	3	4	5	6	7
$f(x)$	−12	−5	5	12	7.5	−2.5	−7.5	−7.5

Table 2.4.24

x	1	2	3	4
$f'(x)$	8	10	2.5	10

(c) (i) $f''(0.5) > 0$, since f is concave up at $x = 0.5$.
(ii) $f''(2.5) < 0$, since f is concave down at $x = 2.5$.
(iii) $f''(4.5) = 0$, since f is a straight line at $x = 4.5$.

2. Suppose the graph of f is in Figure 2.4.58. Are the following quantities positive, negative, or zero?

(a) $f(A)$ (b) $f'(A)$ (c) $f''(A)$
(d) $f(B)$ (e) $f'(B)$ (f) $f''(B)$
(g) $f(C)$ (h) $f'(C)$ (i) $f''(C)$
(j) $f(D)$ (k) $f'(D)$ (l) $f''(D)$

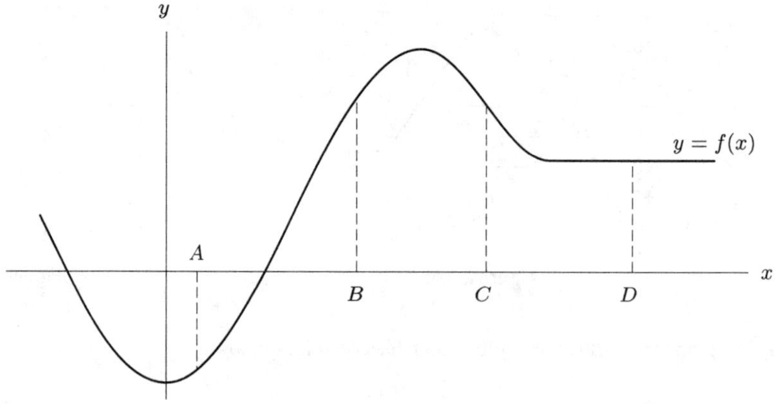

Figure 2.4.58

ANSWER:

(a) $f(A) < 0$ (b) $f'(A) > 0$ (c) $f''(A) > 0$
(d) $f(B) > 0$ (e) $f'(B) > 0$ (f) $f''(B) < 0$
(g) $f(C) > 0$ (h) $f'(C) < 0$ (i) $f''(C) = 0$
(j) $f(D) > 0$ (k) $f'(D) = 0$ (l) $f''(D) = 0$

3. The function f has only one turning point and f'' is shown in Figure 2.4.59. Sketch f and indicate a, b and c on your sketch.

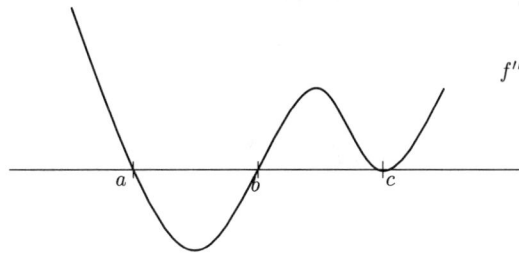

Figure 2.4.59

ANSWER:

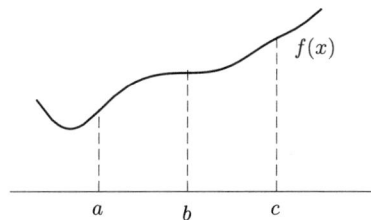

Figure 2.4.60

4. Suppose a function is given by a table of values as follows:

x	1.1	1.3	1.5	1.7	1.9	2.1
$f(x)$	12	15	21	23	24	25

(a) Estimate the instantaneous rate of change of f at $x = 1.7$.

(b) Write an equation for the tangent line to f at $x = 1.7$ using your estimate found in (a).

(c) Use your answer in (b) to predict a value for f at $x = 1.8$. Is your prediction too large or too small? Why?

(d) Is f'' positive or negative at $x = 1.7$? How can you tell? Can you estimate its value?

ANSWER:

(a) We approximate the instantaneous rate of change of $f(x)$ at $x = 1.7$ by the slope of the line joining the points $(1.7, 23)$ and $(1.9, 24)$, which is $\frac{1}{0.2} = 5$.

(b) The equation of a line with slope 5 passing through the point $(1.7, 23)$ is

$$y - 23 = 5(x - 1.7)$$
$$y = 23 + 5x - 8.5$$
$$= 14.5 + 5x.$$

(c) At $x = 1.8$, we predict that

$$y = 14.5 + 5 \cdot (1.8)$$
$$= 23.5.$$

Since the curve appears to be concave down over the interval $x \geq 1.3$, the line joining $(1.7, 23)$ and $(1.9, 24)$ lies *below* the curve, and hence 23.5 is an underestimate.

(d) f'' appears to be negative. To estimate the value of $f''(1.7)$, we first estimate values of $f'(1.6)$ and $f'(1.8)$:

$$f'(1.6) \approx \frac{f(1.7) - f(1.5)}{1.7 - 1.5} = 10 \quad \text{and}$$

$$f'(1.8) \approx \frac{f(1.9) - f(1.7)}{1.9 - 1.7} = 5.$$

Now,

$$f''(1.7) \approx \frac{f'(1.8) - f'(1.6)}{1.8 - 1.6} = \frac{5 - 10}{0.2} = -25.$$

For Problems 5– 6, circle the correct answer(s) or fill in the blanks. No reasons need be given.

5. Each graph in the right-hand column below represents the *second* derivative of some function shown in the left-hand column. Match the functions and their second derivatives.

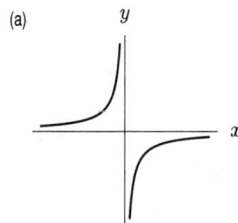

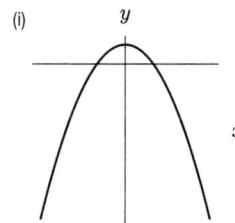

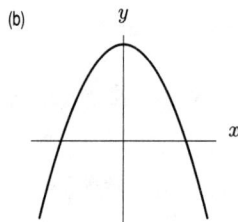

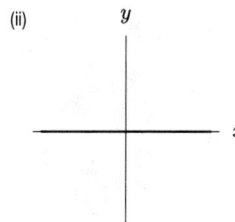

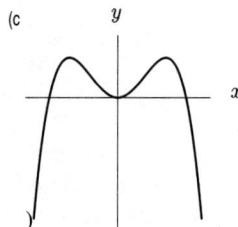

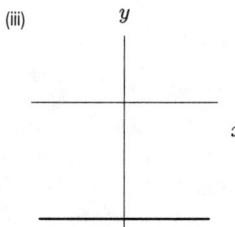

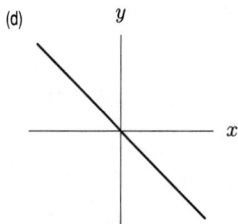

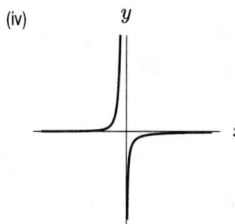

Functions   Second Derivatives

Function (a) has second derivative_____.
Function (b) has second derivative_____.
Function (c) has second derivative_____.
Function (d) has second derivative_____.

ANSWER:

(a) (iv)

(b) (iii)

(c) (i)

(d) (ii)

6. The cost of mining a ton of coal is rising faster every year. Suppose $C(t)$ is the cost of mining a ton of coal at time t.

   (a) Which of the following must be positive? (Circle those which are.)

   (i) $C(t)$

   (ii) $C'(t)$

   (iii) $C''(t)$

   (b) Which of the following must be increasing? (Circle those which are.)

   (i) $C(t)$

   (ii) $C'(t)$

   (iii) $C''(t)$

   (c) Which of the following must be concave up? (Circle those which are.)

   (i) $C(t)$

   (ii) $C'(t)$

   (iii) $C''(t)$

   ANSWER:

   (a) $C(t), C'(t), C''(t)$ positive

   (b) $C(t), C'(t)$ increasing

   (c) $C(t)$ concave up.

7.

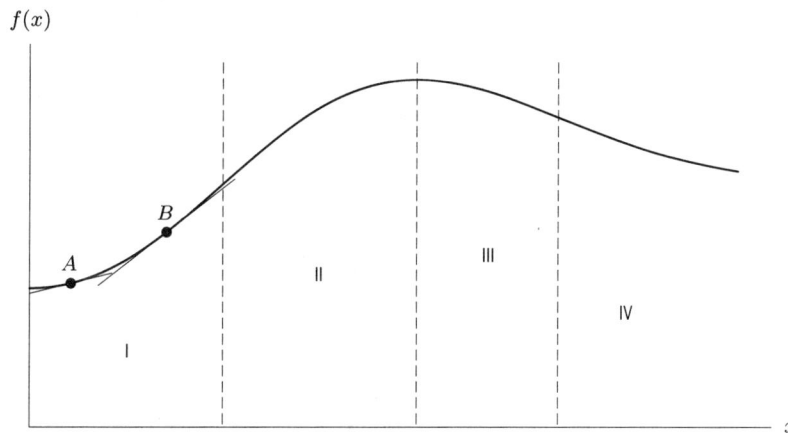

Figure 2.4.61

(a) Explain clearly why $f''(x) > 0$ for the x in region I of the Figure 2.4.61. Your answer should include a discussion of how $f'(x)$ changes as you For each of the three regions indicated, give the signs (positive or negative) of $f'(x)$ and $f''(x)$.

(b) region II: $f'(x)$ is _____ and $f''(x)$ is _____.

(c) region III: $f'(x)$ is _____ and $f''(x)$ is _____.

(d) region IV: $f'(x)$ is _____ and $f''(x)$ is _____.

   ANSWER:

(a) Looking at point A and B, we see that the slope at B is greater. Thus $f'(x)$, which is the slope, is increasing from A to B. Since $f'(x)$ is increasing, the derivative of $f'(x)$, which tells how $f'(x)$ is changing, will be positive. The derivative of $f'(x)$ is $f''(x)$, so $f''(x)$ is positive.

(b) $f'(x)$ is positive, $f''(x)$ is negative.

(c) $f'(x)$ is negative, $f''(x)$ is negative.

(d) $f'(x)$ is negative, $f''(x)$ is positive.

8. Sketch the graph of a function $f(x)$ such that $f'(x) > 0$ at all points and such that $f'(b) > f'(a)$ whenever $b > a$.

   ANSWER:

$f(x)$

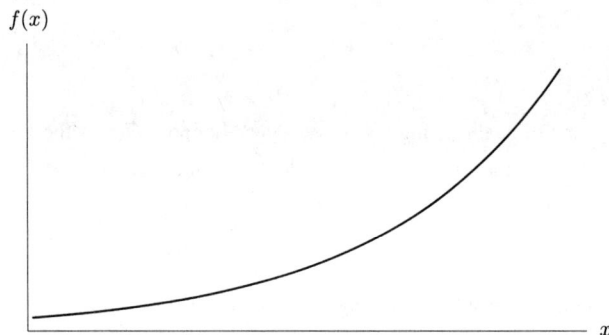

9. Consider the function f sketched in Figure 2.4.62. Decide whether each of the following is positive, negative or zero, and indicate each of these expressions graphically on the sketch. **(a)** $\dfrac{f(4) - f(2)}{2}$ **(b)** $f'(2)$ **(c)** $f''(3)$

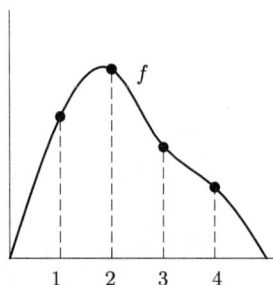

Figure 2.4.62

ANSWER:

slope$= \lim_{h \to 0} \dfrac{f(2+h) - f(2)}{h}$

slope$= \dfrac{f(4) - f(2)}{2}$

$f''(3)$ indicates the concavity at $(3, f(3))$

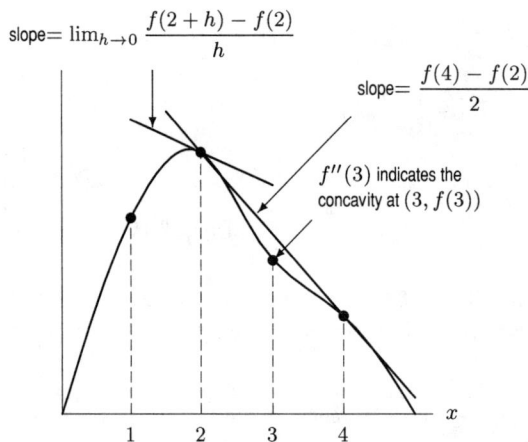

Figure 2.4.63

(a) $\dfrac{f(4) - f(2)}{h}$ is negative.

(b) $\lim_{h \to 0} \dfrac{f(2+h) - f(2)}{h}$ is negative.

(c) $f''(3)$ is positive.

10. Sketch a function f with the following two properties:

(a) $f'(x) > 0$ and f' increasing for $x > 0$.

(b) $f'(x) > 0$ and f' decreasing for $x < 0$.

ANSWER:

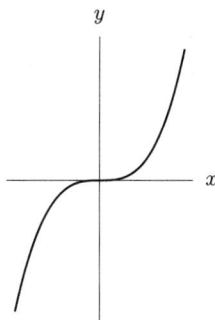

Figure 2.4.64

11. Let $S(t)$ represent the number of students enrolled in school in the year t. What do each of the following tell us about the signs of the first and second derivatives of $S(t)$?

   (a) The number of students enrolling is increasing faster and faster.
   (b) The enrollment is getting close to reaching its maximum.
   (c) Enrollment decreased steadily.

   ANSWER:

   (a) $ds/dt > 0$ and $d^2s/dt^2 > 0$.
   (b) $ds/dt > 0$ and $d^2s/dt^2 < 0$ (but ds/dt is close to zero).
   (c) $ds/dt < 0$ and d^2s/dt^2 is constant.

12. A driver obeys the speed limit as she travels past different towns in the order A, B, C. In town A, the speed limit is 50 mph. In town B, the speed limit is 35 mph and in town C, speed limit is 65mph.

   (a) If $S(t)$ represents the driver's position at time t, compare $S'(t)$ when she is passing town A to $S'(t)$ when she is passing town C.
   (b) If it always takes her two minutes to reach the new speed limit when she passes by a new town, what can you say about $S''(t)$ for the first two minutes she travels by towns B and C?

   ANSWER:

   (a) $S'(t)$ represents the velocity or rate of change in position. $S'(t)$ at town A is less than $S'(t)$ at town C.
   (b) $S''(t)$ represents acceleration. When she gets to town B she has to slow down from 50 mph to 35 mph. When she gets to town C she has to speed up to reach 65 mph. So $S''(t)$ at town B is negative and $S''(t)$ at town C is positive. The magnitude of $S''(t)$ at town B is less than the magnitude of $S''(t)$ at town C.

13. A company graphs $C'(t)$, the derivative of the number of pints of ice cream sold over the past ten years.

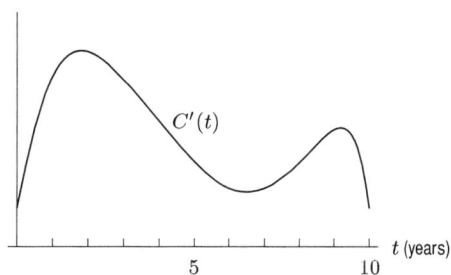

Figure 2.4.65

   At what year was

   (a) $C(t)$ greatest?   **(b)** $C'(t)$ greatest?

(c) $C''(t)$ greatest?   (d) $C(t)$ least?

ANSWER:

Since $C'(t)$ is positive everywhere, $C(t)$ is increasing everywhere. Hence $C(t)$ is greatest at year ten and least at year zero. We see $C'(t)$ is greatest around year two. $C''(t)$ is greatest where $C'(t)$ is rising most rapidly, namely before year 2.

14. Sketch a graph of a continuous function $f(x)$ with the following properties:

(a) $f''(x) < 0$ for $x < 4$

(b) $f''(x) > 0$ for $x > 4$

(c) $f''(4)$ is undefined

ANSWER:

Many possible.

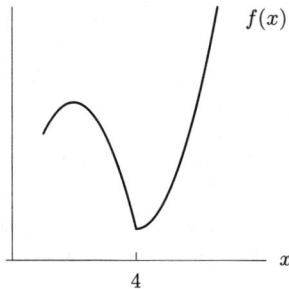

Figure 2.4.66

Problems and Solutions for Section 2.5

1. Cost and revenue functions for a certain chemical manufacturer are given in Figure 2.5.67. Fill in the blanks in the following statements.

(a) It costs _____ to produce 10 tons.

(b) Revenue from sale of 10 tons is _____.

(c) Break even points are at _____ and _____.

(d) Marginal cost at 20 tons is _____ dollars/ton.

(e) Marginal revenue at 20 tons is _____ dollars/ton.

(f) Sale price is _____ dollars/ton.

(g) Should the company increase production beyond 20 tons? Explain your reasoning.

(h) To maximize profit, the company should produce and sell _____ tons, and then its profit will be _____ dollars.

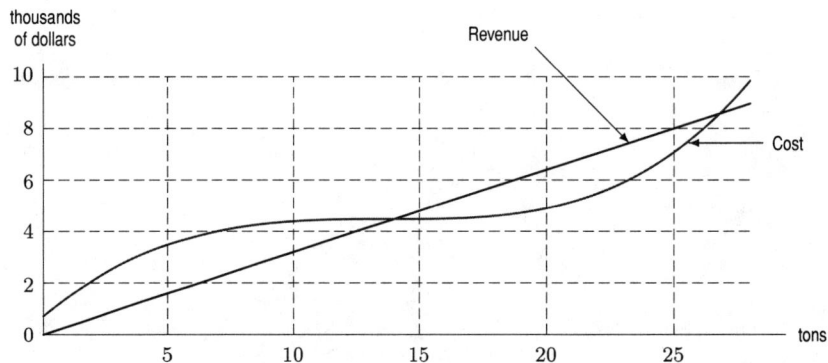

Figure 2.5.67

ANSWER:

(a) It costs \$4,500 to produce 10 tons.
(b) Revenue from sale of 10 tons is \$3,200.
(c) Break even points are at 14 and 27 tons.
(d) Marginal cost at 20 tons is about $5000/25 = \$200$/ton.
(e) Marginal revenue at 20 tons is about $6500/25 = \$160$/ton.
(f) Sale price is \$320/ton.
(g) Yes, since the marginal cost is less than the marginal revenue.
(h) To maximize profit, the company should produce and sell 22 tons, and then its profit will be $22 \times 320 - 5000 = \$2{,}040$.

2. The world's only manufacturer of left-handed widgets has determined that if q left-handed widgets are manufactured and sold per year at price p, then the cost function is $C = 8000 + 40q$, and the manufacturer's revenue function is $R = pq$. The manufacturer also knows that the demand function for left-handed widgets is $q = 2000 - 25p$.

(a) Using the demand function, rewrite the cost and the revenue functions in terms of price p.
(b) Write the profit function π in terms of price p and sketch its graph.
(c) For about what price is the profit largest? How many left-handed widgets should be produced at that price?

ANSWER:

(a) $C = 8000 + 40(2000 - 25p) = 88{,}000 - 1000p$.
$R = p(2000 - 25p) = 2000p - 25p^2$.
(b) $\pi = R - C = 2000p - 25p^2 - 88{,}000 + 1000p = 3000p - 25p^2 - 88{,}000$.

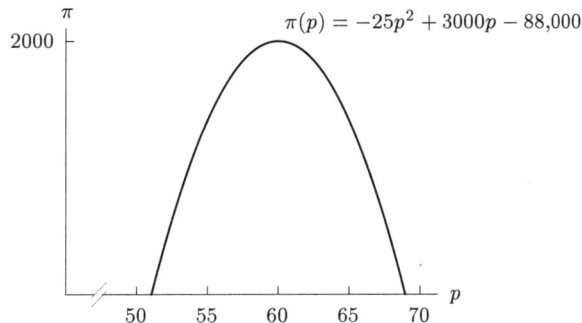

Figure 2.5.68

(c) The profit is largest at $p = 60$ and $q = 2000 - 25 \times 60 = 500$ widgets should be produced.

3. Your friend Herman operates a neighborhood lemonade stand. He asks you to be his financial advisor and wants to know how much lemonade he can make with the \$3.27 he happens to have on hand. The only information he can give you is that once last month he spent \$2 and made 19 glasses of lemonade, and another time he spent \$5 and got 83 glasses of lemonade. You decide to use this data to create a cost function, $C(q)$, giving the cost in dollars of making q glasses of lemonade.

(a) You first decide to create a linear cost function based on this data. What is the linear cost function, and how much lemonade can Herman make according to this model?
(b) You decide to create an exponential cost function. What is the exponential cost function, and how much lemonade can Herman make according to this model?
(c) Herman routinely sells his lemonade for 10¢ per glass. In the case of the linear model, what is his break-even point?
(d) In the case of the exponential model, what can you tell Herman about maximizing his profit? Again assume a sales price of 10¢ per glass.

ANSWER:

(a) Let $C(q) = kq + b$, then $C(19) = 2$ and $C(83) = 5$, implying that $k = 3/64$ and $b = 71/64$. So the linear cost function is

$$C(q) = \frac{3}{64}q + \frac{71}{64}.$$

Let $\dfrac{3}{64}q + \dfrac{71}{64} = 3.27$ and solve for q. We obtain $q = 46.1$; so Herman can make about 46 glasses according to this model.

(b) Let $C(q) = Ae^{kq}$, then $C(19) = 2$ and $C(83) = 5$, implying that $k \approx 0.014$ and $A \approx 1.524$. So the exponential cost function is

$$C(q) = 1.524e^{0.014q}.$$

Let $1.524e^{0.014q} = 3.27$ and solve for q; we have

$$q = \ln \frac{3.27}{1.524}/0.014 \approx 55.$$

So 55 glasses can be made according to this model.

(c) Let the revenue $R(q) = 0.1q$, then at break-even point $R(q) = C(q)$, i.e.,

$$0.1q = \frac{3}{64}q + \frac{71}{64}.$$

Solve for q; we have

$$q = \frac{71}{6.4}/(1 - \frac{3}{6.4}) \approx 21.$$

At 10¢ per glass, he will break even by selling 21 glasses.

(d) The profit $\pi(q) = R(q) - C(q) = 0.1q - 1.524e^{0.014q}$. By sketching a graph of this function, we can see that to maximize the profit, he should make about 275 glasses.

4. At a production level of 2000 for a product, marginal revenue is $4 per unit and marginal cost is $3.25 per unit. Do you expect maximum profit to occur at a production level above or below 2000? Explain your answer for credit.

ANSWER:

Maximum profit will be at a higher production level. Profit is increased by increasing production. Even producing just 1 more unit brings in $4 additional revenue at a cost of only $3.25 for an additional $0.75 profit.

5. The marginal cost of extracting iron ore from a mine already producing 10,000 tons of ore is $15 per ton.

   (a) In practical terms, what does this statement mean?

   (b) Give a clear explanation using a difference quotient of why the given marginal cost equals the derivative $C'(10,000)$, where $C(x)$ is the price in dollars of extracting x tons of ore.

   ANSWER:

   (a) It will cost $15 more to produce 10,001 tons than to produce 10,000 tons. In fact, the increase in cost of producing a little more than 10,000 tons will be about $15 times the extra tons.

   (b) $C'(10,000) \approx \frac{C(10,001) - C(10,000)}{10,001 - 10,000} = C(10,001) - C(10,000) = $ marginal cost.

6. Sketch a graph for the derivative of an advertising company's revenue with the following properties: First revenue increased steadily. That was followed by a short period of near zero revenue. Then revenue grew faster and faster.

   ANSWER:

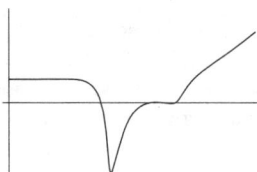

Figure 2.5.69

Problems and Solutions to Review Problems for Chapter 2 —————

1. Two politicians, named A and B, carefully inspect a table of values, x versus y. A claims that the table is linear, while B claims it is exponential.

   (a) You look at the table and agree with A. Explain what you saw in the table.

   (b) You look at the table and agree with B. Explain what you saw in the table.

   (c) You look at the table and realize that neither is *exactly* right, but *both* of them are *approximately* correct. Explain why this can be so. Referring to the derivative might be appropriate.

ANSWER:

(a) For each change in x, the corresponding change in y is proportional with the same constant of proportionality.

(b) Each time x changes by a certain amount, the ratio of the corresponding y-values is the same.

(c) The table could be approximately linear and approximately exponential if the exponential function used were approximately linear. This could happen if we used a portion of an exponential graph and had zoomed in sufficiently so that the exponential looked almost straight.

2. A table of values is given in Table 2.5.25 for $f(x)$.

(a) Is $f'(x)$ positive or negative?

(b) Is $f''(x)$ positive or negative?

(c) Approximate $f'(4)$.

Table 2.5.25

x	3	3.5	4	4.5	5	5.5	6
$f(x)$	17	27	34	38	41	43	44

ANSWER:

(a) $f'(x)$ is positive, since as x increases, $f(x)$ increases.

(b) $f''(x)$ is negative, since as x increases, $f(x)$ increases at a decreasing rate (thus it is concave down).

(c) We look at the slope $\dfrac{\Delta f(x)}{\Delta x}$ for the smallest Δx given, 0.5:

$$\frac{\Delta f(x)}{\Delta x} = \frac{38 - 34}{4.5 - 4} = \frac{4}{0.5} = 8$$

$$\text{and } \frac{\Delta f(x)}{\Delta x} = \frac{34 - 27}{4 - 3.5} = \frac{7}{0.5} = 14.$$

The slope for the region $3.5 \le x \le 4$ is 14, and for $4 \le x \le 4.5$ is 8. We approximate $f'(4)$ by averaging these two values: $\dfrac{14 + 8}{2} = 11$. So $f'(4) \approx 11$.

3. The following questions refer to the function $y = f(x)$, whose graph is given in Figure 2.5.70.

(a) Is $f'(1)$ positive or negative?

(b) Is $f''(1)$ positive or negative?

(c) In each pair given below, tell which is larger.

   (i) $f'(1)$ or $f'(2)$

   (ii) $f'(4)$ or $f'(5)$

   (iii) the average rate of change of $f(x)$ from $x = 3$ to $x = 4$ or the average rate of change of $f(x)$ from $x = 4$ to $x = 6$

   (iv) the instantaneous rate of change of $f(x)$ at $x = 1$ or the instantaneous rate of change of $f(x)$ at $x = 5$

   (v) $f''(3)$ or $f''(5)$

   (vi) $f(3)$ or $f(5)$

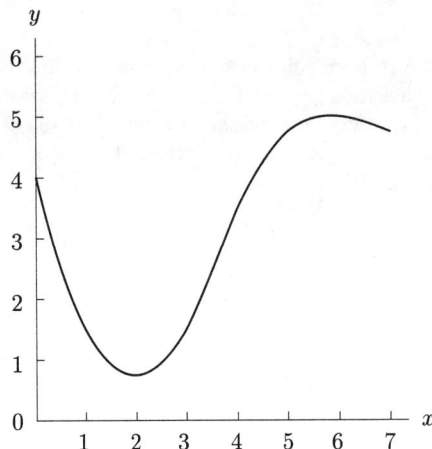

Figure 2.5.70

ANSWER:

(a) f is decreasing at $x = 1$, so $f'(1)$ is negative.

(b) f is concave up at $x = 1$, so $f''(1)$ is positive.

(c) (i) $f'(1)$ is negative and $f'(2) \approx 0$, so $f'(2)$ is larger.

  (ii) f is increasing at $x = 4$ and at $x = 5$, but the graph of f is steeper at $x = 4$, so $f'(4)$ is greater.

  (iii) From $x = 3$ to $x = 4$, the average rate of change is greater than from $x = 4$ to $x = 6$. To see this, imagine two lines, one through the points $(3, f(3))$ and $(4, f(4))$, and the other through the points $(4, f(4))$ and $(6, f(6))$. The slopes of the lines represent the average rates of change given in this question. The first line would be much steeper than the second; thus the average rate of change represented by its slope would be greater.

  (iv) Since f is decreasing at $x = 1$ and increasing at $x = 5$, its instantaneous rate of change is greater at $x = 5$.

  (v) Since f is concave up at $x = 3$ and concave down at $x = 5$, $f''(3)$ is greater.

  (vi) $f(5)$ is clearly greater from the graph.

4. Table 2.5.26 gives the number of passenger cars, in millions, in the United States, C, as a function of years, t. We have $C = f(t)$.

(a) Is $f'(t)$ positive or negative?

(b) Is $f''(t)$ positive or negative?

(c) Estimate $f'(1970)$.

(d) What are the units for your answer to part (c)? Clearly explain in terms of number of cars what information the derivative in part (c) gives us.

Table 2.5.26

t (year)	1940	1950	1960	1970	1980
C (# of cars, in millions)	27.5	40.3	61.7	89.3	121.6

ANSWER:

(a) As t increases, $f(t)$ increases, so $f'(t)$ is positive.

(b) As t increases, $f(t)$ increases at an increasing rate, so $f''(t)$ is also positive.

(c) The smallest Δt we have to work with is 10. We look at $\dfrac{\Delta C}{\Delta t}$ for t around 1970:

$$\frac{121.6 - 89.3}{10} = 3.23.$$

So $f'(1970) \approx 3.23$.

(d) The units are millions of cars per year. The derivative tells us the instantaneous rate of increase of the number of passenger cars in the US, in millions per year, in 1970. One way we can use this derivative is to estimate that there were roughly 3.23 million cars produced between 1970 and 1971.

5. Each of the quantities below can be represented in the picture. For each quantity, state whether it is represented by a length, a slope or an area. Then using the letters on the picture, make clear exactly which length, slope or area represents it. [Note: The letters P, Q, R, etc., represent points.]

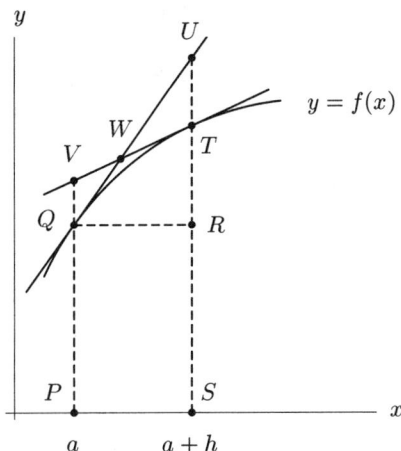

(a) $f(a + h) - f(a)$
(b) $f'(a + h)$
(c) $f'(a)h$
(d) $f(a)h$

ANSWER:

(a) $f(a + h) - f(a)$ is represented by the length TR.
(b) $f'(a + h)$ is the slope of the line TV.
(c) $f'(a)h$ is the length RU.
(d) $f(a)h$ is the area of the rectangle $PQRS$.

6. Given the following data about a function, f,

x	3	3.5	4	4.5	5	5.5	6
$f(x)$	10	8	7	4	2	0	−1

(a) Estimate $f'(4.25)$ and $f'(4.75)$.
(b) Estimate the rate of change of f' at $x = 4.5$.
(c) Find, approximately, an equation of the tangent line at $x = 4.5$.
(d) Use the tangent line to estimate $f(4.75)$.

ANSWER:

(a) $f'(4.25) \approx \frac{f(4.5) - f(4)}{4.5 - 4} = \frac{4 - 7}{0.5} = -6$
$f'(4.75) \approx \frac{f(5) - f(4.5)}{5 - 4.5} = \frac{2 - 4}{0.5} = -4$
(b) $f''(4.5) \approx \frac{f'(4.75) - f'(4.25)}{0.5} = \frac{-4 + 6}{0.5} = 4$
(c) $f'(4.5) \approx \frac{f(5) - f(4.5)}{0.5} = -4$, thus $y - 4 = -4(x - 4.5)$ is the equation of the tangent line.
(d) $f(4.75) \approx f(4.5) + .25 \cdot f'(4.5) \approx 3$

7. There is a population of $P(t)$ thousand bacteria in a culture at time t hours after the beginning of an experiment. You know that $P(10) = 20$, $P'(10) = 0.4$, and $P''(10) = 0.008$.

(a) Give the meaning, in practical terms, of the equation $P'(10) = 0.4$.
(b) Using the value of $P'(10)$, make a prediction for the population at time $t = 10.5$ hours.
(c) Give the meaning, in practical terms, of the equation $P''(10) = 0.008$. Would you expect the population to increase more or less between $t = 10.5$ and $t = 11$ than between $t = 10$ and $t = 10.5$?
(d) Using the value of $P''(10)$, make a prediction for the rate of change of the population at time $t = 10.5$.

(e) Give your best prediction of the population at time $t = 11$ hours.

ANSWER:

(a) 10 hours after the beginning of the experiment, the population is increasing at 400 bacteria per hour.

(b) $P(10.5) \approx P(10) + \frac{1}{2}P'(10) = 20,200$.

(c) At $t = 10$, the increasing rate (thousand/h) is increasing at a rate of 0.008 thousand/h per hour. The population will increase more between $t = 10.5$ and $t = 11$ than between $t = 10$ and $t = 10.5$, since the increasing rate is increasing.

(d) $P'(10.5) = P'(10) + \frac{1}{2}P''(10) = 0.4 + 0.004 = 0.404$ thousand/hour $= 404$ bacteria/hour.

(e) $P(11) = P(10.5) + \frac{1}{2}P'(10.5) = 20.2 + 0.202 = 20.402$. The population will be approximately 20,402 at 11 hours.

8. The graphs below are each the derivative of a function. Sketch a graph of each original function.

(a)

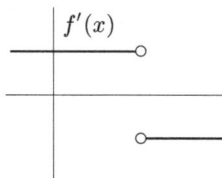

Figure 2.5.71

(b)

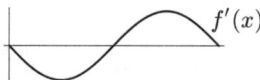

Figure 2.5.72

(c)

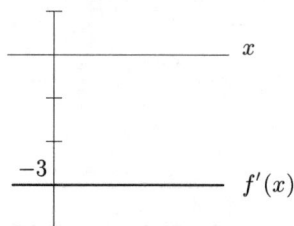

Figure 2.5.73

ANSWER:

(a)

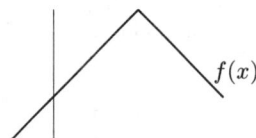

Figure 2.5.74

(b)

Figure 2.5.75

(c)

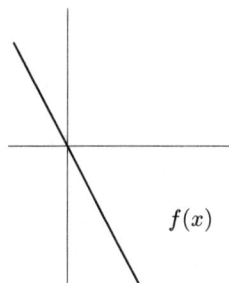

Figure 2.5.76

Problems and Solutions on Limits, Continuity, and the Definition of the Derivative ——

1. There is a function used by statisticians, called the error function, which is written

$$y = \text{erf}(x).$$

Suppose you have a statistical calculator, which has a button for this function. Playing with your calculator, you discover the following:

x	$\text{erf}(x)$
1	0.29793972
0.1	0.03976165
0.01	0.00398929
0	0

   (a) Using this information alone, give an estimate for $\text{erf}'(0)$, the derivative of erf at $x = 0$. Only give as many decimal places as you feel reasonably sure of, and explain why you gave that many decimal places.
   (b) Suppose that you go back to your calculator and find that

$$\text{erf}(0.001) = 0.000398942.$$

   With this extra information, would you refine the answer you gave in (a)? Explain.

   ANSWER:

   (a) Since $\text{erf}'(0) = \lim_{h \to 0} \dfrac{\text{erf}(h) - \text{erf}(0)}{h - 0} = \lim_{h \to 0} \dfrac{\text{erf}(h)}{h}$, we approximate $\text{erf}'(0)$ by $\dfrac{\text{erf}(h)}{h}$ where h is small. As $\dfrac{\text{erf}(0.1)}{0.1}$ and $\dfrac{\text{erf}(0.01)}{0.01}$ agree in the first two decimal places, it seems safe to estimate $\text{erf}'(0) = 0.39$.

   (b) The new value for $\text{erf}(0.001)$ gives us agreement out to four decimal places between $\dfrac{\text{erf}(0.01)}{0.01}$ and $\dfrac{\text{erf}(0.001)}{0.001}$, so we can refine our answer to 0.3989.

2. Estimate the value of $f'(x)$ for the function $f(x) = 10^x$.

   ANSWER:

$$f'(x) = \lim_{h \to 0} \frac{f(x + h) - f(x)}{h} = \lim_{h \to 0} \frac{10^{x+h} - 10^x}{h} = 10^x \lim_{h \to 0} \frac{10^h - 1}{h}.$$

   So far, our calculation is exact. We now estimate the limit by substituting small values of h;

h	$\frac{10^h - 1}{h}$
1	9
0.1	2.589
0.01	2.329
0.001	2.305
0.0001	2.303
0.00001	2.303

   So $f'(x)$ appears to be approximately equal to $(2.303)10^x$.

3. Assume that f and g are differentiable functions defined on all of the real line. Mark the following TRUE or FALSE.

(a) It is possible that $f > 0$ everywhere, $f' > 0$ everywhere, and $f'' < 0$ everywhere.

(b) f can satisfy: $f'' > 0$ everywhere, $f' < 0$ everywhere, and $f > 0$ everywhere.

(c) f and g can satisfy: $f'(x) > g'(x)$ for all x, and $f(x) < g(x)$ for all x.

(d) If $f'(x) = g'(x)$ for all x and if $f(x_0) = g(x_0)$ for some x_0, then $f(x) = g(x)$ for all x.

(e) If $f'' < 0$ everywhere and $f' < 0$ everywhere then $\lim\limits_{x \to +\infty} f(x) = -\infty$.

(f) If $f'(x) > 0$ for all x and $f(x) > 0$ for all x then $\lim\limits_{x \to +\infty} f(x) = \infty$.

ANSWER:

(a) FALSE.

(b) TRUE.

(c) TRUE.

(d) TRUE.

(e) TRUE.

(f) FALSE.

4. (a) Let $f(x) = x^2 + 3$. Derive an exact formula for the derivative $f'(x)$ by computing algebraically the limit of a difference quotient.

(b) Write an equation for the line tangent to the graph of $y = x^2 + 3$ at the point where $x = 4$.

ANSWER:

(a) $\dfrac{f(x+h) - f(x)}{h} = \dfrac{(x+h)^2 + 3 - x^2 - 3}{h} = \dfrac{2xh + h^2}{h} = 2x + h,$

so $\lim_{h \to 0} \dfrac{f(x+h) - f(h)}{h} = 2x.$

(b) $y' = 2x$. So at $x = 4$ and $y' = 8$, we have $y = 8x + b$. Since $y = 19$ at $x = 4$, $b = -13$. The equation of the line is $y = 8x - 13$.

5. Approximate (with difference quotient and calculator) the derivative of $\sqrt{8x + 1}$ at $x = 1$.

ANSWER:

$$f'(1) \approx \frac{f(1.001) - f(1)}{0.001}$$
$$\approx \frac{\sqrt{9.008} - \sqrt{9}}{0.001}$$
$$\approx 1.333$$

6. Find $f'(x)$ algebraically by using the limit definition if $f(x) = \dfrac{1}{x + 2}$.

ANSWER:

By definition,

$$f'(x) = \lim_{h \to 0} \frac{f(x+h) - f(x)}{h} = \lim_{h \to 0} \frac{\frac{1}{x+h+2} - \frac{1}{x+2}}{h}$$
$$= \lim_{h \to 0} \frac{-1}{(x+2)(x+h+2)} = \frac{-1}{(x+2)^2} \quad.$$

7. Using calculator, estimate the derivative of $f(x) = \cos(x)$ at $x = 0$. (Make sure your calculator is set to radians.)

ANSWER:

$$x = 0, \quad h = 0.1, \quad \frac{\cos(x+h) - \cos(x)}{h} = \frac{\cos(0.1) - \cos(0)}{0.1} = \frac{\cos(0.1) - 1}{0.1} = -0.0499583$$

$$x = 0, \quad h = 0.01, \quad \frac{\cos(x+h) - \cos(x)}{h} = \frac{\cos(0.01) - \cos(0)}{0.01} = \frac{\cos(0.01) - 1}{0.01} = -0.00499999$$

$$x = 0, \quad h = 0.001, \quad \frac{\cos(x+h) - \cos(x)}{h} = \frac{\cos(0.001) - \cos(0)}{0.001} = \frac{\cos(0.001) - 1}{0.001} = -0.00049999$$

$$x = 0, \quad h = 0.0001, \quad \frac{\cos(x+h) - \cos(x)}{h} = \frac{\cos(0.0001) - \cos(0)}{0.0001} = \frac{\cos(0.0001) - 1}{0.0001} = -0.000049999$$

The limit appears to be 0.

8. Using a calculator, estimate the derivative of $f(x) = \sin(x)$, at $x = 0$, with the calculator set to degrees. Explain what you get.

   ANSWER:

$$x = 0, \quad h = 1, \quad \frac{\sin(x+h) - \sin(x)}{h} = \frac{\sin(1) - \sin(0)}{1} = \frac{\sin(1)}{1} = 0.0174524$$

$$x = 0, \quad h = 0.1, \quad \frac{\sin(x+h) - \sin(x)}{h} = \frac{\sin(0.1) - \sin(0)}{0.1} = \frac{\sin(0.1)}{0.1} = 0.0174533$$

$$x = 0, \quad h = 0.01, \quad \frac{\sin(x+h) - \sin(x)}{h} = \frac{\sin(0.01) - \sin(0)}{0.01} = \frac{\sin(0.01)}{0.01} = 0.0174533$$

The limit appears to be 0.0174533 but the derivative if $\sin(x)$ is $\cos(x)$ and when $x = 0$ this gives 1. The discrepancy is caused by the fact that we have used degrees rather than radians. Recall that 180 degrees $= 2\pi$ radians so that 1 degree $= \dfrac{\pi}{180}$ radians $= 0.0174533$ radians so if you want to avoid this factor always use radian measure.

9. Give a difference quotient approximation of $f'(10)$ where $f(x) = \sqrt{x^3 + 5}$.

   (a) The two points you will use for the difference quotient are:
   (b) $f'(10) \approx$

   ANSWER:

   (a)

$$(10, \sqrt{10^3 + 5}) = (10, \sqrt{1005}) = (10, 31.701735)$$

$$(10.001, \sqrt{10.001^3 + 5}) = (10.001, 31.706467)$$

   (b) $f'(10) \approx \dfrac{\sqrt{10.001^3 + 5} - \sqrt{10^3 + 5}}{0.001} = 4.73$

10. Give a difference quotient approximation of $f'(4)$ where $f(x) = \sqrt{x^2 + 9}$.

    (a) The two points you will use for the difference quotient are:
    (b) $f'(4) \approx$ _____.
    (c) Find the equation of the line tangent to the graph of $f(x)$ at the point where $x = 4$.

    ANSWER:

    (a)

$$(4, \sqrt{4^2 + 9}) = (4, 5)$$

$$(4.001, \sqrt{(4.001)^2 + 9}) = (4.001, 5.0008)$$

    (b)

$$\frac{\sqrt{4.001^2 + 9} - 5}{4.001 - 4} \approx 0.8.$$

    (c) Point online $(4, 5)$, so the line is $y = 5 + 0.8(x - 4) = 1.8 + 0.8x$.

11. Give the definition for the instantaneous rate of change of a function f at a point x.

    ANSWER:

    The instantaneous rate of change of a function f at a point x is defined by the following limit:

$$f'(x) = \lim_{h \to 0} \frac{f(x+h) - f(x)}{h}.$$

12. Find the derivative of the following functions algebraically:

    (a) $g(x) = 3x^2 + 2x - 4$ at $x = 3$.

(b) $m(x) = 2x^3$ at $x = 2$.
(c) $p(x) = g(x) \cdot m(x)$ at $x = 2$.

ANSWER:

(a) Using the definition of the derivative, we have

$$
\begin{aligned}
g'(3) &= \lim_{h \to 0} \frac{g(3+h) - g(3)}{h} \\
&= \lim_{h \to 0} \frac{(3(3+h)^2 + 2(3+h) - 4) - (3(3)^2 + 2(3) - 4)}{h} \\
&= \lim_{h \to 0} \frac{(3h^2 + 20h + 29) - (29)}{h} \\
&= \lim_{h \to 0} \frac{3h^2 + 20h}{h} = \lim_{h \to 0} 3h + 20 = 20.
\end{aligned}
$$

So $g'(3) = 20$.

(b) Using the definition of the derivative, we have

$$
\begin{aligned}
m'(2) &= \lim_{h \to 0} \frac{2(h+2)^3 - 2h^3}{h} \\
&= \lim_{h \to 0} \frac{4h^3 + 12h^2 + 24h}{h} \\
&= \lim_{h \to 0} 4h^2 + 12h + 24 = 24.
\end{aligned}
$$

So $m'(2) = 24$.

(c) $p(x) = (3x^2 + 2x - 4)(2x^3) = 6x^5 + 4x^4 - 8x^3$

$$
\begin{aligned}
p'(2) &= \lim_{h \to 0} \frac{(6(h+2)^5 + 4(h+2)^4 - 8(h+2)^3) - (6(2)^5 + 4(2)^4 - 8(2)^3)}{h} \\
&= \lim_{h \to 0} \frac{6h^5 + 64h^4 + 264h^3 + 528h^2 + 512h}{h} = 512.
\end{aligned}
$$

So $p'(2) = 512$.

Chapter 3 Exam Questions

Problems and Solutions   for Section 3.1

1. If $y = x^3 + 4x^2 - 7x + 3$, find the following derivatives.

   (a) $\dfrac{dy}{dx}$ (b) $\dfrac{d^2y}{dx^2}$ (c) $\dfrac{d^3y}{dx^3}$

   ANSWER:

   (a) $\dfrac{dy}{dx} = 3x^2 + 8x - 7$.

   (b) $\dfrac{d^2y}{dx^2} = 6x + 8$.

   (c) $\dfrac{d^3y}{dx^3} = 6$.

2. Find the first derivative of the following functions.

   (a) $y = \sqrt{x}$

   (b) $x = y^3 + \dfrac{5}{y}$

   (c) $t = z^2 + \dfrac{2}{z} + 2z^{-2}$

   (d) $s = t^\pi + \sqrt{2}t$

   (e) $w = x^2 + ax$

   ANSWER:

   (a) Since $y = \sqrt{x}$, then the first derivative of this function is

   $$\frac{dy}{dx} = \frac{1}{2\sqrt{x}}$$

   (b) Since $x = y^3 + \dfrac{5}{y}$, then the first derivative of this function is

   $$\frac{dx}{dy} = \frac{-5}{y^2} + 3y^2$$

   (c) Since $t = z^2 + \dfrac{2}{z} + 2z^{-2}$, then the first derivative of this function is

   $$\frac{dt}{dz} = \frac{-4}{z^3} - \frac{2}{z^2} + 2z$$

   (d) Since $s = t^\pi + \sqrt{2}t$, then the first derivative of this function is

   $$\frac{ds}{dt} = \sqrt{2} + \pi t^{\pi-1}$$

   (e) $w = x^2 + ax$, then the first derivative of this function is

   $$\frac{dw}{dx} = 2x + a$$

3. Find the first derivative of the following functions.

   (a) $y = x^2 + 3x^3 - \dfrac{1}{x}$

   (b) $t = z^3 + \dfrac{2}{z^2} + 2z^{-3}$

   (c) $s = t^{\sqrt{2}} + 3t$

   (d) $w(x) = bx^2 + x^t$

ANSWER:

(a) Since $y = x^2 + 3x^3 - \dfrac{1}{x}$, then the first derivative of this function is

$$\frac{dy}{dx} = x^{-2} + 2x + 9x^2$$

(b) Since $t = z^3 + \dfrac{2}{z^2} + 2z^{-3}$, then the first derivative of this function is

$$\frac{dt}{dz} = \frac{-6}{z^4} - \frac{4}{z^3} + 3z^2$$

(c) Since $s = t^{\sqrt{2}} + 3t$, then the first derivative of this function is

$$\frac{ds}{dt} = 3 + \sqrt{2}t^{\sqrt{2}-1}$$

(d) Since $w(x) = bx^2 + x^t$, then the first derivative of this function is

$$\frac{dw}{dx} = 2bx + tx^{t-1}$$

4. Find the equation for the tangent line to the curve $x^2 + y = 4$ when $x = 1$. Determine where the tangent line meets the axes.

ANSWER:

From $x^2 + y = 4$, we get that $y = 4 - x^2$. So

$$\frac{dy}{dx} = -2x$$

When $x = 1, y = 3, \dfrac{dy}{dx} = -2$

$$y - 3 = -2(x - 1)$$

or

$$y = -2x + 5$$

This line meets the axes at $x = 0, y = 5$ and $y = 0, x = \dfrac{5}{2}$.

5. Given the function $f(x) = x^4 - 9x^2 + 2$

   (a) Find the slope of the tangent line at $x = 3$.
   (b) What is the equation of this line?
   (c) Where does the tangent line meet the y axis?
   (d) Find all points where the curve has a horizontal tangent.

   ANSWER:

   (a) The slope of the curve is $f' = 4x^3 - 18x$ so that $f'(3) = 54$. The slope of the tangent at this point will also be 54.
   (b) When $x = 3, f(3) = 3^4 - 9 3^2 + 2 = 2$ so that the equation of the tangent is $(y - 2) = 54(x - 3)$ or $y = 54x + 166$.
   (c) The tangent line meets the y axis when $x = 0$ so that $y = 166$.

   (d) The slope of the curve is $f' = 4x^3 - 18x$ which vnishes when $x = 0$ or $x = \pm\sqrt{\dfrac{18}{4}} = \pm\dfrac{3}{\sqrt{2}}$.

6. Use the definition of the derivative to justify the power rule for $n = 4$: Show $\dfrac{d}{dx}(2x^4) = 8x^3$.

   ANSWER:

$$\frac{d}{dx}(2x^4) = \lim_{h \to 0}\left(\frac{2(x+h)^4 - 2x^4}{h}\right) = \lim_{h \to 0}\frac{2}{h}\left(x^4 + 4x^3h + 6x^2h^2 + 4xh^3 + h^4 - x^4\right)$$

$$= \lim_{h \to 0} 2\left(4x^3 + 6x^2h + 4xh^2 + h^3\right) = 8x^3$$

7. Find the derivatives of the given functions.

   (a) $y = 2x^3 - \dfrac{1}{2x}$

   (b) $f(x) = 2\sqrt{x} + x^3$

   (c) $g(t) = \dfrac{2t^3 - t^2 + 6}{t^2}$

   ANSWER:

   (a) $y = 2x^3 - \dfrac{1}{2}x^{-1}$

$$y' = 6x^2 + \frac{1}{2}x^{-2} = 6x^2 + \frac{1}{2x^2}$$

   (b) $f(x) = 2x^{1/2} + x^3$

$$f'(x) = x^{-1/2} + 3x^2 = \frac{1}{\sqrt{x}} + 3x^2$$

   (c) $g(t) = \dfrac{2t^3 - t^2 + 6}{t^2} = 2t - 1 + 6t^{-2}$, so

$$g'(t) = 2 - 12t^{-3}.$$

8. If $g(t) = 4t^3 - t^2 + 3t$

   (a) find $g'(t)$ and $g''(t)$

   (b) If $g(t)$ represents the position of a particle at time t seconds, what do $g'(3)$ and $g''(4)$ represent?

   ANSWER:

   (a) $g'(t) = 12t^2 - 2t + 3$

       $g''(t) = 24t - 2$

   (b) $g'(3)$ represents the velocity of the particle at time 3 seconds.

       $g''(4)$ represents the acceleration of the particle at time 4 seconds.

9. Consider the function $f(x) = 2x^4 - 4x^3 + 2$. Are there values of x for which $f(x)$ has the following properties? If so, indicate the values.

   (a) Increasing

   (b) Decreasing and concave up

   ANSWER:

   (a) Increasing means $f'(x) > 0$.

$$f'(x) = 8x^3 - 12x^2 = 4x^2(2x - 3)$$

   So $f'(x) > 0$ when $(2x - 3) > 0$ and $x \neq 0$. Therefore

$$2x > 3$$

$$x > 3/2$$

   (b) Decreasing means $f'(x) < 0$.

       So $f'(x) < 0$ when $(2x - 3) < 0$ and $x \neq 0$. Therefore

$$2x < 3$$

$$x < 3/2$$

   Concave up means $f''(x) > 0$.

$$f''(x) = 24x^2 - 24x = 24x(x - 1)$$

$$f''(x) > 0 \text{ when } 24x(x - 1) > 0$$

$$x < 0 \text{ or } x > 1$$

   So both conditions hold when $x < 0$ or $1 < x < 3/2$.

10. Given a power function of the form $f(x) = ax^n$, find n and a so that $f'(2) = -1$ and $f'(4) = -1/4$, .

ANSWER:

Since $f(x) = ax^n$, $f'(x) = nax^{n-1}$. We know that $f'(2) = na(2)^{n-1} = -1$ and $f'(4) = na(4)^{n-1} = -1/4$. Therefore,

$$\frac{f'(4)}{f'(2)} = \frac{-\frac{1}{4}}{-1}$$

$$\frac{(na)4^{n-1}}{(na)2^{n-1}} = \left(\frac{4}{2}\right)^{n-1} = \frac{1}{4}$$

$$2^{n-1} = \frac{1}{4} = 2^{-2},$$

and thus $n = -1$.

Substituting $n = -1$ into the expression for $f'(2)$, we get

$$(-1)a(2)^{-2} = -1$$

$$\frac{a}{4} = 1, a = 4$$

11. Given $f(x) = 3x^2 - x$ and $g(x) = x^3 + 3x^2 - 3$

(a) find $\dfrac{d}{dx}[f(x) + g(x)]$

(b) find $\dfrac{d}{dx}[g(x) - 2f(x)]$

ANSWER:

(a) $\dfrac{d}{dx}[f(x) + g(x)] = \dfrac{d}{dx}[3x^2 - x + x^3 + 3x^2 - 3] = \dfrac{d}{dx}[x^3 + 6x^2 - x - 3] = 3x^2 + 12x - 1$

(b) $\dfrac{d}{dx}[g(x) - 2f(x)] = \dfrac{d}{dx}[x^3 + 3x^2 - 3 - 6x^2 + 2x] = \dfrac{d}{dx}[x^3 - 3x^2 + 2x - 3] = 3x^2 - 6x + 2$

12. Use the definition of the derivative to justify the formula $\dfrac{d(x^n)}{dx} = nx^{n-1}$ for $n = -2$.

ANSWER:

$$\frac{d(x^{-2})}{dx} = \lim_{h \to 0} \frac{(x+h)^{-2} - x^{-2}}{h} = \lim_{h \to 0} \frac{1}{h}\left[\frac{1}{(x+h)^2} - \frac{1}{x^2}\right]$$

$$= \lim_{h \to 0} \frac{1}{h}\left[\frac{x^2 - (x+h)^2}{x^2(x+h)^2}\right] = \lim_{h \to 0} \frac{1}{h}\left[\frac{-2xh - h^2}{x^2(x+h)^2}\right]$$

$$= \lim_{h \to 0} \frac{-2x - h}{x^2(x+h)^2} = \frac{-2x}{x^4} = -2x^{-3}.$$

Problems and Solutions for Section 3.2

1. (a) What is the slope of the graph of $y = e^x$ at $x = a$?

(b) Find the equation of the line tangent to the graph of $y = e^x$ at the point (a, e^a).

(c) Find the x- and y-intercepts of the tangent line in (b).

(d) Explain why the y-intercept of the tangent line in (b) can never be higher than 1.

ANSWER:

(a) Since $y' = y = e^x$, the slope at $x = a$ is e^a.

(b) Let $y = kx + b$, then $k = e^a$ and $e^a = e^a \cdot a + b$, i.e. $b = e^a(1 - a)$. So we have

$$y = e^a x + e^a(1 - a).$$

(c) The x-intercept is $x = a - 1$; the y-intercept is $y = e^a(1 - a)$.

(d) Since for any real number a, we have

$$e^a \geq 1 + a \quad \text{or equivalently} \quad e^{-a} \geq 1 - a,$$

which is the same as $e^a(1 - a) \leq 1$. So the y-intercept can never be higher than 1.

2. The value of a car is falling at 10% per year so that if C_0 is the purchase price of the car in dollars its value after t years is given by
$$V(t) = C_0(0.9)^t.$$

(a) At what rate is its value falling when its is driven out of the show room?

(b) How fast is the car depreciating after 1 year?

ANSWER:

(a) The rate at which the car's value is changing is given by $V' = C_0 \ln(0.9)0.9^x$ so that as soon as the car is driven out of the show room its value is changing at a rate of $\ln(0.9)C_0$ dollars per year. Since $\ln(0.9) < 0$ this indicates that the car's value is of course falling so the answer to this part of the question is $-C_0 \ln(0.9)$ dollars per year.

(b) As in (a) the car's value is changing at $V' = C_0 \ln(0.9)0.9^x$ so that after 1 year it is depreciating at $-C_0 0.9 \ln(0.9)$ dollars per year.

3. Consider the graphs of $s = \ln(t)$ and $t = e^s, t > 0$.

(a) Find the slope of the curves when $t = 3$.

(b) Determine the angle between the tangent lines at this point. (Plot the graphs of $s = \ln(t)$ and $t = e^s$ on the same axis but note that for the second graph the roles of the axes have interchanged.)

(c) Is this result true for any other values of t? Explain.

ANSWER:

(a) If $s = \ln t$ then $\dfrac{ds}{dt} = \dfrac{1}{t}$ so that when $t = 3$, $\dfrac{ds}{dt} = \dfrac{1}{3}$. If $t = e^s$ then $\dfrac{dt}{ds} = e^s$ so that when $t = 3$ so that $\dfrac{dt}{ds} = 3$

(b) The tangent lines are the same.

(c) It is true for all values of t since the functions are inverses of each other.

4. Find the equation of the tangent to the curve $y = e^x$ which passes through the origin.

ANSWER:

Let x_0 be the point of contact of the required tangent. The slope of the curve $y = e^x$ is $y' = e^x$ which is equal to e^{x_0} at the point of contact. The slope of the line pasing through the origin and (x_0, e^{x_0}) is $\dfrac{e^{x_0}}{x_0}$ therefore we require

$$\frac{e^{x_0}}{x_0} = e^{x_0}$$

which gives $x_0 = 1$. The required tangent line is therefore $y = ex$.

5. Differentiate $e^x + x^e$.

ANSWER:
$$\frac{d}{dx}(e^x + x^e) = e^x + ex^{e-1}$$

6. Find the derivative of the functions below:

(a) $y = 3^x - 3$

(b) $f(x) = 6e^x - 5^x$

(c) $g(x) = 2e^{\pi x} = 2(e^\pi)^x$

ANSWER:

(a) $y' = (\ln 3)3^x$

(b) $f'(x) = 6e^x - \ln 5(5^x)$

(c) $g(x) = 2e^{\pi x} = 2(e^\pi)^x$ so
$$g'(x) = 2(e^\pi)^x \ln(e^\pi) = 2\pi(e^\pi)^x = 2\pi e^{\pi x}.$$

7. Find the derivative:

(a) $f(x) = (\ln 2)x^2 + (\ln 5)e^x$

(b) $g(x) = 3x - \dfrac{1}{\sqrt[3]{x}} + 3^x$

ANSWER:

(a) $f'(x) = (2 \ln 2)x + (\ln 5)e^x$

(b) $g'(x) = \dfrac{d}{dx}[3x - x^{-1/3} + 3^x] = 3 + \dfrac{1}{3x^{4/3}} + \ln 3(3^x)$

8. Find the derivative:

(a) $h(t) = t^{\pi^3} + (\pi^3)^t + \pi t^3$

(b) $g(t) = (1/e)^t + e^t + e$

ANSWER:

(a) $h'(t) = \pi^3 t^{(\pi^3 - 1)} + (\pi^3)^t \ln(\pi^3) + 3\pi t^2$

(b) $g'(t) = \ln(1/e)(1/e)^t + e^t = -(1/e)^t + e^t$

9. Consider the function $g(x) = 3x^3 + 3^x$. Give the equation of the tangent lines at $x = 0$ and $x = 2$.

ANSWER:

$$g'(x) = 9x^2 + 3^x \ln 3$$
$$g'(0) = 9(0)^2 + 3^0 \ln 3 = \ln 3$$
$$g(0) = 3(0)^3 + 3^0 = 1$$
$$y = (\ln 3)x + 1.$$
$$g'(3) = 9(3)^2 + 3^3 \ln 3 = 81 + 27 \ln 3$$
$$y = (81 + 27 \ln 3)x + 1.$$

10. With a yearly inflation rate of 3%, prices are described by $P = P_0(1.03)^t$, where P_0 is the price in dollars when $t = 0$ and t is time in years. If $P_0 = 1.2$, how fast (in cents/year) are prices rising when $t = 15$?

ANSWER:

Since $P = (1.2)(1.03)^t$, $\dfrac{dP}{dt} = (1.2) \ln(1.03)(1.03)^t$. When $t = 15$, $\dfrac{dP}{dt} = (1.2) \ln(1.03)(1.03)^{15} \approx 0.055$ dollars or 5.5 cents/year.

11. (a) Find the slope of the graph of $g(x) = x - 2e^x$ at the point where it crosses the y-axis.

(b) Find the equation of the tangent line to the curve at the point from part (a).

(c) Find the equation of the line perpendicular to the line from part (a) that goes through $(1, 1)$.

ANSWER:

(a) $g'(x) = 1 - 2e^x$.

The graph crosses the y-axis when $x = 0$, so the point is $(0, -2)$.

Slope at $(0, -2)$ is $1 - 2e^0 = -1$

(b) $y + 2 = -1(x)$, $y = -x - 2$

(c) $y - 1 = 1(x - 1)$

$$y = x$$

12. Given $y = 4^x + 3^x$, find y', y'', and y'''.

ANSWER:

$$y' = \ln 4(4^x) + \ln 3(3^x)$$

$$y'' = (\ln 4)^2(4^x) + (\ln 3)^2(3^x)$$

$$y''' = (\ln 4)^3(4^x) + (\ln 3)^3(3^x)$$

Problems and Solutions for Section 3.3

1. If $100 is invested at $r\%$ interest per year, compounded yearly, then the yield after 15 years is given by $F = 100\left(1 + \dfrac{r}{100}\right)^{15}$.

(a) Find $\left.\dfrac{dF}{dr}\right|_{r=5}$.

(b) Interpret your answer for part (a).

ANSWER:

(a)

$$\frac{dF}{dr} = (15)(100)(1 + \frac{r}{100})^{14} \cdot \frac{1}{100}$$

$$= 15(1 + \frac{r}{100})^{14}$$

So $\left.\dfrac{dF}{dr}\right|_{r=5} = 15(1 + \dfrac{5}{100})^{14} \approx 29.70$.

(b) For a \$100 investment at 5% interest per year, the yield after 15 years will increase by \$29.70 per additional point of interest rate.

2. Find the first derivative of the following functions.

   (a) $y = \ln(x + 1)$
   (b) $s = \ln z^3$
   (c) $t = \ln(x^3 + 4)$
   (d) $y = a^x$

   ANSWER:

   (a) Since $y = \ln x + 1$, the first derivative of this functon is

   $$\frac{dy}{dx} = \frac{1}{1 + x}$$

   (b) Since $s = \ln z^3 = 3\ln z$, the first derivative of this function is

   $$\frac{ds}{dz} = \frac{3}{z}$$

   (c) Since $t = \ln(x^3 + 4)$, the first derivative of this function is

   $$\frac{dt}{dx} = \frac{3x^2}{(x^3 + 4)}$$

   (d) We have $y = a^x$; taking logs of both sides gives $\ln y = x \ln a$. Now set $t = \ln y$ then $t = x \ln a$. Differentiating the first gives $\dfrac{dt}{dy} = \dfrac{1}{y}$ and the second gives $\dfrac{dt}{dx} = \ln a$ so that $\dfrac{dy}{dx} = y \ln a = a^x \ln a$.

3. Find the first derivative of the following functions.

   (a) $y = e^{-x^2}$
   (b) $s = e^x + x^e$
   (c) $t = e^{x+2}$
   (d) $y = \ln(e^x + 1)$

   ANSWER:

   (a) Since $y = e^{-x^2}$, the first derivative of this function is

   $$\frac{dy}{dx} = -2xe^{-x^2}$$

   (b) Since $s = e^x + x^e$, the first derivative of this function is

   $$\frac{ds}{dx} = e^x + ex^{e-1}$$

   (c) Since $t = e^{x+2}$, the first derivative of this function is

   $$\frac{dt}{dx} = e^{x+2}$$

   (d) Since $y = \ln(e^x + 1)$, the first derivative of this function is

   $$\frac{dy}{dx} = \frac{e^x}{(e^x + 1)}$$

4. The population of Mexico in millions is described by the formula $P(t) = 67e^{0.027t}$, where t is the number of years after 1980.

(a) The population in the year 2000 will be _____.
(b) In the year 2000 the population will be increasing at a rate of _____.
(c) How long will it take for the population to double?
(d) When will the population be increasing at a rate of 5 million persons per year?

ANSWER:

(a) $67e^{(0.027)(20)} = 114.97 \approx 115$ million
(b) $67(0.027)e^{(0.027)(20)} = 3.104$ million persons/year
(c)

$$(2)(67) = 67e^{0.027t}$$
$$t = \ln 2/0.027 = 25.67 \text{ years}$$

(d)

$$67(0.027)e^{0.027t} = 5$$
$$t = \frac{\ln(5/(67 \cdot 0.027))}{0.027} = 37.65 \text{ years.}$$

In the year 2017 or 2018.

5. Find the equation of the line tangent to the curve $y = \sqrt{3x + 7}$ at the point above $x = 3$.

ANSWER:
$y' = \frac{1}{2}\frac{3}{\sqrt{3x+7}}$, so $y'(3) = \frac{3}{8}$. We have $y = \frac{3}{8}x + b$. Since $y(3) = 4$, we have $4 = \frac{9}{8} + b$, or $b = \frac{23}{8}$, and the equation is

$$y = \frac{3}{8}x + \frac{23}{8}.$$

6. The price in dollars of a house during a period of mild inflation is described by the formula $P(t) = (80,000)e^{0.05t}$, where t is the number of years after 1990.

(a) The value of the house in the year 2000 will be _____.
(b) In the year 2000 the value will be increasing at a rate of _____.
(c) How long will it take for the house to double in value?
(d) When will the house be increasing in value at a rate of $10,000 per year?

ANSWER:

(a) $131,898.
(b) $6595 per year.
(c)

$$160,000 = 80,000e^{0.05t}$$
$$2 = e^{0.05t}$$
$$\ln 2 = 0.05t$$
$$t = 13.86.$$

Every 13.86 years.

(d)

$$P'(t) = 10,000$$
$$(80,000)(0.05)e^{0.05t} = 10,000$$
$$e^{0.05t} = 2.5$$
$$t = \frac{\ln 2.5}{0.05} = 18.32.$$

In the year 2008.

7. The cost to produce q aircraft in South Africa is given by the function $C(q) = 2.5q^{0.848}$, with C in millions of rands. Find the marginal cost for $q = 50$ and interpret your answer.

   ANSWER:

   The marginal cost is given by

   $$C'(q) = (0.848)(2.5)q^{-0.152} = 2.12q^{-0.152}.$$

   For $q = 50$, we have $C'(50) = 2.12(50)^{-0.152} \approx 1.17$. At the production level of producing 50 aircraft, the cost of producing one more aircraft is 1.17 million rands.

8. Sand falls from an overhead hopper to form a right circular cone. If the cone formed has angle α find the rate of change of volume with respect to height. If the height is changing at 2 cms per minute what volume of sand is falling from the hopper when the height of the cone is 3 meters? (The volume of a cone is $\frac{1}{3}\pi r^2 h$ where h is the height and r is the base radius.)

   ANSWER:

   If the angle of the cone is α then $r = h \tan \alpha$ so that the volume is

   $$V(h) = \frac{1}{3}\pi h^3 \tan^2 \alpha$$

   Then by the chain rule the rate at which sand is falling is

   $$\frac{dV}{dt} = \frac{dV}{dh}\frac{dh}{dt} = (\pi h^2 \tan^2 \alpha)(2) \quad \text{cc/min}$$

   so that when $h = 300cm$ the volume of sand falling per minute is $2(300^2)\pi \tan^2 \alpha$ cubic centimeters .

9. Find the derivatives
   (a) $h(x) = (2x^3 + e^x)^3$
   (b) $g(x) = \sqrt{2x^3 + e^x}$
   (c) $k(x) = \dfrac{(2x^3 + e^x)^3}{\sqrt{2x^3 + e^x}}$

   ANSWER:

   (a) $h'(x) = 3(2x^3 + e^x)^2(6x^2 + e^x)$
   (b) $g(x) = (2x^3 + e^x)^{1/2}$
   $$g'(x) = \frac{1}{2}(2x^3 + e^x)^{-1/2}(6x^2 + e^x)$$
   (c) $k(x) = (2x^3 + e^x)^{5/2}$
   $$k'(x) = (2x^3 + e^x)^3 \left(-\frac{1}{2}\right)(2x^3 + e^x)^{-3/2}(6x^2 + e^x) + (2x^3 + e^x)^{-1/2}(3)(2x^3 + e^x)^2(6x + e^x)$$
   $$= \frac{5}{2}(6x^2 + e^x)(2x^3 + e^x)^{3/2}.$$

10. Find the derivatives:
    (a) $f(x) = 2^{(4x-5)}$
    (b) $g(x) = \sqrt{e^x + e^{x^2}}$

    ANSWER:

    (a) $f'(x) = 4\ln 2 \cdot 2^{4x-5}$
    (b) $g(x) = \left(e^x + e^{x^2}\right)^{1/2}$
    $$g'(x) = \frac{1}{2}\left(e^x + e^{x^2}\right)^{-1/2}\left(e^x + 2xe^{x^2}\right)$$

11. Find the first three derivatives of $g(x) = (ax^2 + b)^2$. Assume that a and b are constants.

    ANSWER:

    $$g'(x) = 2(ax^2 + b)(2ax) = 4ax(ax^2 + b)$$
    $$g''(x) = 4ax(2ax) + (ax^2 + b)4a = 12a^2x^2 + 4ab$$
    $$g'''(x) = 24a^2x$$

Problems and Solutions for Section 3.4 ━━━━━━

1. Complete:

   (a) $\dfrac{d}{dx}(3^x \cdot \ln x) =$

   (b) $\dfrac{d}{dx}\left(\dfrac{10}{1 + 9e^{-x}}\right) =$

   ANSWER:

   (a) $\dfrac{d}{dx}(3^x \cdot \ln x) = \ln 3 \cdot 3^x \ln x + \dfrac{3^x}{x}$.

   (b) $\dfrac{d}{dx}\left(\dfrac{10}{1 + 9e^{-x}}\right) = \dfrac{-10(-9e^{-x})}{(1 + 9e^{-x})^2} = \dfrac{90e^{-x}}{(1 + 9e^{-x})^2}$.

2. Compute the derivatives of the following functions.

   (a) $y = 2x^4 + \dfrac{x^2 - 10x}{x^3}$ (b) $y = 3.5^x + \dfrac{2}{\sqrt{x}}$ (c) $y = 8e^{-3x}$

   ANSWER:

   (a) $y' = 8x^3 + \dfrac{2x^4 - 10x^3 - 3x^4 + 30x^3}{x^6} = 8x^3 + 20x^{-3} - x^{-2}$.

   (b) $y' = (\ln 3.5)(3.5)^x - \dfrac{1}{(\sqrt{x})^3}$.

   (c) $y' = -24e^{-3x}$.

3. Find $\dfrac{dy}{dx}$ for the following functions.

   (a) $y = x^e \cdot (1.012)^x$ (b) $y = \ln\left(\dfrac{x}{2} + \dfrac{2}{x}\right)$ (c) $y = \sqrt{e^x + e^{-x}}$

   ANSWER:

   (a) $\dfrac{d}{dx}(x^e \cdot 1.012^x) = ex^{e-1} \cdot 1.012^x + x^e \cdot \ln(1.012)(1.012)^x$.

   (b) $\dfrac{d}{dx}[\ln(x/2 + 2/x)] = \dfrac{1}{x/2 + 2/x} \cdot \left(\dfrac{1}{2} - \dfrac{2}{x^2}\right) = \dfrac{1/2 - 2/x^2}{x/2 + 2/x}$.

   (c) $\dfrac{d}{dx}\sqrt{e^x + e^{-x}} = \dfrac{1}{2}\dfrac{e^x - e^{-x}}{\sqrt{e^x + e^{-x}}}$.

4. Find a linear approximation for the function $f(x) = \dfrac{2x + 8}{x - 2}$ valid for x near 3.

   $\dfrac{2x + 8}{x - 2} \approx$ _____ for x near 3.

   ANSWER:

   A linear approximation for f valid for x near 3 is given by $f(3) + f'(3)(x - 3)$. Since

   $$f(x) = \dfrac{2x + 8}{x - 2},$$

   we have

   $$f(3) = 14.$$

   Since

   $$f'(x) = \dfrac{-12}{(x - 2)^2},$$

   we have

   $$f'(3) = -12.$$

   Therefore,

   $$f(x) \simeq 14 - 12(x - 3)$$
   $$f(x) \simeq -12x + 50$$

5. Table 3.4.27 gives values for two functions f and g and their derivatives.

Table 3.4.27

x	-1	0	1	2	3
f	3	3	1	0	1
g	1	2	2.5	3	4
f'	-3	-2	-1.5	-1	1
g'	2	3	2	2.5	3

Find:

(a) $\left. \dfrac{d}{dx}[f(x)g(x)]\right|_{x=1}$ **(b)** $\left. \dfrac{d}{dx}\left(\dfrac{f(x)}{g(x)}\right)\right|_{x=2}$ **(c)** $\left. \dfrac{d}{dx}[f(g(x))]\right|_{x=0}$

ANSWER:

(a) $\left. \dfrac{d}{dx}(f(x)g(x))\right|_{x=1} = \left.\big(f'(x)g(x) + f(x)g'(x)\big)\right|_{x=1} = -3.75 + 2 = -1.75.$

(b) $\left. \dfrac{d}{dx}\left(\dfrac{f(x)}{g(x)}\right)\right|_{x=2} = \left.\dfrac{f'(x)g(x) - f(x)g'(x)}{(g(x))^2}\right|_{x=2} = \dfrac{-3-0}{9} = -\dfrac{1}{3}.$

(c) $\left. \dfrac{d}{dx}(f(g(x)))\right|_{x=0} = \left.f'(g(x))g'(x)\right|_{x=0} = \left.f'(2)g'(x)\right|_{x=0} = (-1)\cdot 3 = -3.$

6. (a) Find the derivative of $f(x) = x^{-3}(2x^4 + 2)$ using the product rule and simplify the answer.
(b) Rewrite $f(x)$ as a quotient and use the quotient rule to find $f'(x)$.

ANSWER:

(a)

$$f'(x) = x^{-3}(8x^3) + (2x^4 + 2)(-3x^{-4}) = 8 - 6 - 6x^{-4} = 2 - 6x^{-4}$$

(b)

$$f(x) = \dfrac{2x^4 + 2}{x^3}$$

$$f'(x) = \dfrac{x^3(8x^3) - (2x^4 + 2)(3x^2)}{(x^3)^2} = \dfrac{8x^6 - 6x^6 - 6x^2}{x^6}$$

$$= \dfrac{2x^6 - 6x^2}{x^6} = 2 - 6x^{-4}$$

7. Find the equation of the tangent line to the graph of $g(x) = \dfrac{x^2 - 2}{x + 1}$ at the point at which $x = 1$.

ANSWER:

Since $g(1) = \dfrac{1^2 - 2}{1 + 1} = -\dfrac{1}{2}$, the tangent line passes through the point $(1, -1/2)$.

We find the derivative of $g(x)$ using the quotient rule:

$$g'(x) = \dfrac{(x + 1)(2x) - (x^2 - 2)(1)}{(x + 1)^2} = \dfrac{2x^2 + 2x - x^2 + 2}{(x + 1)^2} = \dfrac{x^2 + 2x + 2}{(x + 1)^2}$$

At $x = 1$, the slope of the tangent line is $m = g'(1) = \dfrac{1 + 2 + 2}{(1 + 1)^2} = \dfrac{5}{4}$. The equation of the tangent line is

$$y + \dfrac{1}{2} = \dfrac{5}{4}(x - 1)$$

$$y = \dfrac{5}{4}x - \dfrac{7}{4}$$

8. Differentiate $g(t) = e^{-t} + 2e^{-2t}$ by writing each term as a quotient before finding the derivative.

ANSWER:

$$g(t) = \frac{1}{e^t} + \frac{2}{e^{2t}}$$

$$g'(t) = \frac{e^t(0) - 1(e^t)}{(e^t)^2} + \frac{e^{2t}(0) - 2(2e^{2t})}{(e^{2t})^2}$$

$$= \frac{-e^t}{e^{2t}} - \frac{4e^{2t}}{e^{4t}} = -e^{-t} - 4e^{-2t}$$

9. Given $f(x) = e^x$, $g(x) = 2^x$, $h(x) = f(x)g(x)$, and $j(x) = \dfrac{g(x)}{f(x)}$

(a) Find $h'(x)$ and $h''(x)$.

(b) Find $j'(x)$ and $j''(x)$

ANSWER:

(a)

$$h(x) = e^x 2^x$$

$$h'(x) = e^x(\ln 2)2^x + 2^x e^x$$

$$h''(x) = e^x(\ln 2)^2 2^x + (\ln 2)2^x e^x + 2^x e^x + e^x(\ln 2)2^x$$

$$= 2^x e^x ((\ln 2)^2 + 2\ln 2 + 1)$$

(b)

$$j(x) = \frac{2^x}{e^x}$$

$$j'(x) = \frac{e^x(\ln 2)2^x - 2^x e^x}{(e^x)^2} = \frac{2^x e^x(\ln 2 - 1)}{e^{2x}} = \frac{2^x}{e^x}(\ln 2 - 1)$$

$$j''(x) = (\ln 2 - 1)(\frac{2^x}{e^x})'$$

$$= (\ln 2 - 1)(\ln 2 - 1)\frac{2^x}{e^x}$$

$$= (\ln 2 - 1)^2 \frac{2^x}{e^x}.$$

10. Given $f(x) = \dfrac{x^3}{2x + 1}$ and $g(x) = \dfrac{x^2 + 2}{3x^2}$ and $h(x) = f(x)g(x)$, find $h'(2)$.

ANSWER:

$$h'(x) = f(x)g'(x) + g(x)f'(x)$$

$$= \frac{x^3}{2x + 1}\left(\frac{3x^2(2x) - (x^2 + 2)(6x)}{(3x^2)^2}\right) + \frac{x^2 + 2}{3x^2}\left(\frac{(2x + 1)(3x^2) - x^3(2)}{(2x + 1)^2}\right)$$

$$= \frac{x^3}{2x + 1}\left(\frac{6x^3 - 6x^3 - 12x}{9x^4}\right) + \frac{x^2 + 2}{3x^2}\left(\frac{6x^3 + 3x^2 - 2x^3}{4x^2 + 4x + 1}\right)$$

$$h'(2) = \frac{2^3}{2(2) + 1}\left(\frac{-12(2)}{9(2)^4}\right) + \frac{2^2 + 2}{3(2)^2}\left(\frac{4(2)^3 + 3(2)^2}{4(2)^2 + 4(2) + 1}\right)$$

$$= \frac{8}{5}\left(\frac{-24}{9 \cdot 2^4}\right) + \frac{6}{12}\left(\frac{32 + 12}{16 + 8 + 1}\right)$$

$$= -\frac{4}{15} + \frac{22}{25} = \frac{46}{75}$$

11. If $j(x) = g(x)h(x)$ and $j'(x) = 2x^3 + 3x^2(2x + 1)$, what are $g(x)$ and $h(x)$?

ANSWER:

$$j'(x) = g(x)h'(x) + h(x)g'(x)$$

$$g(x) = x^3 \text{ and } h(x) = 2x + 1$$

12. Differentiate $g(x) = 3e^{2x}$ by first writing it as a product.
ANSWER:

$$g(x) = 3e^x \cdot e^x$$
$$g'(x) = 3e^x \cdot e^x + e^x \cdot 3e^x = 3e^{2x} + 3e^{2x} = 6e^{2x}$$

13. Differentiate $\dfrac{6x^2}{x^3 + 1}$.
ANSWER:

$$\frac{(x^3 + 1)(12x) - 6x^2(3x^2)}{(x^3 + 1)^2} = \frac{12x^4 + 12x - 18x^4}{(x^3 + 1)^2} = \frac{-6x^4 + 12x}{(x^3 + 1)^2} = \frac{6x(-x^3 + 2)}{(x^3 + 1)^2}$$

Problems and Solutions   for Section 3.5

1. Find $\dfrac{dy}{dx}$ if: **(a)** $y = e^{-0.01x} + \ln x$ **(b)** $y = 4x^2 \sin(\pi x)$ **(c)** $y = (2^x + \sqrt{x})^5$
ANSWER:

(a) $dy/dx = -0.01e^{-0.01x} + \dfrac{1}{x}$.

(b) $dy/dx = 8x \sin(\pi x) + 4\pi x^2 \cos(\pi x)$.

(c) $dy/dx = 5(2^x + \sqrt{x})^4 \cdot (\ln 2 \cdot 2^x + \dfrac{1}{2\sqrt{x}})$.

2. Find the derivatives of the functions given below.

(a) $y = (x^2 + 3\sqrt{x})^5$

(b) $y = 8 \sin \left(2^{-x} \right)$

(c) $y = x \cos(3x)$

(d) $y = \dfrac{\ln x + 5}{x^2 + 7}$

ANSWER:

(a) $\dfrac{dy}{dx} = 5(x^2 + 3\sqrt{x})^4 \cdot \left(2x + \dfrac{3}{2} \left(\dfrac{1}{\sqrt{x}} \right) \right)$.

(b) $\dfrac{dy}{dx} = 8 \cos(2^{-x}) \cdot (-\ln 2 \cdot 2^{-x}) = -8 \ln 2 \cdot 2^{-x} \cdot \cos(2^{-x})$.

(c) $\dfrac{dy}{dx} = \cos(3x) - 3x \sin(3x)$.

(d) $\dfrac{dy}{dx} = \dfrac{x + \frac{7}{x} - 2x \ln x - 10x}{(x^2 + 7)^2} = \dfrac{\frac{7}{x} - 2x \ln x - 9x}{(x^2 + 7)^2}$.

3. Compute $\dfrac{dy}{dx}$ for the following functions: **(a)** $y = x \cos(\ln x)$ **(b)** $y = \dfrac{x^2 + e^{3x}}{\sin(2x) + 8}$
ANSWER:

(a) $\dfrac{dy}{dx} = \cos(\ln x) - \sin(\ln x)$.

(b) $\dfrac{dy}{dx} = \dfrac{(2x + 3e^{3x})(\sin(2x) + 8) - (x^2 + e^{3x})(2\cos(2x))}{(\sin(2x) + 8)^2}$.

4. The size of an impala population is represented by the function $R(t) = 10 + 2\cos(\pi t/6)$, with t in months since the beginning of the year and where $R(t)$ is measured in thousands. At what rate does the population change after 3 months?
ANSWER:
$R'(t) = -2\sin(\dfrac{\pi}{6}t) \cdot \dfrac{\pi}{6} = -\dfrac{\pi}{3} \sin(\dfrac{\pi}{6}t)$, so $R'(3) = -\dfrac{\pi}{3} \sin \dfrac{\pi}{2} = -\dfrac{\pi}{3} \approx -1.05$ thousand/month. After 3 months the population decreases at the rate of 1.05 thousand, or 1050, per month.

5. The number of hours, H, of daylight in Madrid as a function of the date is given by the formula

$$H = 12 + 2.4 \sin(0.0172(t - 80))$$

where t is the number of days since the beginning of the year.

(a) What are the units of $\dfrac{dH}{dt}$? _____

(b) Explain in words the meaning of $\left.\dfrac{dH}{dt}\right|_{t=100}$.

(c) Estimate $\left.\dfrac{dH}{dt}\right|_{t=100} \approx$ _____.

ANSWER:

(a) $\dfrac{dH}{dt}$ is measured in hours/day.

(b) $\left.\dfrac{dH}{dt}\right|_{t=100}$ represents the number of daylight hours gained with the passing of one whole day around April 10 (when 100 days have already passed).

(c)

$$\left.\frac{dH}{dt}\right|_{t=100} \approx \frac{H(100.1) - H(100)}{0.1}$$

$$\left.\frac{dH}{dt}\right|_{t=100} \approx 0.0388 \text{ hours/day}$$

6. Find two different functions $G_1(x)$ and $G_2(x)$ that satisfy $G'(x) = -\cos(3x)$.

ANSWER:

Since $G'(x)$ is of the form $-\cos u$, we can make an initial guess that

$$G(x) = -\sin(3x).$$

Then $G'(x) = -3\cos(3x)$, so we're off by a factor of 3 and get

$$G_1(x) = -\frac{1}{3}\sin(3x)$$

We can also add any constant to get $G_2(x)$. For example, $G_2(x) = -\dfrac{1}{3}\sin(3x) + \pi$.

7. Find the derivative of the following functions:

(a) $f(w) = \sin(2w^2) + \cos(2w^2)$
(b) $g(x) = \cos(\sin 2x)$
(c) $h(z) = e^{\cos z} + e^{\sin z}$

ANSWER:

(a) $f'(w) = 4w\cos(2w^2) - 4w\sin(2w^2)$
(b) $g'(x) = -\sin(\sin 2x)(2\cos 2x) = -2\cos 2x \sin(\sin 2x)$
(c) $h'(z) = -\sin z e^{\cos z} + \cos z e^{\sin z}$

8. Is the graph of $f(w) = \cos(w^6)$ increasing or decreasing when $w = 5$? Is it concave up or concave down?

ANSWER:

We begin by taking the derivative of $f(w)$ and evaluating at $w = 5$;

$$f'(w) = -6w^5 \sin(w^6)$$

Evaluating $\sin(5^6) = \sin(15,625)$ on a calculator, we see that $\sin(15,625) < 0$, so we know that $-6w^5 \sin(w^6) > 0$ when $w = 5$, and therefore the function is increasing.

Next, we take the second derivative and evaluate it at $w = 5$;

$$f''(w) = \underbrace{-30\sin w^6 \cdot w^4}_{\text{negative}} - \underbrace{36\cos w^6 \cdot w^{10}}_{\text{negative}}$$

From this we see $f''(w) < 0$, thus the graph is concave down.

Problems and Solutions to Review Problems for Chapter 3 ━━━━━━━━━━

1. Find $\dfrac{dy}{dx}$ for the following functions.

(a) $y = \sin 3x - \cos 4x$ (b) $y = \ln(x^2 e^{3x})$ (c) $y = \sqrt{x^2 + 9}$

ANSWER:

(a) $\dfrac{dy}{dx} = 3\cos 3x + 4\sin 4x.$

(b) $\dfrac{dy}{dx} = \dfrac{2xe^{3x} + 3x^2 e^{3x}}{x^2 e^{3x}} = \dfrac{2 + 3x}{x}.$

(c) $\dfrac{dy}{dx} = \dfrac{x}{\sqrt{x^2 + 9}}.$

2. Find the first derivative of the following functions.

(a) $y = \cos(x^2)$
(b) $s = \ln(1 + \sin(x))$
(c) $t = \sqrt{(1 + x^2)}$
(d) $s = \sin(1 + x^2)$
(e) $y = e^{\sin(x)}$

ANSWER:

(a) Let $u = x^2$ then $y = \cos u$ so that $\dfrac{dy}{du} = -\sin u$ then $\dfrac{dy}{dx} = \dfrac{dy}{du}\dfrac{du}{dx} = (-\sin u)(2x) = -2x\sin(x^2)$

(b) Let $u = 1 + \sin(x)$ then $s = \ln u$ so that $\dfrac{ds}{du} = \dfrac{1}{u}$ then $\dfrac{ds}{dx} = \dfrac{ds}{du}\dfrac{du}{dx} = (\dfrac{1}{1 + \sin x})(\cos x)$

(c) Let $u = 1 + x^2$ then $t = \sqrt{u}$ so that $\dfrac{dt}{du} = \dfrac{1}{2\sqrt{u}}$ then $\dfrac{dt}{dx} = \dfrac{dt}{du}\dfrac{du}{dx} = (\dfrac{1}{2\sqrt{1 + x^2}})(2x),$

(d) Let $u = 1 + x^2$ then $s = \sin u$ so that $\dfrac{ds}{du} = \cos u$ then $\dfrac{ds}{dx} = \dfrac{ds}{du}\dfrac{du}{dx} = (\sin(1 + x^2))(2x),$

(e) Let $u = \sin(x)$ then $y = e^u$ so that $\dfrac{dy}{du} = e^u$ then $\dfrac{dy}{dx} = \dfrac{dy}{du}\dfrac{du}{dx} = \cos(x)e^{\sin(x)}.$

3. Find the first derivative of the following functions.

(a) $y = x(x + 1)^5$
(b) $s = x\sin(x)$
(c) $t = x^2 \ln(x)$
(d) $s = 3t\cos^2 t$
(e) $x = \dfrac{w^2 - 4}{w}$

ANSWER:

(a) $y = x(x + 1)^5, \dfrac{dy}{dx} = 5\,x\,(1 + x)^4 + (1 + x)^5$

(b) $s = x\sin(x), \dfrac{ds}{dx} = x\,\cos(x) + \sin(x)$

(c) $t = x^2 \ln(x), \dfrac{dt}{dx} = 2\,x\,\ln(x) + x$

(d) $s = 3t\cos^2 t = 3\cos t\cos t, \dfrac{ds}{dt} = 3\cos(t)^2 - 6\,t\,\cos(t)\,\sin(t)$

(e) $x = \dfrac{w^2 - 4}{w} = w - \dfrac{4}{w}, \dfrac{dx}{dw} = 1 - \dfrac{4}{w^2}$

4. Find the first derivative of the following functions.

(a) $y = x(x + a)^4$
(b) $s = x^2 \sin(x)$
(c) $t = x\ln(x)$
(d) $s = 3t\sin^2 t$
(e) $x = (w^2 - 4)\sin w$

ANSWER:

(a) $y = x(x + a)^4, \dfrac{dy}{dx} = 4\,x\,(a + x)^3 + (a + x)^4$

(b) $s = x^2 \sin(x), \dfrac{ds}{dx} = x^2\,\cos(x) + 2\,x\,\sin(x)$

(c) $t = x\ln(x), \dfrac{dt}{dx} = 1 + \ln(x), x > 0$

(d) $s = 3t\sin^2 t = 3t\sin(t)\sin(t), \dfrac{ds}{dt} = 6\,t\,\cos(t)\,\sin(t) + 3\sin(t)^2$

(e) $x = (w^2 - 4)\sin w$ $\dfrac{dz}{dw} = \left(-4 + w^2\right)\cos(w) + 2\,w\,\sin(w)$

5. Differentiate the following functions.

(a) $y = \dfrac{w^2 - 9}{w^2 - 2}$

(b) $s = \dfrac{z^2}{\cos z^2}$

(c) $y = \dfrac{1 + \sin(t)}{1 - \cos(t)}$

(d) $x = \dfrac{\cos(w) - 9}{\sin(w) - 2}$

ANSWER:

(a) Since $y = \dfrac{w^2 - 9}{w^2 - 2}$, we have

$$\frac{dy}{dw} = \frac{14w}{(w^2 - 2)^2}, \qquad w^2 \neq 2$$

(b) Since $s = \dfrac{z^2}{\cos z^2}$, we have

$$\frac{ds}{dz} = 2\,z\,\sec(z^2) + 2\,z^3\,\sec(z^2)\,\tan(z^2)$$

(c) Since $t = \dfrac{1 + \sin(t)}{1 - \cos(t)}$, we have

$$\frac{dy}{dt} = \frac{\cos(t) - \sin(t) - 1}{(1 - \cos(t))^2}$$

(d) Since $x = \dfrac{\cos(w) - 9}{\sin(w) - 2}$, we have

$$\frac{dx}{dw} = \frac{9\,\cos(w) + 2\,\sin(w) - 1}{\left(\sin(w) - 2\right)^2}$$

6. Find the first derivative of the following functions.

(a) $y = \sqrt{x^3 + 1}$

(b) $s = z^2 + \sin z^3$

(c) $t = \ln\sin(x) + 4)$

(d) $x = \dfrac{w - 4}{w^2 + 3}$

ANSWER:

(a) Since $y = \sqrt{x^3 + 1}$, then the first derivative of this function is

$$\frac{dy}{dx} = \frac{3x^2}{2\sqrt{1 + x^3}}$$

(b) Since $s = z^2 + \sin z^3$, then the first derivative of this function is

$$\frac{ds}{dz} = 2z + 3z^2\,\cos(z^3)$$

(c) Since $t = \ln(\sin(x) + 4)$, then the first derivative of this function is

$$\frac{dt}{dx} = \frac{\cos(x)}{(4 + \sin(x))}$$

(d) Since $x = \dfrac{w - 4}{w^2 + 3}$, then the first derivative of this function is

$$\frac{dx}{dw} = \frac{(3 + 8w - w^2)}{(w^2 + 3)^2}$$

7. Find the first derivative of the following functions.

(a) $y = \sin(x^3 + 1)$

(b) $s = z^4 + \sin z^3$

(c) $t = \cos(\sin(x) + 4)$

(d) $x = \dfrac{w^2 - 4}{w + 3}$

ANSWER:

(a) Since $y = \sin(x^3 + 1)$, then the first derivative of this function is

$$\frac{dy}{dx} = 3x^2 \cos(x^3 + 1)$$

(b) Since $s = z^4 + \sin z^3$, then the first derivative of this function is

$$\frac{ds}{dz} = 4z^3 + 3z^2 \cos(z^3)$$

(c) Since $t = \cos(\sin(x) + 4)$, then the first derivative of this function is

$$\frac{dt}{dx} = -\cos(x)\sin(\sin(x) + 4)$$

(d) Since $x = \dfrac{w^2 - 4}{w + 3}$, then the first derivative of this function is

$$\frac{dx}{dw} = \frac{(4 + 6w + w^2)}{(w + 3)^2}, \qquad w \neq -3$$

8. Find the first derivative of the following functions.

(a) $y = e^{\sin(x)}$

(b) $s = e^{5x - 2}$

(c) $t = e^{\sqrt{x}}$

(d) $y = e^{(e^x + 1)}$

ANSWER:

(a) Since $y = e^{\sin(x)}$, then the first derivative of the function is

$$\frac{dy}{dx} = \cos(x)e^{\sin(x)}$$

(b) Since $s = e^{5x - 2}$, then the first derivative of the function is

$$\frac{ds}{dx} = 5e^{5x - 2}$$

(c) Since $t = e^{\sqrt{x}}$, then the first derivative of the function is

$$\frac{dt}{dx} = \frac{1}{(2\sqrt{x})}e^{\sqrt{x}}$$

(d) Since $y = e^{(e^x + 1)}$, then the first derivative of the function is

$$\frac{dy}{dx} = e^x e^{(e^x + 1)}$$

9. Find the first derivative of the following functions.

(a) $y = \sqrt{x}$

(b) $x = y^3 + \dfrac{5}{y}$

(c) $s = \sin t^2 + \sin^2 t$

(d) $w = \dfrac{x^2 + 5}{x - 3}$

ANSWER:

(a) Since $y = \sqrt{x}$, then the first derivative of this function is

$$\frac{dy}{dx} = \frac{1}{2\sqrt{x}}$$

(b) Since $x = y^3 + \dfrac{5}{y}$, then the first derivative of this function is

$$\frac{dx}{dy} = 3y^2 - \frac{5}{y^2}$$

(c) Since $s = \sin t^2 + \sin^2 t$, then the first derivative of this function is

$$\frac{ds}{dt} = 2t\cos(t^2) + 2\cos(t)\sin(t)$$

(d) Since $w = \dfrac{x^2 + 5}{x - 3}$, by the quotient rule we get

$$\frac{dw}{dx} = \frac{(x-3)2x - 1(x^2 + 5)}{(x-3)^2}$$

10. Compute the derivatives:

(a) $f(x) = 8x^3 - 10x^2 + 5x - 17$
$f'(x) =$
(b) $f(x) = 2^x + 5e^{3x}$
$f'(x) =$
(c) $f(x) = (\frac{6}{x} + \sqrt{x})^{15}$
$f'(x) =$

ANSWER:

(a) $f'(x) = 24x^2 - 20x + 5$
(b) $f'(x) = (\ln 2) \cdot 2^x + 15e^{3x}$
(c) $f'(x) = 15(\dfrac{6}{x} + \sqrt{x})^{14} \cdot (-\dfrac{6}{x^2} + \dfrac{1}{2}x^{-1/2})$.

11. The amount, W, of fuel used by an aircraft flying at speed v km min^{-1} is given, in liters, by

$$W = 25v^2 + \frac{400}{v^2}$$

Find the rate of change of W with respect to v when $v = 4$. If the plane now begins to accelerate so that $\dfrac{dv}{dt} = k$ km min^{-2} intepret the quantity

$$k\frac{dW}{dv}.$$

ANSWER:

$$\frac{dW}{dv} = 50v - \frac{800}{v^3}$$

When $v = 4$ this gives

$$\frac{dW}{dv} = \frac{375}{2}$$

If $\dfrac{dv}{dt} = k$ then

$$k\frac{dW}{dv} = \frac{dW}{dv}\frac{dv}{dt} = \frac{dW}{dt}$$

which is the increase in fuel required to accelerate.

12. Compute the derivatives:

(a) $f(x) = 5x^4 - 20x^2 + 18x + 2$
   $f'(x) =$
(b) $f(x) = (10x^2 + 5)^{100}$
   $f'(x) =$
(c) $f(x) = \frac{5}{x} + \sqrt{3x + 1}$
   $f'(x) =$

ANSWER:

(a) $f'(x) = 20x^3 - 40x + 18$
(b) $f'(x) = 100(10x^2 + 5)^{99} \cdot 20x = 2000x(10x^2 + 5)^{99}$
(c) $-\dfrac{5}{x^2} + \dfrac{1}{2}(3x + 1)^{-1/2} \cdot 3 = -\dfrac{5}{x^2} + \dfrac{3}{2}(3x + 1)^{-1/2}$

Problems and Solutions on Establishing the Derivative Formulas

1. Show how to use algebra to derive the formula for the derivative of $f(x) = x^2 + 1$.
ANSWER:

$$\begin{aligned}
f'(x) &= \lim_{h \to 0} \frac{f(x+h) - f(x)}{x + h - x} = \lim_{h \to 0} \frac{(x+h)^2 + 1 - (x^2 + 1)}{h} \\
&= \lim_{h \to 0} \frac{x^2 + 2hx + h^2 + 1 - x^2 - 1}{h} \\
&= \lim_{h \to 0} \frac{2hx + h^2}{h} = \lim_{h \to 0} 2x + h = 2x + 0 = 2x.
\end{aligned}$$

For Problems 2– 4, use the definition of the derivative to obtain the stated results.

2. If $f(x) = 8 - 3x$, then $f'(x) = -3$.
ANSWER:

$$\begin{aligned}
f'(x) &= \lim_{h \to 0} \frac{f(x+h) - f(x)}{h} \\
&= \lim_{h \to 0} \frac{8 - 3(x+h) - (8 - 3x)}{h} \\
&= \lim_{h \to 0} \frac{8 - 3x - 3h - 8 + 3x}{h} \\
&= \lim_{h \to 0} \frac{-3h}{h} = \lim_{h \to 0}(-3) = -3.
\end{aligned}$$

3. If $f(x) = x^2 - 12$, then $f'(x) = 2x$.
ANSWER:

$$\begin{aligned}
f'(x) &= \lim_{h \to 0} \frac{f(x+h) - f(x)}{h} \\
&= \lim_{h \to 0} \frac{(x+h)^2 - 12 - (x^2 - 12)}{h} \\
&= \lim_{h \to 0} \frac{x^2 + 2xh + h^2 - 12 - x^2 + 12}{h} \\
&= \lim_{h \to 0} \frac{2xh + h^2}{h} = \lim_{h \to 0} \frac{h(2x + h)}{h} \\
&= \lim_{h \to 0}(2x + h) = 2x.
\end{aligned}$$

4. If $f(x) = 3x^2 + 2$, then $f'(x) = 6x$.

ANSWER:

$$
\begin{aligned}
f'(x) &= \lim_{h \to 0} \frac{f(x+h) - f(x)}{h} \\
&= \lim_{h \to 0} \frac{3(x+h)^2 + 2 - (3x^2 + 2)}{h} \\
&= \lim_{h \to 0} \frac{3(x^2 + 2xh + h^2) + 2 - 3x^2 - 2}{h} \\
&= \lim_{h \to 0} \frac{3x^2 + 6xh + 3h^2 + 2 - 3x^2 - 2}{h} \\
&= \lim_{h \to 0} \frac{6xh + 3h^2}{h} = \lim_{h \to 0} \frac{h(6x + 3h)}{h} \\
&= \lim_{h \to 0} (6x + 3h) = 6x.
\end{aligned}
$$

Chapter 4 Exam Questions

Problems and Solutions for Section 4.1

1. The sketch in Figure 4.1.77 represents the function f, with $f(x) = e^{-ax} \sin x$, $a > 0$ and $x \geq 0$.

   (a) Determine the x-intercepts of f.

   (b) What must the value of a be so that $x_0 = \dfrac{\pi}{4}$?

   (c) If $x_0 = \dfrac{\pi}{4}$, calculate x_1.

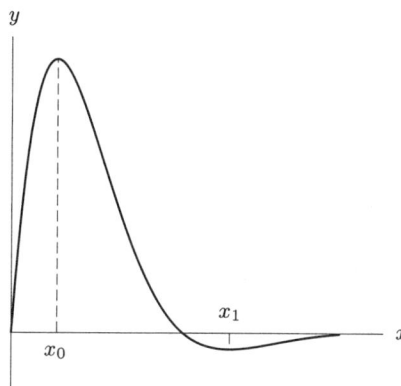

Figure 4.1.77

   ANSWER:

   (a) For $x \geq 0$, $f(x) = e^{-ax} \sin x = 0$ implies that $\sin x = 0$. Therefore, the x-intercepts of f are $x = k\pi$, $k = 0, 1, 2, \ldots$.

   (b) Let $f'(x) = -ae^{-ax} \sin x + e^{-ax} \cos x = 0$. Then $\cos x - a \sin x = 0$ or $\cos x = a \sin x$. So, if $x_0 = \pi/4$, we have $a = 1$.

   (c) If $x_0 = \pi/4$, then $a = 1$ and f has critical points at $\cos x = \sin x$ or at $x_0 = \pi/4$, $x_1 = 5\pi/4, \ldots$.

2. (a) Suppose the graph in Figure 4.1.78 is that of a function $g(x)$, $-1 \leq x \leq 2$. Sketch the graph of the derivative g'.

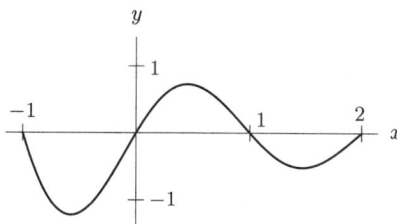

Figure 4.1.78

   (b) On the other hand, suppose the graph above is that of the derivative f' of a function f. For the interval $-1 \leq x \leq 2$, tell where the function f is

   (i) increasing;

   (ii) decreasing.

   (iii) Tell whether f has any extrema, and if so, where they are.

   ANSWER:

   (a) The graph of g' is shown in Figure 4.1.79.

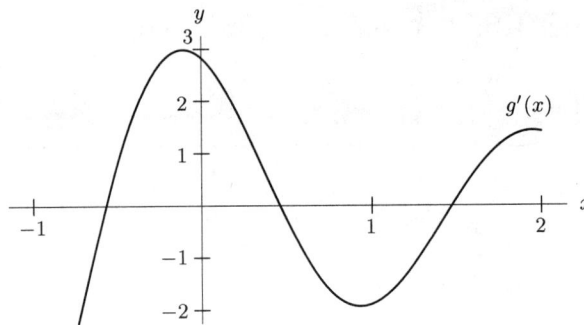

Figure 4.1.79

(b)   (i) f is increasing for $0 \leq x \leq 1$.

      (ii) f is decreasing for $-1 \leq x \leq 0$ and $1 \leq x \leq 2$.

      (iii) f has a local minimum at $x = 0$ and a local maximum at $x = 1$.

3. A brick is heated in an oven and taken out to cool off after a certain time. The temperature T of the brick at any time t is given by $T = 100e^{-(t-1)^2}$ for $t \geq 0$, with T in $°C$ and t in minutes.

    (a) What is the temperature of the brick when it is placed in the oven?

    (b) At what time t is the brick taken out of the oven?

    (c) What will the temperature of the brick eventually be?

    (d) Sketch the temperature/time curve.

    ANSWER:

    (a) $T(0) = 100e^{-1} \approx 36.8°C$ when it is placed in the oven.

    (b) We assume that the brick is taken out of the oven at its maximal temperature. Since $T' = -200e^{-(t-1)} \cdot (t - 1)$, $t = 1$ is a critical point and T reaches maximum at $t = 1$. So the brick is taken out after 1 minute.

    (c) As $t \to \infty$, $T \to 0$. So eventually the temperature will be 0.

    (d)

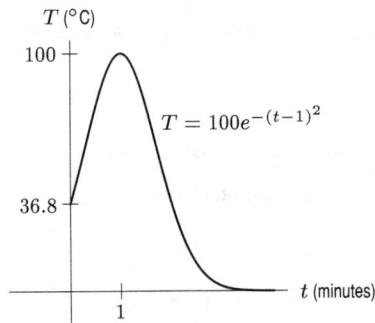

Figure 4.1.80

4. A stone is thrown vertically upwards so that its height, measured in feet, after t seconds is given by $s(t) = 100t - 16t^2$. What is its initial velocity? Determine the maximum height of the stone and the time of flight of the stone.

    ANSWER:

    The velocity of the stone is $s' = 100 - 32t$ and its acceleration is $s'' = -32$. The initial velocity is therefore 100 feet per second upwards.

    The stone reaches its maximum height when $s' = 0$, i.e. $t = \dfrac{100}{32}$ seconds. The time of flight is twice this. Note that this can also be found by setting $s = 0$ and solving $100t - 16t^2 = 0$.

5. The graph of $f(x)$ is given in Figure 4.1.81.

    (a) Sketch the graph of $f'(x)$ on the same axes.

    (b) Where does $f'(x)$ change its sign?

    (c) Where does $f'(x)$ have a local maximum or minimum?

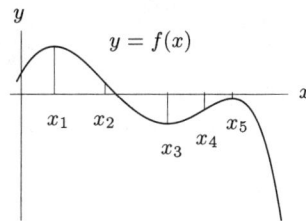

Figure 4.1.81

ANSWER:

(a)

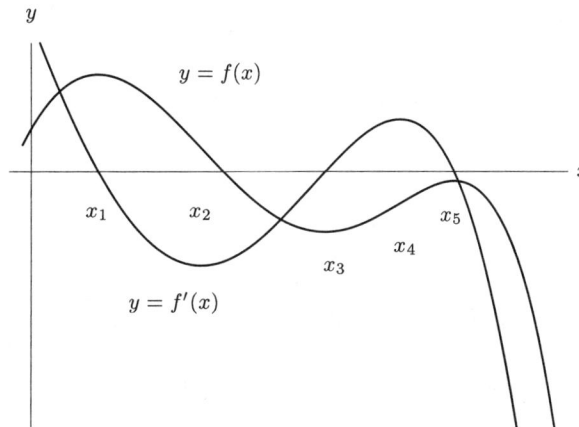

(b) At x_1, x_3, x_5.

(c) At x_2, x_4.

6. Given the curve

$$y = ax^3 + bx^2 + cx + d, \quad a \neq 0,$$

find the relation between the parameters $a, b,$ and c that will ensure that the curve:

(a) has only one turning point.

(b) has no turning points.

ANSWER:

Let $y' = 3ax^2 + 2bx + c = 0$. Then $x = \dfrac{-2b \pm \sqrt{4b^2 - 12ac}}{6a}$.

(a) If $4b^2 - 12ac = 0$, i.e. $b^2 - 3ac = 0$, then the curve has one critical point and it has only one turning point.

(b) If $b^2 - 3ac < 0$, then the curve has no critical point and it has no turning point.

7. Find all critical points of $f(x) = 4x^3 + 7x^2 + 4x$.

ANSWER:

$$f'(x) = 12x^2 + 14x + 4$$
$$= 2(6x^2 + 7x + 2)$$
$$= 2(2x + 1)(3x + 2)$$

$$2x + 1 = 0, \quad 3x + 2 = 0$$
$$x = -\frac{1}{2} \text{ or } x = -\frac{2}{3}$$

Problems and Solutions for Section 4.2

1. (a) Sketch the graph of $y = 2x^3 + 3x^2 - 36x + 100$ on the axes in Figure 4.2.82.

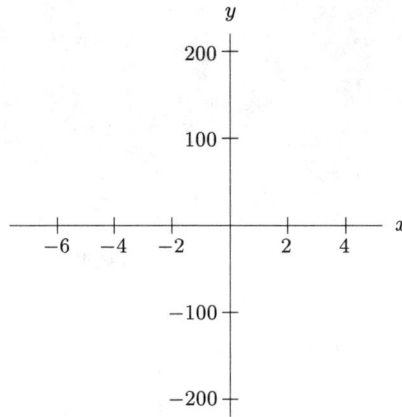

Figure 4.2.82

(b) The graph is:

    (i) Increasing and concave up on the interval _____ .

    (ii) Increasing and concave down on the interval _____ .

    (iii) Decreasing and concave up on the interval _____ .

    (iv) Decreasing and concave down on the interval _____ .

    ANSWER:

(a) The graph of $y = 2x^3 + 3x^2 - 36x + 100$ is shown in Figure 4.2.83.

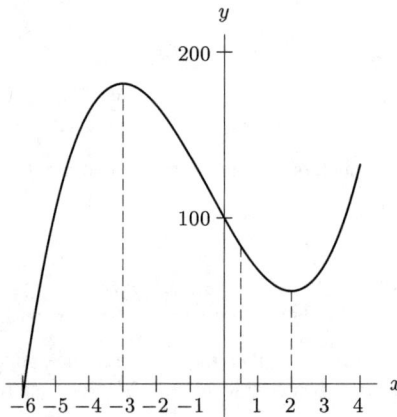

Figure 4.2.83

(b) Let $y' = 6x^2 + 6x - 36 = 6(x - 2)(x + 3) = 0$. The graph of y' is shown in Figure 4.2.84.

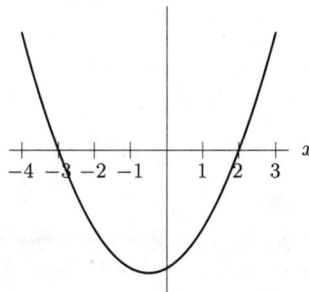

Figure 4.2.84

Let $y'' = 12x + 6 = 0$; then $x = 1/2$.

   (i) Increasing and concave up on $(2, \infty)$.

   (ii) Increasing and concave down on $(-\infty, -3)$

   (iii) Decreasing and concave up on $(-\frac{1}{2}, 2)$

   (iv) Decreasing and concave down on $(-3, -\frac{1}{2})$

2. Use the first and second derivatives to identify all critical points and inflection points of the following functions. Then graph the functions.

  (a) $f(x) = x^2 - 3x - 10$

  (b) $f(x) = 2x^3 - 3x^2 - 36x + 7$

  (c) $f(x) = x^4 - 12x^3 + 2$

  ANSWER:

(a) First find $f'(x)$ and set it equal to zero to find the critical points:

$$f'(x) = 2x - 3 = 0$$

$$x = 3/2$$

$f(3/2) = (3/2)^2 - 3(3/2) - 10 = -49/4$, so the point is (3/2,-49/4). Now, checking the second derivative,

$$f''(x) = 2$$

Thus there is no inflection point. Since $f''(3/2) > 0$, the point $(3/2, -49/4)$ is a local minimum.

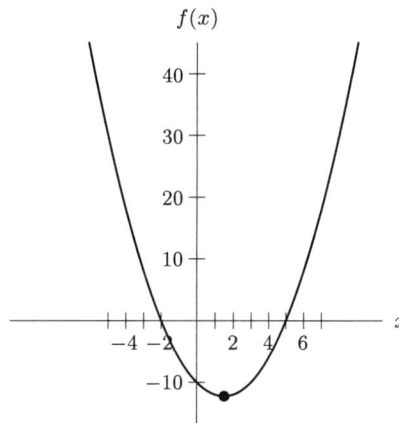

Figure 4.2.85

(b) Setting $f'(x)$ equal to zero to find the critical points:

$$f'(x) = 6x^2 - 6x - 36 = 0$$

$$6(x - 3)(x + 2) = 0$$

$$x = -2, 3$$

$f(-2) = 2(-2)^3 - 3(-2)^2 - 36(-2) + 7 = 51$, and $f(3) = 2(3)^3 - 3(3)^2 - 36(3) + 7 = -74$. Thus, the points are (-2,51), and (3,-74). Now, checking the second derivative,

$$f''(x) = 12x - 6 = 0$$

$$x = 2$$

$f(2) = 2(2)^3 - 3(2)^2 - 36(2) + 7 = -61$, so the point is (2,-61). Since $f''(-2) < 0$, (-2,51) is a local maximum. Since $f''(3) > 0$, (3,-74) is a local minimum. (2,-61) is an inflection point.

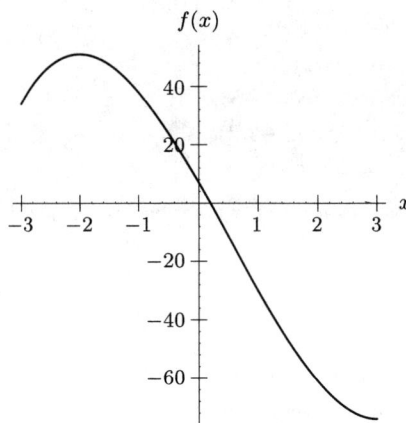

$$f(x)$$

Figure 4.2.86

(c) Setting $f'(x)$ equal to zero to find the critical points:

$$f'(x) = 4x^3 - 36x^2 = 0$$

$$4x^2(x - 9) = 0$$

$$x = 0, 9$$

$f(0) = (0)^4 - 12(0)^3 + 2 = 2$, and $f(9) = (9)^4 - 12(9)^3 + 2 = -2185$, so the points are (0,2) and (9,-2185). Now, checking the second derivative,

$$f''(x) = 12x^2 - 72x = 0$$

$$12x(x - 6) = 0$$

$$x = 0, 6$$

$f(6) = (6)^4 - 12(6)^3 + 2 = -1294$, so the point is (6,-1294). Thus, (0,2) and (6,-1294) are inflection points. Since $f''(9) > 0$, (9,-2185) is a local minimum.

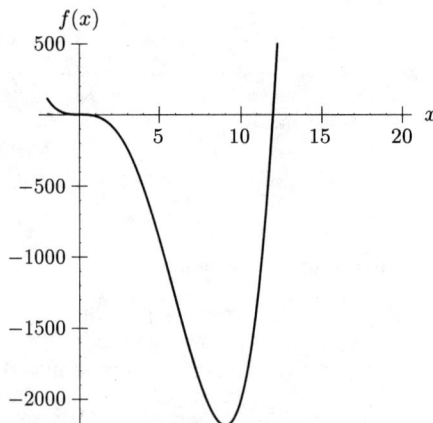

Figure 4.2.87

3. Estimate the inflection point(s) of $f(x)$ if the following figure is the graph of

(a) $f(x)$ (b) $f'(x)$ (c) $f''(x)$

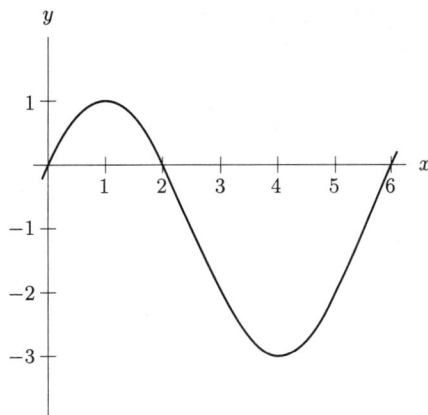

Figure 4.2.88

ANSWER:

(a) The inflection point is where the graph changes concavity: at $(2.5, -1)$.

(b) The inflection points are where the slope of the graph is equal to zero, or at the local minima and maxima: $(1, 1), (4, -3)$.

(c) The inflection points are where the values on the graph are equal to zero, or at the intercepts: $(0, 0), (2, 0), (6, 0)$.

4. If water is flowing at a constant rate (i.e. constant volume per unit of time) into the container shown in Figure 4.2.88, sketch a graph of the depth of the water against time. Mark on the graph where the water reaches points A and B.

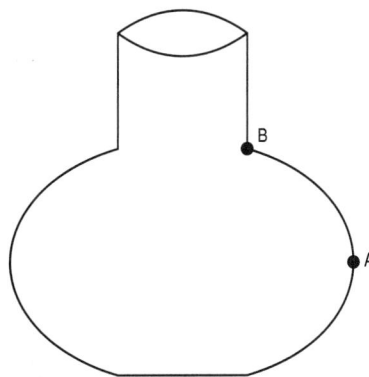

Figure 4.2.89

ANSWER:

At first the water level rises quite rapidly, because the base of the container is narrow. Then, as the container becomes wider, the rate at which the water rises decreases. This means that initially, until point A, the graph is concave down. At point A, the container starts to narrow, so the rate at which the water is rising begins increasing. Thus the graph begins to be concave up, until point B. At this point the width of the container becomes fixed, so the rate at which the water is rising becomes constant. Thus the graph becomes a straight line.

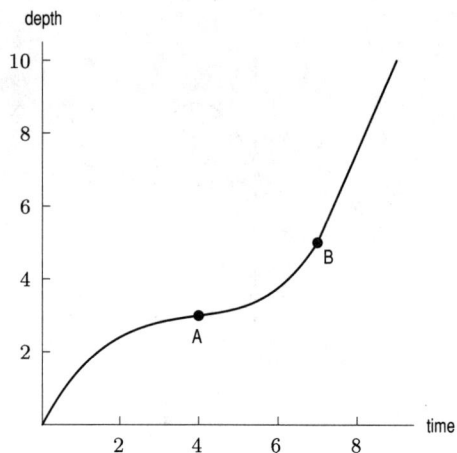

Figure 4.2.90

5. Sketch $g'(x)$ and $f(x)$, assuming $f(0) = 0$.

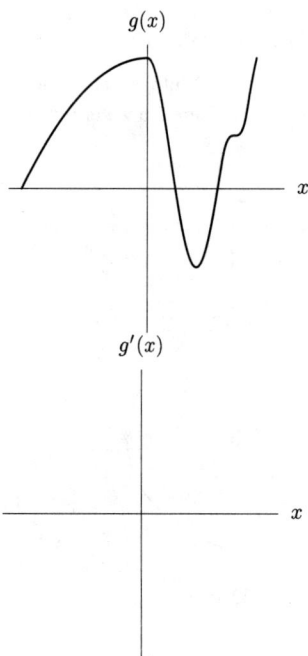

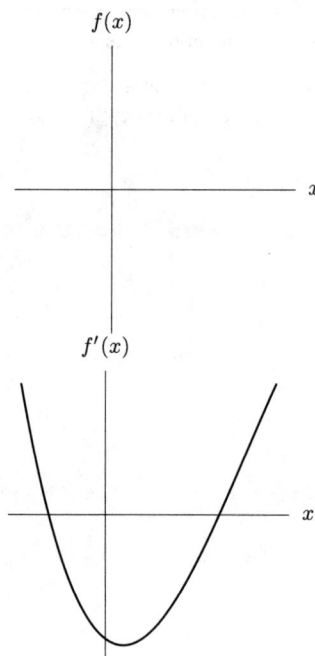

ANSWER:

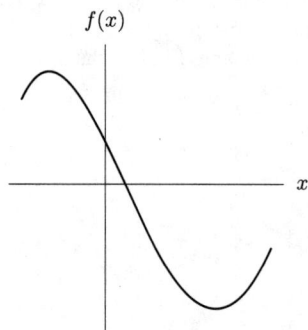

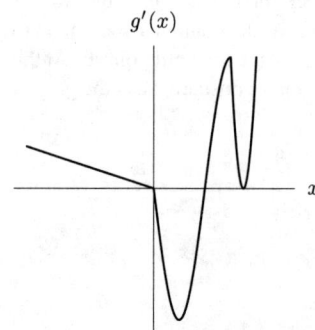

6. Sketch the curve $y = 10(1 - e^{-t})$.

ANSWER:

$y' = 10e^{-t} > 0$ for all t, so y increases for all t. $y'' = -10e^{-t} < 0$ for all t, so y is concave down for all t. $y \to -\infty$ as $t \to -\infty$ and $y \to 10$ as $t \to \infty$. In addition, $y(0) = 0$, so we have the graph given in Figure 4.2.91.

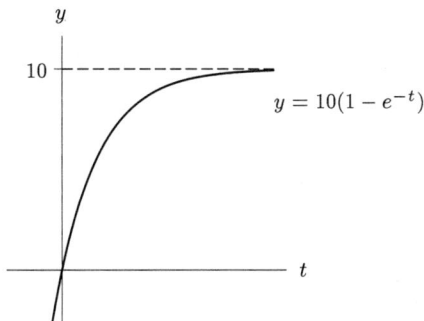

Figure 4.2.91:

7. Given $f(x) = e^{-x} \cos x$, for $0 \le x \le 2\pi$,

(a) find f' and f''.

(b) find the critical points of f.

(c) evaluate f at the critical points and at the endpoints.

(d) find the inflection points.

(e) sketch the graph of f.

(f) identify the local and global extrema of f.

ANSWER:

(a) $f' = -e^{-x} \cos x - e^{-x} \sin x$.

$f'' = e^{-x} \cos x + e^{-x} \sin x + e^{-x} \sin x - e^{-x} \cos x = 2e^{-x} \sin x$.

(b) Let $f' = -e^{-x} \cos x - e^{-x} \sin x = 0$, for $0 \le x \le 2\pi$. We have $\sin x = -\cos x$, or $x = 3\pi/4$ and $7\pi/4$.

(c) $f(0) = 1$.

$f\left(\frac{3}{4}\pi\right) = -e^{-3\pi/4} \cdot \frac{\sqrt{2}}{2} \approx -0.066$.

$f\left(\frac{7}{4}\pi\right) = e^{-7\pi/4} \cdot \frac{\sqrt{2}}{2} \approx 0.0029$.

$f(2\pi) = e^{-2\pi} \approx 0.0019$.

(d) Let $f'' = 2e^{-x} \sin x = 0$, $0 \le x \le 2\pi$. Then $x = 0$, $x = \pi$, $x = 2\pi$, so $x = \pi$ is the inflection point and f is concave up for $0 \le x < \pi$ and concave down for $\pi < x < 2\pi$.

(e) Since $f' = -e^{-x}(\sin x + \cos x)$, $0 \le x \le 2\pi$, we have $f' < 0$ for $0 < x < \frac{3}{4}\pi$ and $\frac{7}{4}\pi < x < 2\pi$; $f' > 0$ for $\frac{3}{4}\pi < x < \frac{7}{4}\pi$. So from (c) and (d) we have the graph in Figure 4.2.92.

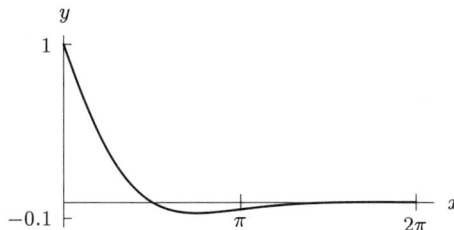

Figure 4.2.92

(f) f has a local maximum at $x = 7\pi/4$, a local and global minimum at $3\pi/4$, and a global maximum at $x = 0$.

8. I am thinking of a function f such that:

the only critical points are at $x = 200$ and $x = 600$;

the only inflection point is at $x = 400$;

$f'(x) < 0$ for $x < 200$;
$f'(x) > 0$ for $200 < x < 600$;
$f'(x) < 0$ for $600 < x$.
Sketch the graph, marking the critical points and inflection points.

ANSWER:

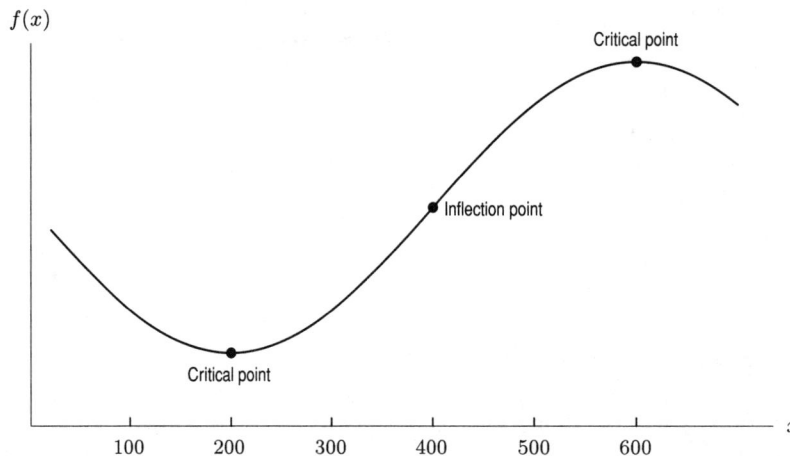

Figure 4.2.93

9. Given the function $y = f(x) = e^{-x^2/2}$:

   (a) Devise functions $g(x)$ and $h(x)$ so that $f(x) = g(h(x))$.

   (b) Graph the function $f(x)$; set the range of your calculator to $-3 \le x \le 3$. Copy your graph onto graph paper. Estimate the extrema (the maximum and/or minimum) point(s) and the inflection point(s) for this function.

   (c) The function $N(x) = \dfrac{1}{\sqrt{2\pi}} \cdot e^{-\frac{x^2}{2}}$ is one of the cornerstones of statistics. In a sentence, briefly describe what $\dfrac{1}{\sqrt{2\pi}}$ does to $f(x)$, i.e., briefly describe the curve $N(x)$ in terms of that of $f(x)$.

   ANSWER:

   (a) $h(x) = -\dfrac{x^2}{2}$, the "inside" function and $g(x) = e^x$.

   Then $f(x) = g(h(x)) = g\left(-\dfrac{x^2}{2}\right) = e^{-\frac{x^2}{2}}$.

   (b)

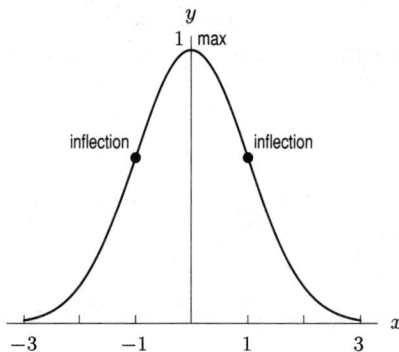

   (c) The coefficient $\dfrac{1}{\sqrt{2\pi}}$ "dilates" the curve; since $\dfrac{1}{\sqrt{2\pi}} < 1$, the effect is to "squash the curve down" a bit.

   This curve is the *Gaussian* or *Normal distribution*. The $\dfrac{1}{\sqrt{2\pi}}$ is chosen so that the area under the whole curve is 1.

   Also note that the x-axis is a horizontal asymptote.

Problems and Solutions for Section 4.3

1. Graph a function that has local minima at $x = -2$ and $x = 5$, with $x = -2$ also a global minimum; a local maximum at $x = 3$ but no global maximum; and inflection points at $x = 1$ and $x = 4$.
Which of the following could be the equation for such a function?

(a) $ax^3 + bx^2 + c$
(b) $ax^4 + bx^3 + cx^2 + dx + e$
(c) $ax^5 + bx^2 + cx + d$
(d) $ax^6 + bx^5 + cx^2 + d$

ANSWER:

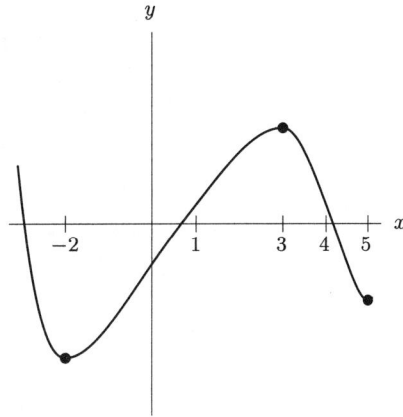

Figure 4.3.94

Both (b) and (d) are possible equations.

2. Indicate all critical points on the following closed graph. Determine which correspond to local minima, local maxima, global minima, global maxima, or none of these.

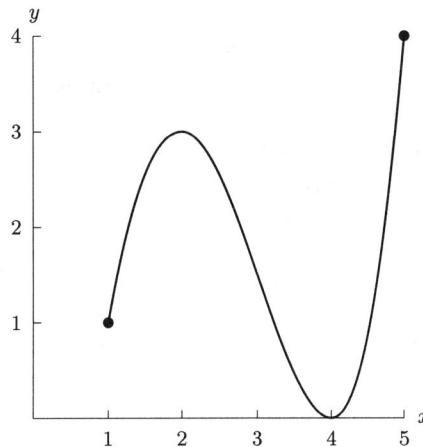

Figure 4.3.95

ANSWER:

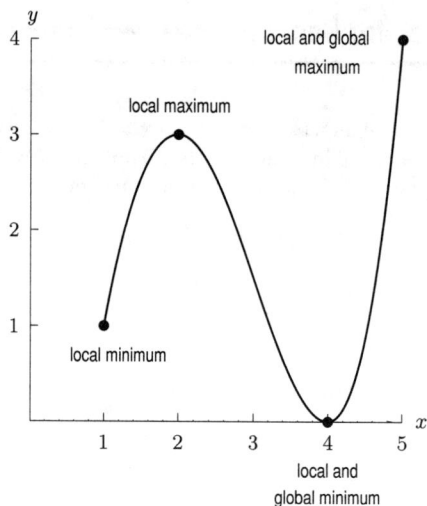

Figure 4.3.96

3. Graph a function which has a local minimum at $x = 2$, a local maximum and global maximum at $x = 4$, but no global minimum.

ANSWER:

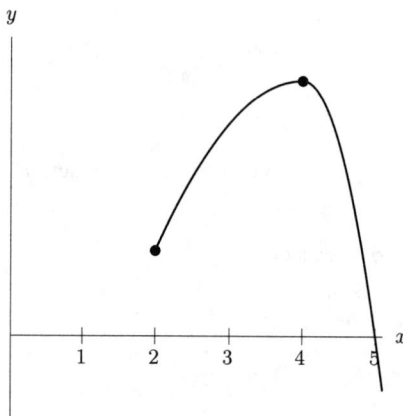

Figure 4.3.97

4. For $f(x) = x^2 - e^{x^2}$, and $-0.5 \le x \le 1.5$, find the value(s) of x for which

(a) $f(x)$ has a local minimum or local maximum.

(b) $f(x)$ has a global minimum or global maximum.

ANSWER:

(a) We first find the critical point(s) by differentiating, setting equal to zero, and solving for x :

$$f'(x) = 2x - 2xe^{x^2} = 0$$
$$2x(1 - e^{x^2}) = 0$$

Thus, $x = 0$ or $1 - e^{x^2} = 0$, which means $1 = e^{x^2}$, or $x = 0$. We now evaluate $f(x)$ at the critical point and at both endpoints.

$$f(-0.5) \approx -1.03$$
$$f(0) = -1$$
$$f(1.5) \approx -7.24$$

Thus $f(x)$ has local minima at $x = -0.5$ and $x = 1.5$, and $f(x)$ has a local maximum at $x = 0$.

(b) $f(x)$ has a global minimum at $x = 1.5$ and a global maximum at $x = 0$.

5. The function $y = p(x)$ is positive and continuous for all real numbers and has a global maximum at the point $(0, 4)$. Draw a possible graph for $p(x)$ if $p'(x)$ is positive for $x < 0$, $p'(x)$ is negative for $x > 0$, $p''(x)$ is negative for $-1 < x < 1$, and $p''(x)$ is positive for $x < -1$ or $x > 1$.

ANSWER:

One possible graph:

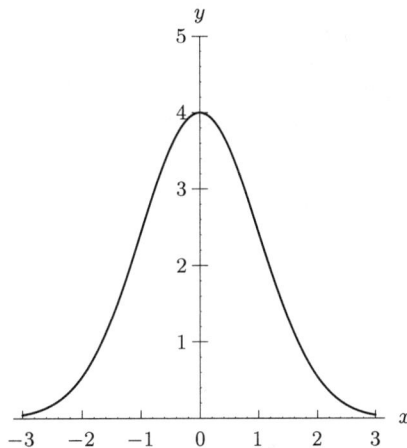

Figure 4.3.98

6. The distance, s, traveled by a runner in a 20 mile race is given in Figure 4.3.99, where time, t, is in hours.

(a) Suggest a reason for the shape of the graph in terms of the runner's speed.

(b) Estimate the point(s) where the runner's speed is slowest and fastest.

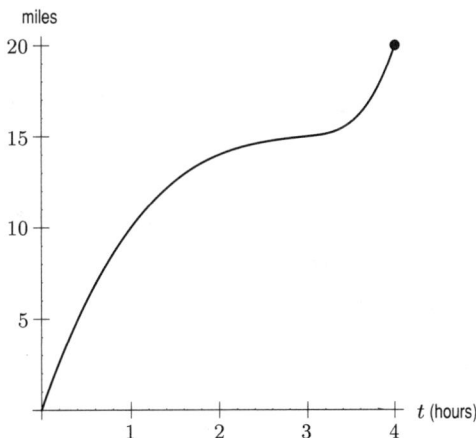

Figure 4.3.99

ANSWER:

(a) The runner started of quickly, then slowed down as he tired, so the slope of the graph got smaller. Near the end, the runner applied a burst of speed to get to the finish, so the slope got steeper again.

(b) The runner is slowest when the slope of the tangent line is least, at about $t = 3$ hours. The runner is fastest right at the beginning of the race, just after $t = 0$, and right at the end of the race, at about $t = 4$.

7. Given that

$$f(x) = \frac{x}{(x+1)^2} \qquad f'(x) = \frac{1-x}{(x+1)^3} \qquad f''(x) = \frac{2x-4}{(x+1)^4} \quad ,$$

do the following:

(a) Find the intercepts and the asymptotes of f.

(b) Find the critical points, the results of the first-derivative test, and where f is increasing or decreasing.

(c) Find the results of the second-derivative test, and the points of inflection of f.

(d) Sketch the graph of f.

ANSWER:

(a) $f(x) = 0$ implies that $x = 0$, and f is not defined at $x = -1$. As $x \to -\infty$, $f(x) \to 0^-$; as $x \to +\infty$, $f(x) \to 0^+$; and as $x \to -1$, $f(x) \to -\infty$.

(b) $f'(x) >$ implies that $x < 1$ and $x > -1$,
$f'(x) < 0$ implies that $x > 1$ and $x < -1$,
and $f'(x) = 0$ implies that $x = 1$.
So f increases for $-1 < x < 1$ and decreases for $x > 1$ and $x < -1$. $f(1) = 1/4$.

(c) $f''(x) > 0$ implies that $x > 2$,
$f''(x) < 0$ implies that $x < 2$.
So f is concave up for $x > 2$ and concave down for $x < 2$.

(d)

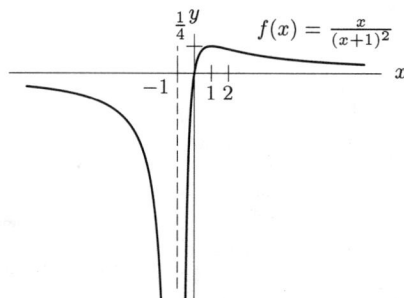

Figure 4.3.100

8. For $f(x) = 2\cos^2 x - \sin x$ and $0 \le x \le \pi$, find, to two decimal places, the value(s) of x for which $f(x)$ has a global maximum or global minimum.

ANSWER:

Find the critical points:

$$f'(x) = -4\cos x \sin x - \cos x = \cos x(1 - 4\sin x)$$

$$x = \pi/2 \text{ or } x = \arcsin(-1/4) \approx -0.25.$$

Since -0.25 is not in the domain $0 \le x \le \pi$, we need only check the first three answers.

$$f(\pi/2) = -1 \qquad f(0) = 2 \qquad f(\pi) = 2$$

Global maxima are at $x = 0$ and $x = \pi$.

Global minimum is at $x = \pi/2$.

9. Sketch a graph of a function with two local minima, no global maximum, but a global minimum. Indicate the domain of your function.

ANSWER:

Answers will vary. One example:

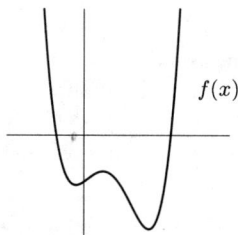

Figure 4.3.101

10. Daily production levels in a plant can be modeled by the function $G(t) = -3t^2 + 12t - 12$ which gives units produced at t, the number of hours since the factory opened at 8am. At what time during the day is factory productivity a maximum?

ANSWER:

To find the maximum productivity, first find the critical points of $G(t)$:

$$G'(t) = -6t + 12 = 0$$
$$t = 2$$

Since $G''(t) = -6$, productivity is a maximum at 10am, 2 hours after the factory opens.

11. The number of plants in a terrarium is given by the function $P(c) = -1.2c^2 + 4c + 10$, where c is the number of mg of plant food added to the terrarium. Find the amount of plant food that produces the highest number of plants.

ANSWER:

Find the critical points:

$$P'(c) = -2.4c + 4 = 0$$
$$c \approx 1.67$$

Since $P''(c) = -2.4$, $c \approx 1.67$ is where the maximum occurs.

12. The function $y = .2(x + 2)^2 - 5x + 2$ gives the population of a town (in 1000's of people) at time x where x is the number of years since 1980. When was the population a minimum?

ANSWER:

$$y' = .4(x + 2) - 5$$
$$= .4x - 4.2$$
$$x = 10.5$$

Since $y'' = .4$, the population was a minimum halfway through 1990.

Problems and Solutions   for Section 4.4

1. With x people on board, a South African airline makes a profit of $(900 - 3x)$ rands per person for a specific flight.

  (a) How many people would the airline prefer to have on board?

  (b) What is the maximum number of passengers that can board such that the airline still profits?

  ANSWER:

  (a) The total profit is $x(900 - 3x) = 900x - 3x^2$, which reaches a maximum at $x = 150$. So the airline would prefer to have 150 people on board for that flight.

  (b) Let $900x - 3x^2 = 0$. We have $x = 0$ or $x = 300$, so the maximum number of passengers such that the airline still profits is 299.

2. **(a)** What quantity of gadgets should be produced to maximize profit if the total revenue and total cost (in dollars) are given by

$$R(x) = 5x - 0.003x^2$$
$$C(x) = 300 + 1.1x$$

  where x is the number of units produced, and $0 \leq x \leq 1000$.

  (b) What is the maximum profit obtainable with conditions as in part (a)?

  (c) Sketch and label a graph showing the cost and revenue curves, and indicating the optimal production level and maximum profit.

  ANSWER:

  (a)

$$\text{Marg rev} = \text{Marg cost}$$
$$R'(x) = C'(x)$$
$$5 - 0.006x = 1.1$$
$$3.9 = 0.006x$$
$$x = \frac{3.9}{0.006} = 650 \text{ units}$$

(b)

$$R(650) - C(650) = \left(5(650) - 0.003(650)^2\right) - (300 + 1.1(650)) = \$967.50$$

(c)

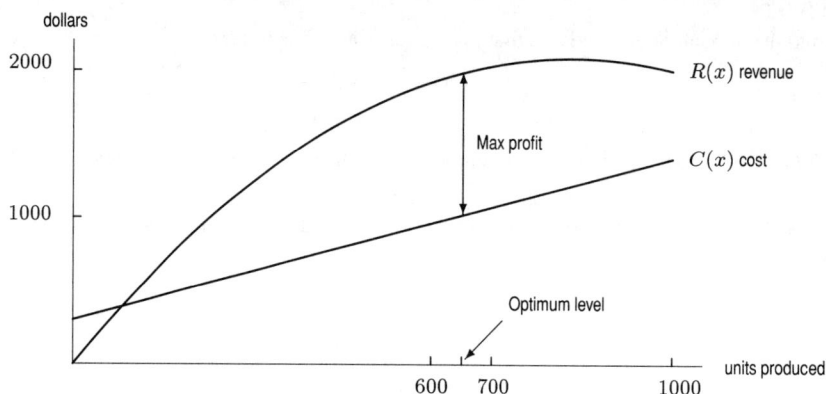

Figure 4.4.102

3. **(a)** What quantity of gadgets should be produced to maximize profit if they sell for $450 per unit and the total cost (in dollars) of producing x units is given by $C(x) = 10,000 + 3x^2$?
   (b) What is the maximum profit obtainable with conditions as in part (a)?
   ANSWER:
   (a)

$$R(x) = 450x$$
$$R'(x) = 450 = 6x = C'(x)$$
$$x = 75 \text{ units.}$$

   (b) $450(75) - (10,000 + 3(75)^2) = \$6875$

4. Total cost and revenue are approximated by the functions $C = 1500 + 3.7q$ and $R = 5q$, both in dollars. Identify the fixed cost, marginal cost per item, and profit.
   ANSWER:
   Fixed cost =1500, marginal cost per item =3.7, and profit $= R - C = 5q - (3.7q + 1500) = 1.3q - 1500$.

5. Write a formula for total cost as a function of quantity r when fixed costs are $50,000 and variable costs are $1,200 per item.
   ANSWER:
   Total cost $= 50,000 + 1,200r$.

6. The revenue for selling q items is $R(q) = 400q - 2q^2$ and the total cost is $C(q) = 100 + 40q$. Write a function that gives the total profit earned, and find the quantity which maximizes profit.
   ANSWER:

$$P = 400q - 2q^2 - (100 + 40q)$$
$$= 360q - 2q^2 - 100$$

   To find the maximum profit we find the critical points of P :

$$P'(q) = 360 - 4q$$

$$P'(q) = 0 \text{ when } q = 90.$$

   $P''(q) = -4$, so profit will be a maximum when $q = 90$.

7. The function $C(r) = 15r^2 - 50$ gives cost in dollars of producing r items. What is the marginal cost of increasing r by 1 item from the current production level of $r = 5$?
   ANSWER:
   Marginal cost $= C'(r) = 30r$. For $r = 5, C'(r) = 150$ dollars.

8. Find the marginal cost and marginal revenue for $q = 100$ when the fixed costs in dollars are 3,000 and the variable costs are 200 per item and each sells for $400.

ANSWER:

Marginal cost $= 200$

Marginal revenue $= 400$.

9. Find the quantity q which maximizes profit if the total revenue, $R(q)$, and the total cost, $C(q)$, are given in dollars by

$$R(q) = 7q - 0.02q^2$$

$C(q) = 400 + 1.5q$, where $0 \leq q \leq 600$ units.

ANSWER:

We look for production levels that give marginal revenue=marginal cost:

$$MR = R'(q) = 7 - 0.04q$$

$$MC = C'(q) = 1.5$$

so

$$7 - 0.04q = 1.5$$

$$q = 138 \text{ units}$$

Does this value of q represent a local maximum or minimum of profit? We can tell by looking at production levels of 137 units and 139 units.

When $q = 137$ we have $MR = 1.52$.

When $q = 139$, $MR = 1.44$.

Therefore, $q = 138$ is a local maximum for the profit function.

Problems and Solutions   for Section 4.5

1. Figure 4.5.103 shows the cost C of producing quantity q of a commodity and the revenue R from the sale of quantity q. Make a copy of the graph on your paper.

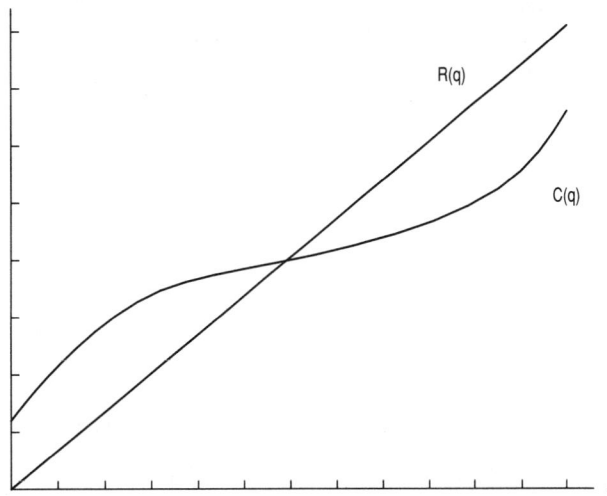

Figure 4.5.103

(a) Label the point F representing the fixed costs of the operation.

(b) Label the point B representing the break-even production level.

(c) Label the point M at which marginal cost is a minimum.

(d) Label the point A at which average cost $a(q) = \dfrac{C(q)}{q}$ is a minimum.

(e) Label the point P at which profit is a maximum.

ANSWER:

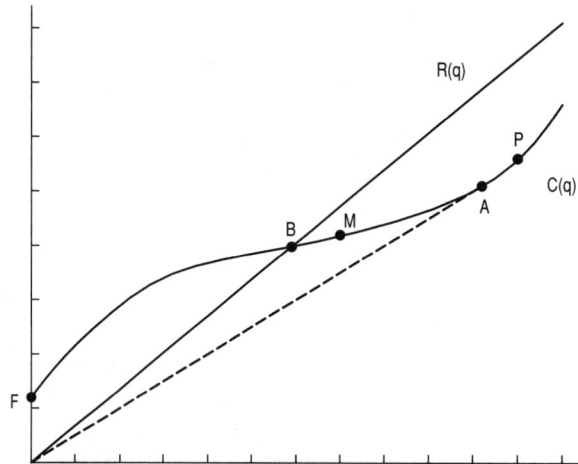

Figure 4.5.104

2. The cost of producing q items is $C(q) = 1000 + 7q$ dollars.

(a) What is the marginal cost of producing the 50th item? The 500th item?

(b) What is the average cost of producing 50 items? 500 items?

ANSWER:

(a) $C'(q) = 7$, so the marginal cost is 7 for both $q = 50$ and $q = 500$.

(b) $a(q) = C(q)/q$ so

$$a(50) = \frac{C(50)}{50} = \frac{1000 + 7(50)}{50} = 27$$

$$a(500) = \frac{C(500)}{500} = \frac{1000 + 7(500)}{500} = 9.$$

3. A factory produces a product that sells for $10. They currently produce 2000 items per month, at an average cost of $4 per item. The marginal cost at this production level is $3. Assume that the factory can sell all items that it produces.

(a) What is the profit at this production level?

(b) How would increasing production affect average cost? Profit?

ANSWER:

(a)

$$P = R - C = 10(2000) - 4(2000) = \$12,000.$$

(b) Look at the effect of increasing production by 1 item:

$$a(2000) = \frac{C(2000)}{2000} = \frac{4(2000)}{2000} = 4$$

$$a(2001) = \frac{C(2001)}{2001} = \frac{4(2000) + 3}{2001} = 3.9995$$

Thus average cost goes down. Now, determine the effect of increasing production on profit. We already have P(2000)=$12,000.

$$P(2001) = 10(2001) - (4(2000) + 3) = \$12,007.$$

Thus profit increases. This makes sense because average cost has gone down.

4. The graph of a cost function is given in Figure 4.5.105. Estimate the value of q at which average cost is minimized. Explain your answer.

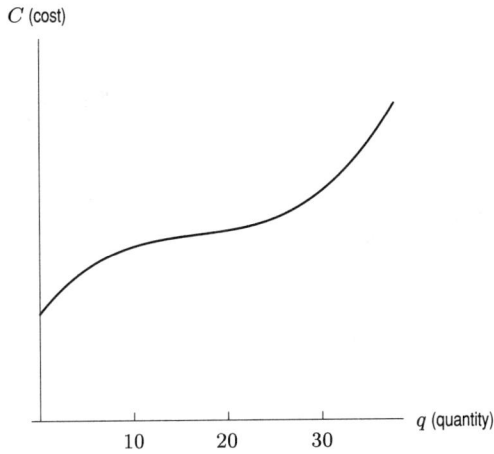

Figure 4.5.105

ANSWER:

The average cost is given by the slope of the line from the origin to the graph of $C(q)$ at q. Figure 4.5.105 shows that this slope is smallest at approximately $q = 30$.

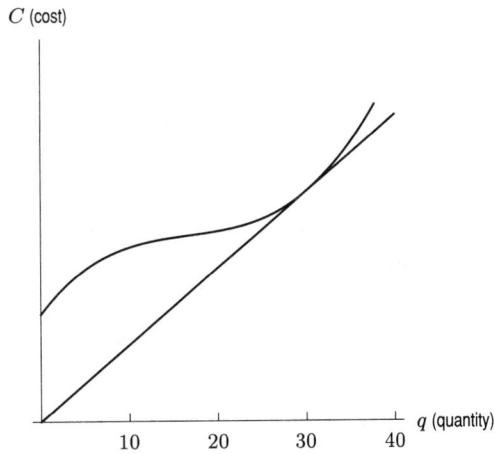

Figure 4.5.106

5. Show analytically that if the derivative of average cost is negative, i.e. $a'(q) < 0$, the marginal cost is less than average cost. Show similarly that if the derivative of average cost is positive, i.e. $a'(q) > 0$, the marginal cost is greater than average cost.

ANSWER:

Suppose $a'(q) < 0$. Then

$$a'(q) = \frac{qC'(q) - C(q)}{q^2} < 0$$
$$qC'(q) - C(q) < 0$$
$$qC'(q) < C(q)$$
$$C'(q) < \frac{C(q)}{q}$$

i.e. marginal cost is less than average cost.

Similarly, if $a'(q) > 0$,

$$a'(q) = \frac{qC'(q) - C(q)}{q^2} > 0$$
$$qC'(q) - C(q) > 0$$
$$qC'(q) > C(q)$$
$$C'(q) > \frac{C(q)}{q}$$

i.e. marginal cost is greater than average cost.

6. Given the cost function $C(q) = 4000 + 50q + 0.002q^2$ and the demand function $p = 80 - 0.025q$, find the value of q for which:

(a) average cost is a minimum.

(b) revenue is a maximum.

(c) profit is a maximum.

ANSWER:

(a) The average cost is

$$\frac{C(q)}{q} = \frac{4000}{q} + 50 + 0.02q.$$

Let $(C(q)/q)' = 0$; then we have $q = 1414$. So the average cost is minimum at $q = 1414$.

(b) Revenue is

$$pq = 80q - 0.025q^2.$$

Let $(pq)'_q = 80 - 0.05q = 0$; then we have $q = 1600$. So the revenue is maximum at $q = 1600$.

(c) Profit is

$$pq - C(q) = 80q - 0.025q^2 - 4000 - 50q - 0.002q^2.$$

Let $(pq - C(q))'_q = 0$; then we have $q = 556$. So the profit is maximum at $q = 556$.

Problems and Solutions for Section 4.6

1. The demand curve for a product is $q = 2000 - 3p^2$. Find the elasticity of demand at a price of $p = 8$. Interpret your answer. Is demand elastic or inelastic at this price?

ANSWER:

Since $q = 2000 - 3p^2$, the derivative is $\frac{dq}{dp} = -6p$. At a price of $p = 8$, we have $q = 2000 - 3(8^2) = 1808$ and $\frac{dq}{dp} = -6(8) = -48$. Therefore, elasticity is

$$E = \left| \frac{p}{q} \cdot \frac{dq}{dp} \right| = \left| \frac{8}{1808} \cdot (-48) \right| = 0.212.$$

This means that if the price goes up 1%, demand will drop by 0.212%.

Demand is inelastic at this price.

2. An amusement park finds that when it charges $15 for an all-day pass, attendance is about 3000 per day. When it charges $18, attendance is about 2700 per day.

(a) Estimate elasticity for this amusement park. Is demand elastic or inelastic?

(b) Is daily revenue higher at a price of $15 or at a price of $18? Show your work to justify your answer.

ANSWER:

(a) The percent change in price is

$$\frac{\Delta p}{p} = \frac{3}{15} = 0.20 = 20\%.$$

The percent change in demand is

$$\frac{\Delta q}{q} = \frac{-300}{3000} = -0.10 = -10\%.$$

The elasticity is the ratio $E = \dfrac{0.10}{0.20} = 0.50$. The elasticity is less than 1 because the percent change in demand is less than the percent change in price.

(b) Revenue = price times quantity sold, so at a price of $15, revenue = $(15)(3000) = 45{,}000$, and at a price of $18, revenue = $(18)(2700) = 48{,}600$. Revenue is higher at a price of $18.

3. Would you expect elasticity to be higher for bread or donuts? Explain your reasoning.

   ANSWER:

   We would expect elasticity to be higher for donuts, since people most likely consider donuts a luxury and bread more of of a necessity.

4. Suppose the elasticity for gasoline in a certain community is 0.25. What does this number tell you about the effect of price increases on the demand for gasoline? Is the demand for gasoline elastic or inelastic? Is this what you would expect? Explain.

   ANSWER:

   The number 0.25 means that a 1% increase in price would cause a 0.25% decrease in demand for gasoline. The demand is inelastic, which is what we would expect because gasoline is a necessity for most people, and few substitutes exist.

5. A youth group wishes to hold a car wash as a fundraiser. Through past experience with car washes, they have constructed the following table, which shows the price, p, charged for a car wash and the quantity, q, of cars washed at that price. Estimate elasticity and revenue at each of the prices. At what price is revenue maximized? What is the elasticity at that price? Is this what you would expect?

Table 4.6.28

p	$2.00	$2.50	$3.00	$3.50	$4.00
q	230	210	190	160	120

   ANSWER:

   At $p = \$2.00$, the percent change in quantity is

$$\frac{\Delta q}{q} = \frac{20}{230} = 0.087$$

and the percent change in price is

$$\frac{\Delta p}{p} = \frac{0.50}{2.00} = 0.25.$$

Thus, the elasticity is given by

$$E = \frac{0.087}{0.25} \approx 0.35.$$

At this price, the revenue is given by $R = pq = (2.00)(230) = \$460$.

   Computing elasticity and revenue in a similar manner for each of the other prices gives us the following table:

Table 4.6.29

p	$2.00	$2.50	$3.00	$3.50	$4.00
q	230	210	190	160	120
elasticity	0.35	0.48	0.95	2.33	2.67
revenue	$460	$525	$570	$560	$480

   Revenue is maximized at a price of $3.00, where the elasticity is 0.95. This is as we would expect, because revenue is maximized when E is approximately one.

Problems and Solutions for Section 4.7

1. A biologist found that the number of Drosophila fruit flies, $N(t)$, assumes the following growth pattern if the food source is limited:

$$N(t) = \frac{400}{1 + 39e^{-0.4t}}.$$

   (a) How many fruit flies were there in the beginning?

(b) At what time was the population increasing most rapidly?

(c) At what rate does the number of fruit flies increase after three days?

ANSWER:

(a) There were $N(0) = \dfrac{400}{1 + 39} = 10$ fruit flies in the begining.

(b) The population is increasing most rapidly when $N(t)$ is half the carrying capacity, or 200.

Let $\dfrac{400}{1 + 39e^{-0.4t}} = 200$ and solve for t. We have

$$e^{0.4t} = 39$$
$$0.4t = \ln 39$$
$$t = \frac{\ln 39}{0.4} \approx 9.16.$$

The population increases most rapidly after 9 days.

(c)

$$N'(t) = \frac{-400(-0.4)(39)e^{-0.4t}}{(1 + 39e^{-0.4t})^2}$$
$$= \frac{6240e^{-0.4t}}{(1 + 39e^{-0.4t})^2}$$

So $N'(3) = 1879.45/161.29 = 11.65$flies/day. After 3 days, the population increases at about 12 flies per day.

2. The following table gives the number of students who have joined a new school club t days after it was formed.

(a) Explain why a logistic model is reasonable for this data.

(b) Find the point where concavity changes in this function. Use it to estimate the maximum membership in the club, L.

(c) Graph $P(t)$.

Table 4.7.30

t (days)	1	2	3	4	5	6	7	8	9	10
P (number of students)	4	9	18	36	70	126	205	296	375	431

ANSWER:

(a) At first, growth in membership in the club is small. As word of the new club spreads, however, the rate of growth increases. This continues until most of the interested students have heard about the club, at which point the rate of growth decreases.

(b) The rate of change increases until $t = 7$ and decreases after that, so the inflection point is approximately at $t = 7$. At this point, $P = 205$, so $L/2 \approx 205$. Thus $L \approx 410$ members.

(c)

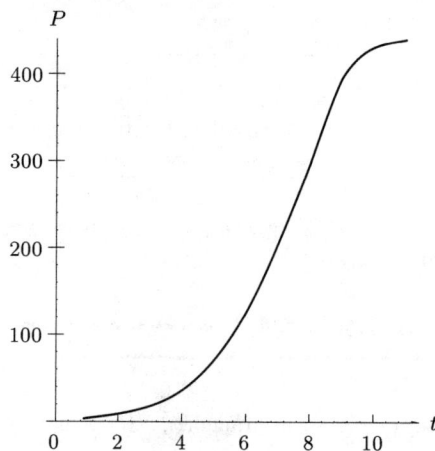

Figure 4.7.107

3. Figure 4.7.108 shows dose-response curves for 3 different allergy drugs. Discuss the advantages and disadvantages of each.

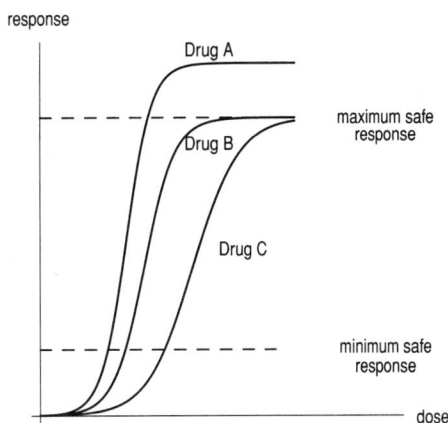

Figure 4.7.108

ANSWER:

Drug A has the highest maximum response, which exceeds the maximum safe response. Drugs B and C have the same maximum response, which is right at the maximum safe response. All three achieve the minimum desired response. Although Drug A reaches the minimum desired response the fastest, it can exceed the maximum safe response, making it a possibly unsafe choice. Of Drugs B and C, Drug B reaches the minimum desired response the fastest and would be good for quick relief. It has a much narrower safe and effective dosage than Drug C, though. Drug C may therefore be the best choice for general use.

4. In Wilson Corners, population 2000, a rumor spreads according to the logistic model. If 5 people know the rumor at 4 PM and 160 people have heard it by 5 PM, how many will have heard the rumor by 6 PM?

ANSWER:

Let $P = \dfrac{2000}{1 + Ae^{-kt}}$, t hours after 4 pm, and $P(0) = 2000/(1 + A) = 5$ implies that $A = 399$, $P(1) = 2000/(1 + 399e^{-k}) = 160$ implies that $k = 3.5466$. Therefore by 6 pm,

$$P(2) = \frac{2000}{1 + 399e^{-2 \times 3.5466}} = 1502 \text{ people}$$

will have heard the rumor.

5. A flu epidemic spreads amongst a group of people according to the formula

$$P(t) = \frac{1000}{1 + 199e^{-0.8t}},$$

where $P(t)$ represents the number of people that are infected by the end of day t.

(a) How many people are infected by the end of the fifth day?

(b) At what rate do the people become infected on day 5?

ANSWER:

(a) $P(5) = \dfrac{1000}{1 + 199e^{-4}} \approx 215$. So 215 people will be infected by the end of 5th day.

(b) $P'(t) = \dfrac{-(1000)(199)(-0.8)e^{-0.8t}}{(1 + 199e^{-0.8t})^2}$. So $P'(5) = \dfrac{2915.8497}{21.5742798} \approx 135$ people/day.

The rate of infection is 135 people per day on day 5.

6. Let $f(t) = \dfrac{a}{1 + 100e^{-bt}}$, for $t \geq 0$.

(a) Give a rough sketch of the general shape of this graph.

(b) On your graph, label any critical points, and label approximate locations of any inflection points.

(c) What is the effect of the parameter a on the graph? Fix b at $b = 1$, and try various values for a. Explain the effect of a using both a graph and words. Assume $a > 0$.

(d) What is the effect of the parameter b on the graph? Fix a at $a = 1$, and try various values for b. Explain the effect of b using both a graph and words. Assume $b > 0$.

ANSWER:

(a)

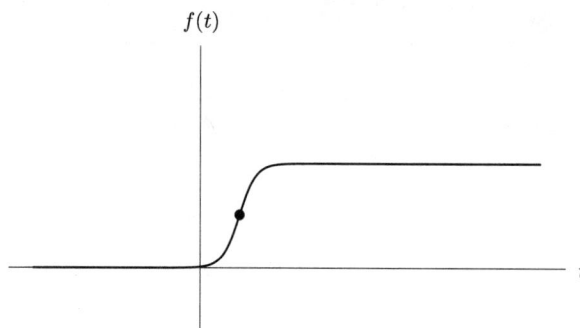

$f(t)$

t

Figure 4.7.109

(b) See Figure 4.7.109. There are no critical points, and the approximate location of the one inflection point is marked on the graph.

(c) The graph of $f(t) = \dfrac{1}{1 + 100e^{-t}}$ has $y = 0$ as an asymptote for $t < 0$, and $y = 1$ as an asymptote for $t > 10$ or so. Adjusting the parameter a while keeping $b = 1$ has no effect on the first asymptote, but it moves the second asymptote to $y = a$. This has the visual effect of "stretching" the graph of $f(t)$ vertically for $t > 0$, while keeping it fixed for $t < 0$. (Actually, the graph is "stretched" at all points, as when any graph is multiplied by a constant, but it approaches 0 so quickly for $t < 0$ that it is hard to see the effect there.) Figure 4.7.110 illustrates these effects of different values of a, with $b = 1$ in all cases.

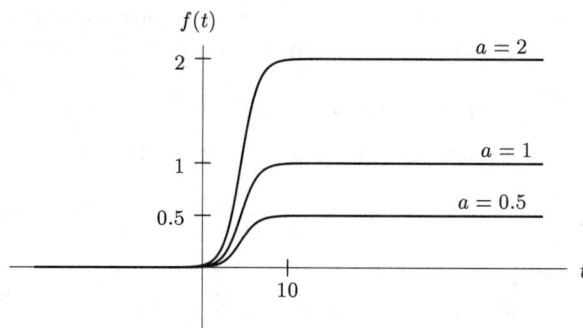

$f(t)$

$a = 2$

$a = 1$

$a = 0.5$

2

1

0.5

10

t

Figure 4.7.110

(d) Adjusting the parameter b while keeping $a = 1$ has no effect on the asymptotes of $f(t)$ described in part (c), but it does affect the rate at which the graph of $f(t)$ approaches the asymptote $y = 1$. Larger values of b cause $f(t)$ to approach 1 more quickly as $t \to +\infty$, while smaller values of b cause it to approach 1 more slowly. This has the visual effect of "stretching" the graph of $f(t)$ horizontally for $t > 0$, while keeping it fixed for $t < 0$. Figure 4.7.111 illustrates these effects of different values of b, with $a = 1$ in all cases.

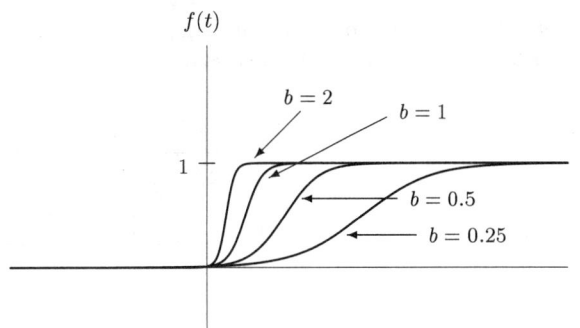

Figure 4.7.111

Problems and Solutions for Section 4.8

1. If time, t, is in hours, and concentration, C, is in ng/ml, the drug concentration curve for a drug is given by

$$C(t) = 7.8te^{-0.3t}.$$

   (a) Graph this curve.
   (b) How many hours does it take for the drug to reach its peak concentration? What is the concentration at this time?
   (c) If the minimum effective concentration is 8 ng/ml, during what time period is the drug effective?
   (d) Suppose scientists wish to alter the drug so that it is effective for 6 hours. Would the coefficient of 7.8 in its concentration curve equation increase or decrease?

   ANSWER:

   (a)

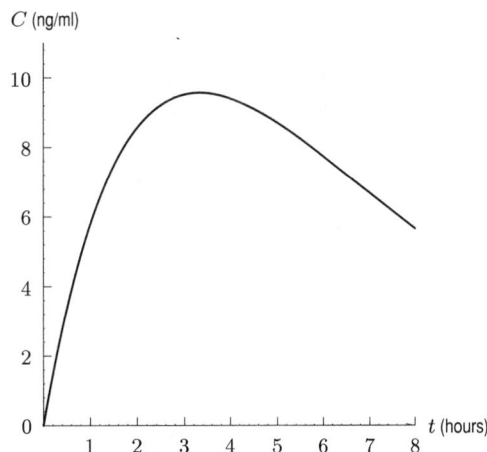

Figure 4.8.112

   (b) It reaches peak concentration just after $t = 3$ hours. The concentration is approximately 9.5 ng/ml at that time.
   (c) The drug is effective just before $t = 2$ hours until just before $t = 6$ hours.
   (d) To make the effective time increase, the curve would need to rise to higher values for a longer period of time. Thus the coefficient would become greater.

2. Table 4.8.31 gives the concentration, C, of a drug, in ng/ml, at time, t, in hours, after it is administered to 2 different people.
   (a) Plot these points on a graph of concentration against time.
   (b) If the concentration for Person A is given by $C = ate^{-b_A t}$, and the concentration for Person B is given by $C = ate^{-b_B t}$, would you expect b_A or b_B to be larger?

(c) If the minimum effective concentreation is 5 ng/ml, during what time period is the drug effective for each person?

(d) What factors might have contributed to the differences in each person's drug concentration?

Table 4.8.31

t (hours)	1	2	3	4	5
Person A	15	10	5	4	3
Person B	20	22	8	5	4

ANSWER:

(a)

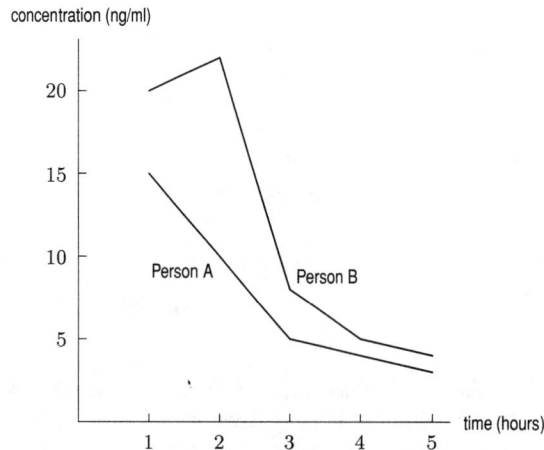

Figure 4.8.113

(b) b_A would be larger.

(c) For Person A, the drug is effective until $t = 3$ hours. For Person B, the drug is effective until $t = 4$ hours.

(d) The differences might be caused by differences in each person's body weight and metabolism, the amount of food or water in each person's stomach at the time the drug was given, or the interaction of any other drugs each person may be taking. (Answers may vary)

3. Consider the two concentration curves, A and B, shown in Figure 4.8.114.

(a) For which value of C (approximately) will A maintain a minimum concentration of C ng/ml for exactly two hours?

(b) Determine the equation of curve B.

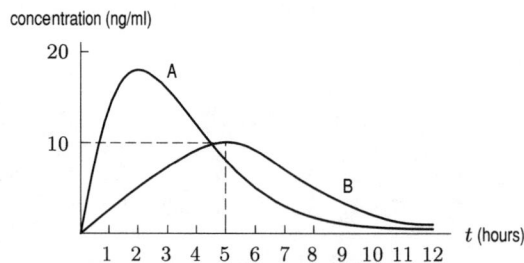

Figure 4.8.114

ANSWER:

(a) $c = 15$

(b) Let $B = ate^{-bt}$. Then $b = \frac{1}{5} = 0.2$ and $10 = a(5)e^{-0.2(5)}$ implies $a = 2e$, so

$$B(t) = 2ete^{-0.2t}.$$

4. Supose $Q = te^{-bt}$, with $t \geq 0$ and b a positive constant.

  (a) For which values of t will Q increase, and for which values of t will Q decrease?
  (b) Determine the coordinates of the point of inflection of Q.
  (c) What happens to Q as t increases indefinitely?
  (d) Sketch the graph of $Q = te^{-bt}$ and show how the value of the constant b influences the shape of the graph.
  (e) If Q represents the level of alcohol in a person's blood, what could the symbols b and t represent?

  ANSWER:

  (a) Since $Q'(t) = e^{-bt} - bte^{-bt} = (1 - bt)e^{-bt}$, we have that Q increases for $t < 1/b$ and decreases for $t > 1/b$.
  (b) Let $Q'' = -be^{-bt} - b(1 - bt)e^{-bt} = (b^2 t - 2b)e^{-bt} = 0$, and solve for t. We have that the inflection point is

$$\left(\frac{2}{b}, \frac{2}{b}e^{-b\frac{2}{b}} \right) = \left(\frac{2}{b}, \frac{2}{be^2} \right).$$

  (c) $Q > 0$ for all $t > 0$. Q increases rapidly at first, and then decreases, approaching zero as t increases indefinitely.
  (d)

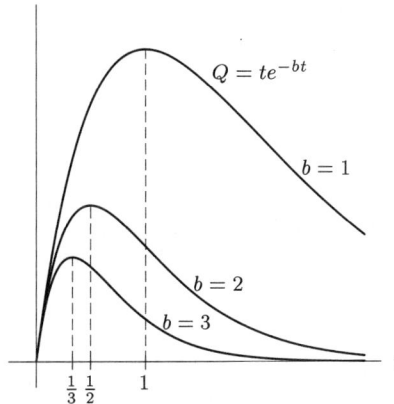

Figure 4.8.115

The graph changes from increasing to decreasing at $t = 1/b$.

  (e) t could represent the time elapsed since the person started to drink alcohol, and $1/b$ could represent the time at which the alcohol level reaches a maximum.

5. Let $f(x) = axe^{-bx}$, for $x \geq 0$.

  (a) Give a rough sketch of the general shape of this graph.
  (b) On your graph, label any critical points, and label approximate locations of any inflection points.
  (c) What is the effect of the parameter a on the graph? Fix b at $b = 1$, and try various values for a. Explain the effect of a using both a graph and words. Assume $a > 0$.
  (d) What is the effect of the parameter b on the graph? Fix a at $a = 1$, and try various values for b. Explain the effect of b using both a graph and words. Assume $b > 0$.
  (e) Find all critical points in terms of a and/or b.
  (f) Find values of a and b so that the function has a local maximum at the point $(3, 7)$.

  ANSWER:

  (a)

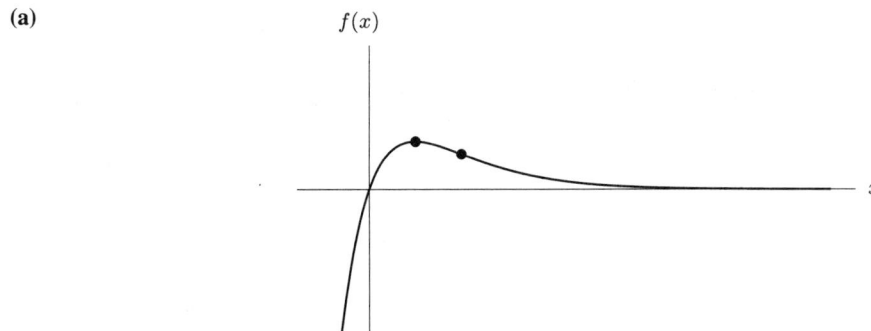

Figure 4.8.116

(b) See Figure 4.8.116. There is one critical point and one inflection point.

(c) Adjusting the parameter a while keeping b fixed has the effect of vertically stretching the graph up or down. (This is the standard effect of multiplying a function by a constant.) Figure 4.8.117 illustrates the effects of adjusting a, with $b = 1$ in all cases.

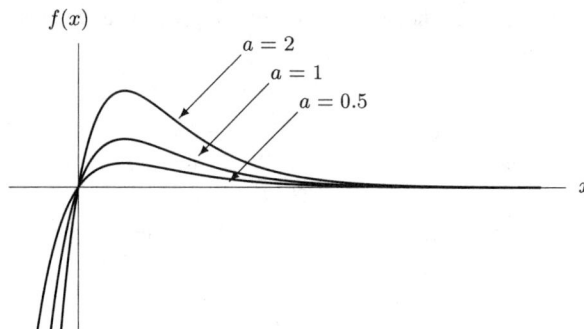

Figure 4.8.117

(d) Adjusting the parameter b while keeping a fixed has the effect of moving the peak point (the only critical point). If b increases, the point moves up and to the right; if b decreases, it moves down and to the left. The general shape of the graph is preserved. Figure 4.8.118 illustrates the effects of adjusting b, with $a = 1$ in all cases.

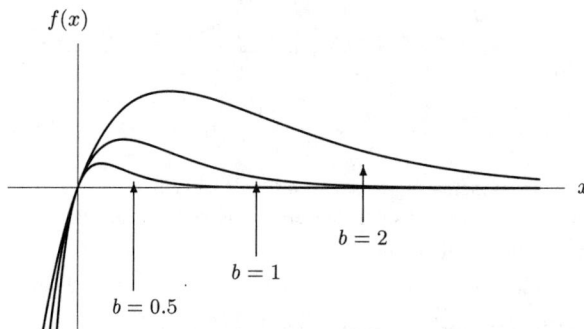

Figure 4.8.118

(e) To find the critical points, we set $f'(x) = 0$. We have $ax(-b)e^{-bx} + ae^{-bx} = 0$, or $ae^{-bx} = abxe^{-bx}$, and so $x = 1/b$. So $y = a(1/b)e^{-b(1/b)} = \dfrac{a}{be}$, and our only critical point is $\left(\dfrac{1}{b}, \dfrac{a}{be}\right)$.

(f) From (e) we know that if $(3, 7)$ is the local maximum, then $x = \dfrac{1}{b}$, so $b = \dfrac{1}{x} = \dfrac{1}{3}$. We also know that $y = \dfrac{a}{be}$, so we can find a: $a = ybe = \dfrac{7e}{3} \approx 6.34$.

Problems and Solutions to Review Problems for Chapter 4 ━━━━━━━━

1. Indicate all critical points on the given closed graph. Label them as local minima, local maxima, global minima, global maxima, or neither. Also, approximate and label any inflection points.

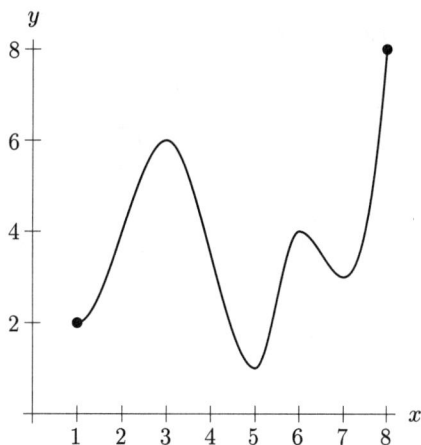

Figure 4.8.119

ANSWER:

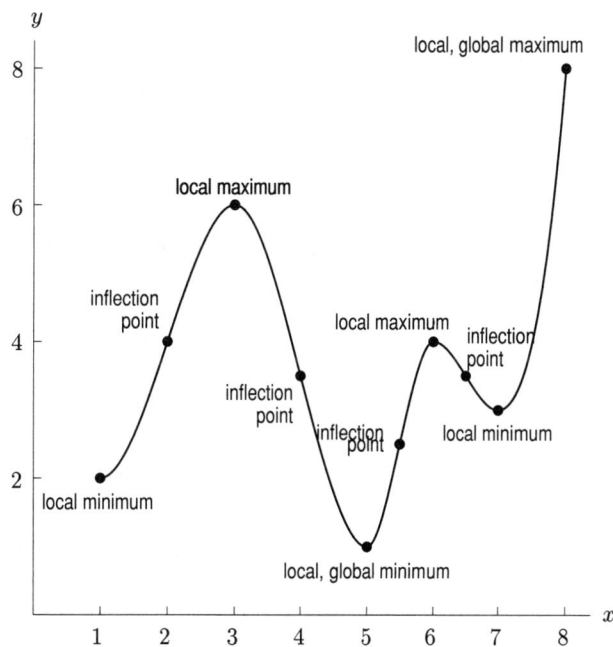

Figure 4.8.120

2. Given $f(x) = e^{-x} + x$.

    (a) Find $f'(x)$ and $f''(x)$.

    (b) Use derivatives to find any critical points of f.

    (c) Use derivatives to find any inflection points of f.

    (d) Identify all local minima, local maxima, global minima, and global maxima.

    (e) Graph $f(x)$.

      ANSWER:

    (a)

$$f'(x) = -e^{-x} + 1$$
$$f''(x) = e^{-x}$$

(b) Set the first derivative equal to zero and solve:

$$f'(x) = -e^{-x} + 1 = 0$$

$$e^{-x} = 1$$

$$x = 0$$

Since $f(0) = e^0 + 0 = 1$, the point $(0, 1)$ is the only critical point.

(c) Since $f''(x) = e^{-x} \neq 0$ for all x, there are no inflection points on the graph.

(d) The critical point $(0, 1)$ is a global minimum.

(e)

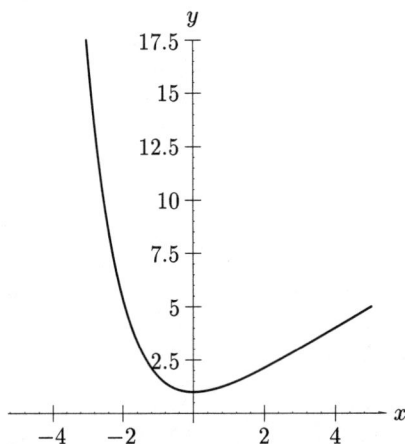

Figure 4.8.121

3. Sketch the graph of $y = ax^2 - 8$ for $a = 1, 2, 3,$ and 4 on the same axes. Discuss the effect of the parameter a on the graph. What if a were negative?

ANSWER:

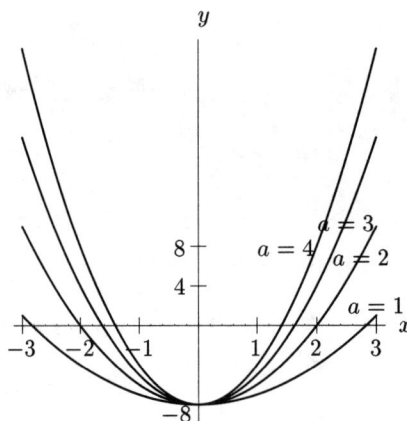

Figure 4.8.122

All of the graphs are parabolas with the same vertex (in this case, (0,-8)) but as a gets larger, the parabola gets narrower. If a were negative, the parabola would have the same vertex, but would open downward instead of upward.

4. The total revenue and cost curves for a product are shown in the following figure.

(a) Estimate the production level that maximizes profit.

(b) Graph marginal cost and marginal revenue for this product on the same axes. Label the point on this graph that maximizes profit.

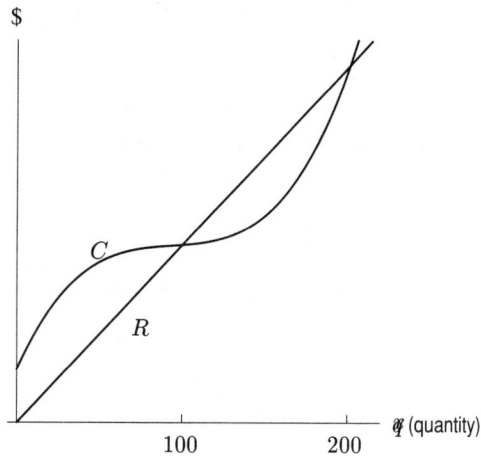

Figure 4.8.123

ANSWER:

(a) Since $P = R - C$, profit is maximized when $R > C$, and the distance between them is greatest, at approximately $q = 150$.

(b) Since R is a straight line, MR, which is the derivative of R, is a horizontal line. Since C is always increasing, MC, which is the derivative of C, is always positive. Since C has an inflection point at approximately $Q = 100$, MC has a minimum there. The point P, where MR and MC cross, shows where maximum profit occurs.

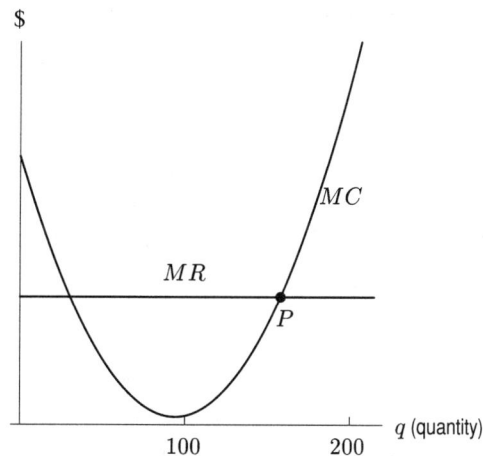

Figure 4.8.124

5. The average cost per item to produce q items is given by

$$a(q) = 0.2q^2 - 0.5q + 11 \text{ for } q > 0.$$

(a) What is the total cost, C, of producing q goods?
(b) What is the marginal cost, MC, of producing q goods?
(c) At what point does marginal cost equal average cost? What is the significance of this point in terms of average cost? Verify this using the formula for average cost.

ANSWER:

(a) Since $a(q) = C(q)/q$, we get $C(q) = q \times a(q) = 0.2q^3 - 0.5q^2 + 11q$.
(b) $MC = C'(q) = 0.6q^2 - q + 11$.
(c) Setting marginal cost equal to average cost, we get

$$0.2q^2 - 0.5q + 11 = 0.6q^2 - q + 11$$

$$0.5q - 0.4q^2 = 0$$

$$q(0.5 - 0.4q) = 0$$

$$q = 0, 5/4$$

At $q = 5/4$, average cost is minimized. To verify this, take the derivative of $a(q)$ and set it equal to zero:

$$a'(q) = 0.4q - 0.5 = 0$$

$$q = 5/4.$$

6. Rank the following products according to their elasticity, with the highest elasticity first. Explain your reasoning.
High performance automobile
Cellular phone
Laundry detergent
Movie theater tickets

ANSWER:

The highest elasticity would go with the product that is either the largest luxury or has close substitutes–in this case, the automobile. The lowest elasticity would go with the product that either is the biggest necessity or has no close substitutes-in this case, the detergent. To rank the other two, the cell phone or the tickets, consider which is more of a necessity or has no close substitutes. It would be the cell phone because substitutes exist for going to a movie theater (i.e. watching videos or DVD's, or attending other forms of entertainment). Thus the final ranking would be:
High performance automobile
Movie theater tickets
Cellular phone
Laundry detergent

7. Classify the following graphs as probable logistic functions, polynomial functions, or surge functions. Write a possible equation for each.

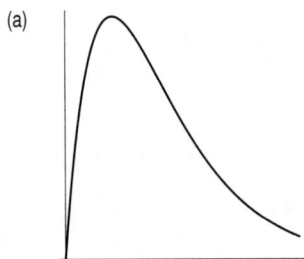

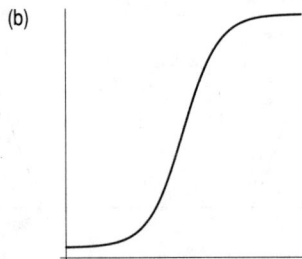

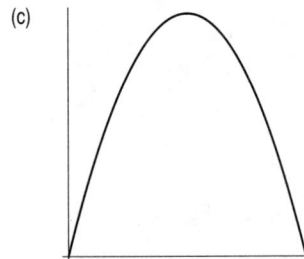

ANSWER:

(a) A surge function. Equations may vary, but have the form

$$y = ate^{-bt}.$$

(b) A logistic function. Equations may vary, but have the form

$$P = \frac{L}{1 + Ce^{-kt}}.$$

(c) A polynomial function. Equations may vary, but are likely to have the form

$$y = -ax^2 + bx + c, \text{ with } a > 0.$$

8. Suppose the spread of a cold virus in an elementary school can be modeled by the equation

$$P = \frac{100}{1 + 45e^{-t}},$$

where P is the number of children with the virus and t is time in days.

(a) Graph P against t.

(b) How many children will eventually catch the virus?

(c) What is the significance of the inflection point in the graph?

(d) If a different strain of the virus is found to fit the equation

$$P = \frac{100}{1 + 45e^{-1.5t}},$$

how would it affect the children compared to the first virus?

ANSWER:

(a)

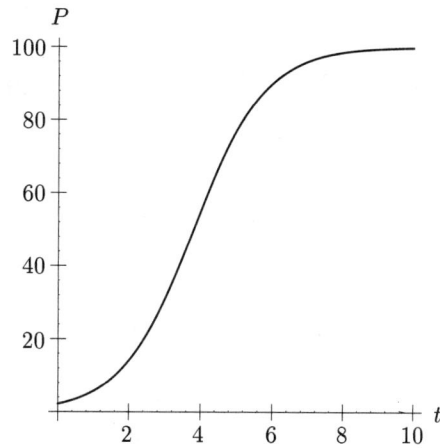

Figure 4.8.125

(b) 100 children.

(c) At the inflection point, approximately $t = 3.8$ days, half of the children who will catch the virus have already caught it.

(d) The children will catch the new virus much more quickly than the old one, though the same number will eventually be infected.

Chapter 5 Exam Questions

Problems and Solutions for Section 5.1

1. The rate of pollution pouring into a lake is measured every 10 days, with results in the following table. About how much pollution has entered the lake in the first 40 days?

Time in days	0	10	20	30	40
Rate of pollution (tons/day)	5	7	10	9	8

ANSWER:

The amount of pollution can be found by calculating the area under the graph in Figure 5.1.126.

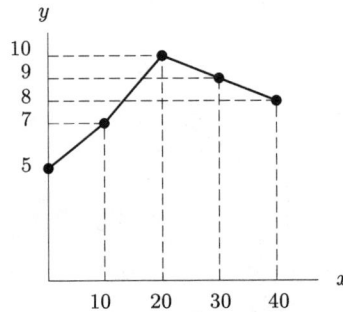

Figure 5.1.126

Using the left-hand value of each interval, we obtain

$$5 \cdot 10 + 7 \cdot 10 + 10 \cdot 10 + 9 \cdot 10 = 310.$$

Using the right-hand value of each interval, we obtain

$$7 \cdot 10 + 10 \cdot 10 + 9 \cdot 10 + 8 \cdot 10 = 310.$$

Averaging these, we obtain an estimate of 325 tons of pollution during the 40-day period.

2. Consider a sports car which accelerates from 0 ft/sec to 88 ft/sec in 5 seconds (88 ft/sec = 60 mph). The car's velocity is given in the table below.

t	0	1	2	3	4	5
$V(t)$	0	30	52	68	80	88

  (a) Find upper and lower bounds for the distance the car travels in 5 seconds.
  (b) In which time interval is the average acceleration greatest? Smallest?

    ANSWER:

  (a) Since $v(t)$ is increasing, a lower bound is given by the left-hand sum, and an upper bound is given by the right-hand sum.

$$\text{lower bound} = 0 + 30 + 52 + 68 + 80 = 230 \text{ feet};$$
$$\text{upper bound} = 30 + 52 + 68 + 80 + 88 = 318 \text{ feet}.$$

  (b) In the first interval it is greatest. In the last interval it is smallest.

3. The graph shown below is that of the velocity of an object (in meters/second).

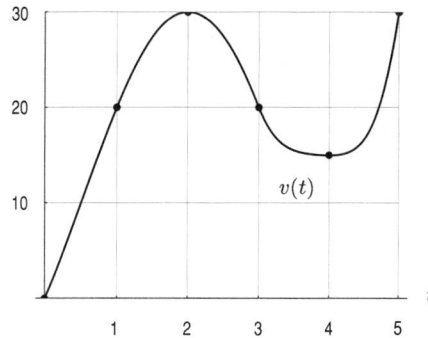

(a) Find an upper and a lower estimate of the total distance traveled from $t = 0$ to $t = 5$ seconds.

(b) At what times is the acceleration zero?

ANSWER:

(a) $v(0) = 0$, $v(1) = 20$, $v(2) = 30$, $v(3) = 20$, $v(4) = 15$, $v(5) = 30$.

Since this function is increasing over $[0, 2]$, decreasing over $[2, 4]$, and increasing over $[4, 5]$, we need to break the function into three parts in order to determine an overestimate and an underestimate of the distance travelled.

$$\text{Over } [0, 2], \text{ lower bound} = 0 + 20 = 20,$$
$$\text{upper bound} = 20 + 30 = 50.$$
$$\text{Over } [2, 4], \text{ lower bound} = 20 + 15 = 35,$$
$$\text{upper bound} = 30 + 20 = 50.$$
$$\text{Over } [4, 5], \text{ lower bound} = 15,$$
$$\text{upper bound} = 30.$$

So, adding the upper and lower bounds for the separate intervals, we get

$$\text{lower bound on distance traveled} = 20 + 35 + 15 = 70 \text{ meters};$$
$$\text{upper bound on distance traveled} = 50 + 50 + 30 = 130 \text{ meters}.$$

(b) $v' = 0$ at $t = 2$ and $t = 4$.

4. At time t, in seconds, the velocity, v, in miles per hour, of a car is given by

$$v(t) = 5 + .5t^2 \text{ for } 0 \le t \le 8.$$

Use $\Delta t = 2$ to estimate the distance traveled during this time. Find the left- and right-hand sums, and the average of the two.

ANSWER:

Using $\Delta t = 2$,

$$\text{Left} - \text{hand sum} = v(0) \cdot 2 + v(2) \cdot 2 + v(4) \cdot 2 + v(6) \cdot 2$$
$$= 2(5) + 2(5 + .5(2)^2) + 2(5 + .5(4)^2) + 2(5 + .5(6)^2)$$
$$= 96$$
$$\text{Right} - \text{hand sum} = v(2) \cdot 2 + v(4) \cdot 2 + v(6) \cdot 2 + v(8) \cdot 2$$
$$= 160$$

The average is $(96 + 160)/2 = 128$.

5. A car travels exactly 20 mph faster than the car from Exercise 4. What are the left- and right-hand estimates of the distance traveled by the new car (also using $\Delta t = 2$)?

ANSWER:

Each term of the new estimates will be $2 \cdot 20$ miles greater for a total of $4 \cdot 2 \cdot 20 = 160$. The left-hand sum will be 256 and the right-hand sum will be 320.

6. Figure 5.1.127 shows the graph of the velocity, v, of an object (in meters/sec.). If the graph were shifted up two units, how would the total distance traveled between $t = 0$ and $t = 6$ change? What would it mean for the motion of the object?

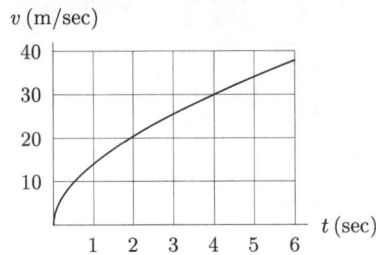

Figure 5.1.127

ANSWER:
The distance would increase by $6 \cdot 2$ or 12 meters. Shifting the graph up 2 units means the velocity at each time would be 2 mph greater.

7. A car is observed to have the following velocities at times $t = 0, 2, 4, 6$:

Table 5.1.32

time(sec)	0	2	4	6
velocity(ft/sec)	0	21	40	66

Give lower and upper estimates for the distance the car traveled.
ANSWER:
Lower estimate $= 0(2) + 21(2) + 40(2) = 122$ feet.
Upper estimate $= 21(2) + 40(2) + 66(2) = 254$ feet.

8. At time t, in seconds, your velocity v, in meters/sec, is given by

$$v(t) = 2 + 2t^2 \text{ for } 0 \leq t \leq 6$$

Use $\Delta t = 2$ to estimate distance during this time.
ANSWER:

$$\text{Left} - \text{hand sum} = v(0) \cdot 2 + v(2) \cdot 2 + v(4) \cdot 2$$
$$= 2 \cdot 2 + 10 \cdot 2 + 34 \cdot 2 = 92 \text{ meters}$$
$$\text{Right} - \text{hand sum} = v(2) \cdot 2 + v(4) \cdot 2 + v(6) \cdot 2$$
$$= 10 \cdot 2 + 34 \cdot 2 + 74 \cdot 2 = 236 \text{ meters}$$

Average $= (92 + 236)/2 = 164$ meters. Distance traveled ≈ 164 meters.

9. Repeat Exercise 8 with $\Delta t = 1$. Compare the accuracy of your two answers.
ANSWER:

$$\text{Left} - \text{hand sum} = v(0) \cdot 1 + v(1) \cdot 1 + v(2) \cdot 1 + v(3) \cdot 1 + v(4) \cdot 1 + v(5) \cdot 1$$
$$= 2 \cdot 1 + 4 \cdot 1 + 10 \cdot 1 + 20 \cdot 1 + 34 \cdot 1 + 52 \cdot 1$$
$$= 122 \text{ meters}$$
$$\text{Right} - \text{hand sum} = v(1) \cdot 1 + v(2) \cdot 1 + v(3) \cdot 1 + v(4) \cdot 1 + v(5) \cdot 1 + v(6) \cdot 1$$
$$= 4 \cdot 1 + 10 \cdot 1 + 20 \cdot 1 + 34 \cdot 1 + 52 \cdot 1 + 74 \cdot 1$$
$$= 194 \text{ meters}$$

Average $= (122 + 194)/2 = 158$ meters. Distance traveled ≈ 158 meters.

Since the estimate of 158 meters was obtained using more and smaller intervals of time, it is more accurate than the estimate obtained by using $\Delta t = 2$.

Problems and Solutions   for Section 5.2

1. Answer the following questions with reference to the graph in Figure 5.2.128.

   (a) $\displaystyle\int_{10}^{70} f(t)\,dt \approx$ _____

     (give three terms of the left Riemann sum).

   (b) Shade the area you computed with the Riemann sum in part (a). Indicate on the graph the error in your estimate of

     $\displaystyle\int_{10}^{70} f(t)\,dt.$

   (c) $\displaystyle\int_{10}^{70} f(t)\,dt \approx$ _____

     (give three terms of the right Riemann sum).

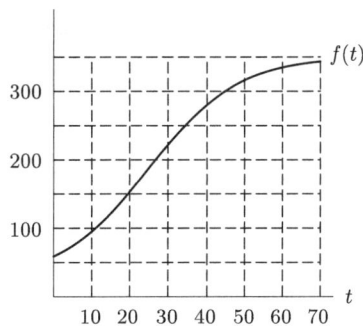

Figure 5.2.128

     ANSWER:

   (a) $\displaystyle\int_{10}^{70} f(t)\,dt \approx 20(100 + 225 + 325) \approx 13{,}000.$

   (b) See Figure 5.2.129.

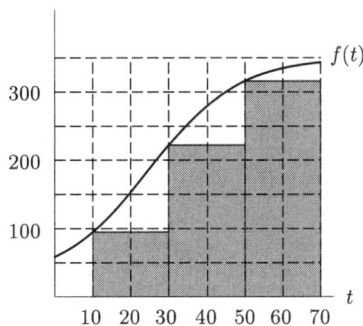

Figure 5.2.129

   (c) $\displaystyle\int_{10}^{70} f(t)\,dt \approx 20(225 + 325 + 350) \approx 16{,}000.$

2. Give a four rectangle Riemann sum approximation for $\int_{0}^{20} f(x)\,dx$ where f is shown in Figure 5.2.130. Draw the rectangles you use on the graph, and give the dimensions of each.

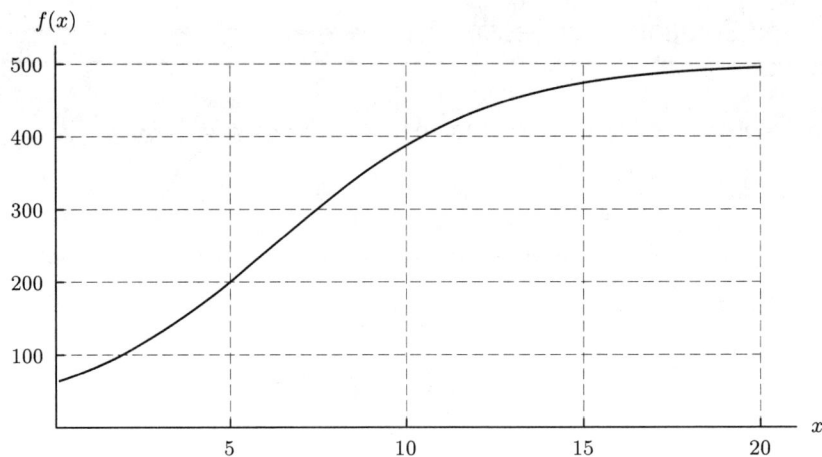

Figure 5.2.130

ANSWER:

Around 6500–each grid rectangle gives area of $5 \times 100 = 500$ and there are roughly 13 such.

Lower sum: $50(5) + 200(5) + 380(5) + 460(5) = 5450$

Upper sum: $200(5) + 380(5) + 460(5) + 500(5) = 7700$

Average: 6575

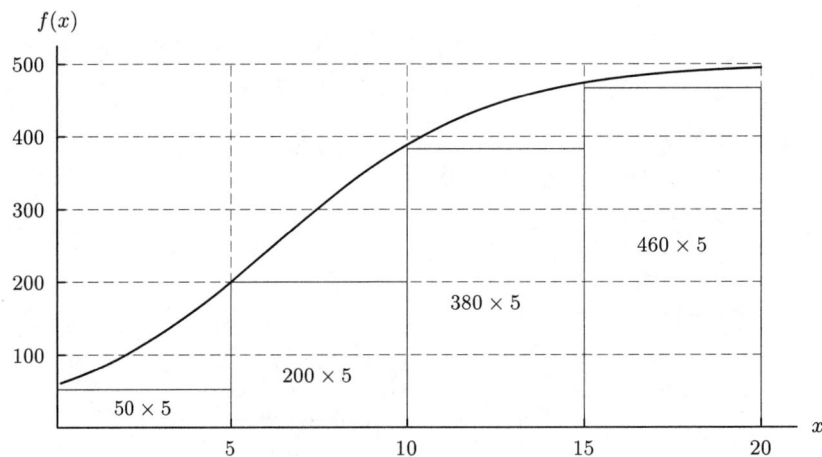

Figure 5.2.131

3. Given the graph of f:

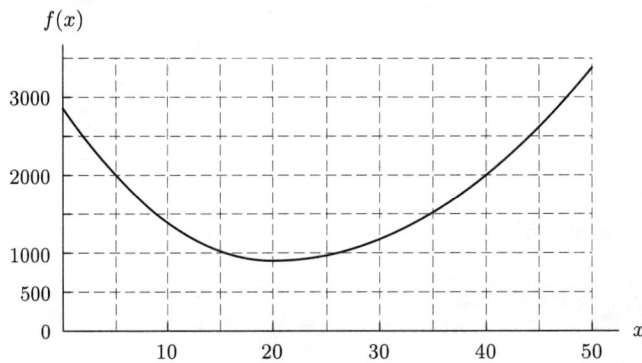

Figure 5.2.132

Give a five rectangle Riemann sum approximation for $\int_0^{50} f(x)dx$ where f is in Figure 5.2.132. Draw the rectangles you use on the graph, and give the dimensions of each.

$$\int_0^{50} f(x)\,dx \approx$$

ANSWER:

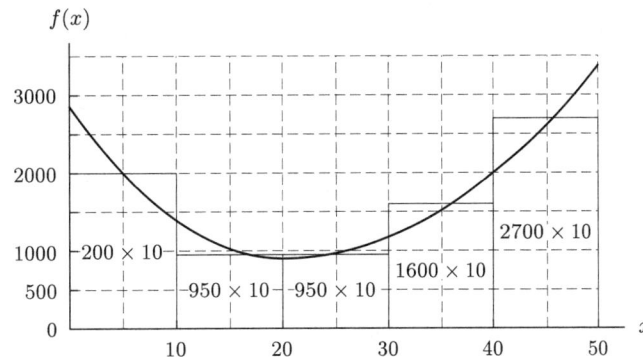

Figure 5.2.133

$$\int_0^{50} f(x)\,dx \approx 82{,}000$$

$2000 \times 10 + 950 \times 10 + 950 \times 10 + 1600 \times 10 + 2700 \times 10 = 82{,}000$

Each square above $5 \times 500 = 2500$. Total number squares= 31.5. Total $(2500)(31.5) = 78750$.

4. Consider the graph of the function $f(x) = e^{-x^2}$ given in Figure 5.2.134.

   (a) Approximate $\int_0^{1.5} e^{-x^2}\,dx$ by using a left-hand sum with 3 subdivisions.

   (b) Is your answer to part (a) a lower estimate or an upper estimate?

   (c) Find $\int_0^{1.5} e^{-x^2}\,dx$ exactly using technology.

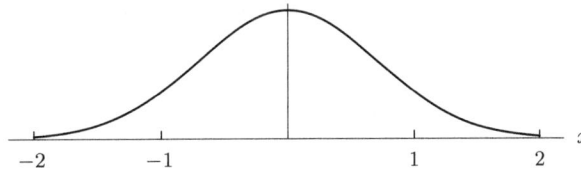

Figure 5.2.134

ANSWER:

(a) $\int_0^{1.5} e^{-x^2}\,dx \approx (0.5)(e^0) + (0.5)(e^{-0.5^2}) + (0.5)(e^{-1}) = 1.07.$

(b) (a) is an upper estimate.

(c) $\int_0^{1.5} e^{-x^2}\,dx = 0.8562.$

5. You plan to approximate the definite integral $\int_5^{10} (2x+3)^2\,dx$ by Riemann sums. Which will be larger, the right Riemann sum or the left Riemann sum, and how do you know?

   ANSWER:

   We can see on a graph that y is increasing for $5 < x < 10$, and so the right Rieman sum is larger.

6. Consider the function $f(x) = \ln x$, as shown in the sketch in Figure 5.2.135. Suppose an approximation for $\int_1^3 \ln x\,dx$ is obtained by using Riemann sums.

   (a) Take $\Delta t = 0.5$, and shade the area that represents the left-hand sum on the sketch in Figure 5.2.135.

(b) Find a lower estimate for $\displaystyle\int_1^3 \ln x\,dx$. (Take $\Delta t = 0.5$).

(c) Find $\displaystyle\int_1^3 \ln x\,dx$ exactly using technology.

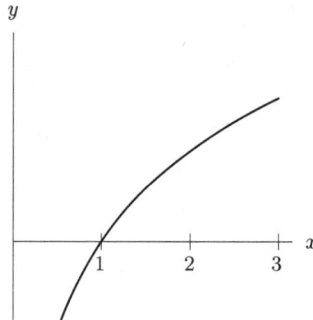

Figure 5.2.135

ANSWER:

(a)

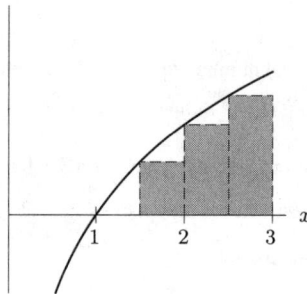

Figure 5.2.136

(b) $\displaystyle\int_1^3 \ln x\,dx \approx 0.5\ln 1.5 + 0.5\ln 2 + 0.5\ln 2.5 = 1.007.$

(c) $\displaystyle\int_1^3 \ln x\,dx = 1.2958.$

7. (a) Give a 4-term left Riemann sum approximation for the integral below.

$$\int_{30}^{34} 3\sqrt{x+1}\,dx \approx \underline{\qquad} + \underline{\qquad} + \underline{\qquad} + \underline{\qquad} = \underline{\qquad}.$$

(b) Is your answer in part (a) an overestimate or an underestimate?

ANSWER:

(a) $\displaystyle\int_{30}^{34} 3\sqrt{x+1}\,dx \approx 3\sqrt{31} + 3\sqrt{32} + 3\sqrt{33} + 3\sqrt{34} \approx 22.8.$

(b) (a) is an underestimate.

8. Using Figure 5.2.137, draw rectangles representing each of the following Riemann sums for the function f on the interval $0 \le t \le 12$. Calculate the value of each sum.

(a) Left-hand sum with $\Delta t = 3$

(b) Right-hand sum with $\Delta t = 3$

(c) Left-hand sum with $\Delta t = 6$

(d) Right-hand sum with $\Delta t = 6$

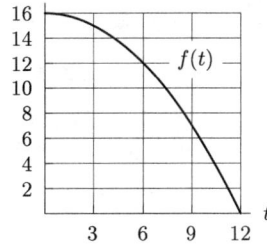

Figure 5.2.137

ANSWER:

(a)

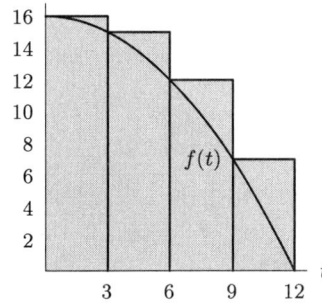

Figure 5.2.138

Left-hand sum $= 16(3) + 15(3) + 12(3) + 7(3) = 150$

(b)

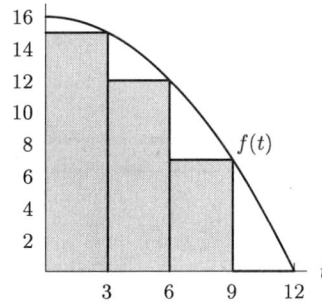

Figure 5.2.139

Right-hand sum $= 15(3) + 12(3) + 7(3) = 108$

(c)

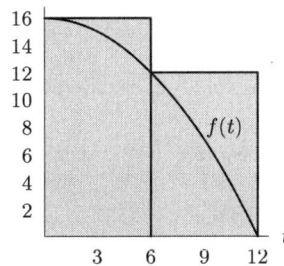

Figure 5.2.140

Left-hand sum $= 16(6) + 12(6) = 168$

(d)

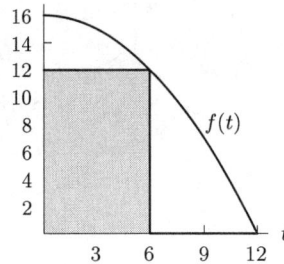

Figure 5.2.141

Right-hand sum $= 12(6) = 72$

9. Use the table to estimate $\displaystyle\int_0^{50} f(x)\,dx$. What values of n and Δx did you use?

Table 5.2.33

x	0	10	20	30	40	50
$f(x)$	30	35	45	50	70	85

ANSWER:

We estimate $\displaystyle\int_0^{50} f(x)\,dx$ using left- and right-hand sums.

Left sum $= 30(10) + 35(10) + 45(10) + 50(10) + 70(10) = 2300$

Right sum $= 35(10) + 45(10) + 50(10) + 70(10) + 85(10) = 2850$

We estimate that $\displaystyle\int_0^{50} f(x)\,dx \approx \dfrac{2300 + 2850}{2} = 2575$. In this estimate, we used $n = 5$ and $\Delta x = 10$.

Problems and Solutions   for Section 5.3

1. Find the area included between the curves $y = \sqrt{x}$ and $y = \dfrac{1}{x}$, from $x = 1$ to $x = 2$.

   ANSWER:

   Area $= \displaystyle\int_1^2 \left(\sqrt{x} - \dfrac{1}{x}\right) dx = 0.54.$

2. Your rich eccentric friend has hired you to cover his back yard with grass and patio stone. Instead of drawing a map, he has given you equations of the boundary lines. If the southwest corner of his yard is taken as the origin, with the x-axis pointing eastward and distances measured in feet, then the boundaries of the yard are the lines $x = 0$, $x = 100$, $y = 0$, and $y = 110 - 0.5x$. The border between grass and stone is $y = 40 + 20\cos(\pi x/40)$, with grass covering all of the yard south of the curve. This border also bounds one side of his pool, the other side of the pool being surrounded by the curve $y = 60 - 0.05(x - 40)^2$. All the rest of the yard is to be covered with stone.

   (a) Sketch a map of your friend's yard.

   (b) Estimate, to the nearest square yard, the area of (i) the grass and (ii) the stone.

   ANSWER:

(a)

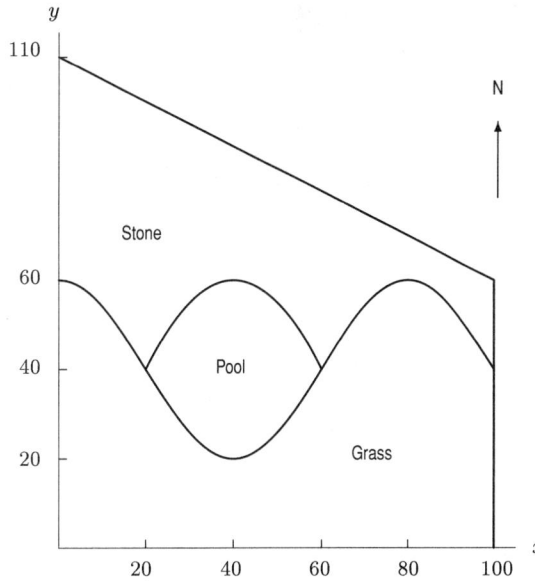

Figure 5.3.142

(b) Let $40 + 20\cos(\pi x/40) = 60 - 0.05(x-40)^2$ and solve for x; we have $x = 20$, $x = 60$. We thus have:

$$\text{Grass area} = \int_0^{100} 40 + 20\cos\left(\frac{\pi x}{40}\right)\, dx = 4254.6 \text{ sq. ft.}$$

$$\text{Pool area} = \int_{20}^{60} \left(60 - 0.05(x-40)^2 - \left(40 + 20\cos\frac{\pi x}{40}\right)\right)\, dx = 1042.62 \text{ sq. ft.}$$

$$\text{Entire yard} = \int_0^{100} (110 - 0.5x)\, dx = 8500 \text{ sq. ft.}$$

So we have the following answers:

(i) Grass area ≈ 4255 square feet ≈ 473 square yards.

(ii) The stone area $\approx 8500 - 4255 - 1043 = 3202$ square feet ≈ 356 square yards.

3. (a) Explain, using words and pictures, how you would decide whether or not the quantity

$$\frac{2}{\pi}\int_0^{\frac{\pi}{2}} \sin t\, dt$$

is greater than, less than, or equal to 0.5 <u>without</u> doing any calculations. (Please be as concise as possible.)

(b) Which of the following best approximates $\frac{2}{\pi}\int_0^{\frac{\pi}{2}} \sin t\, dt$? Circle one. No explanation needed.

0.35   0.4   0.45   0.5   0.55   0.6   0.65
ANSWER:

(a)

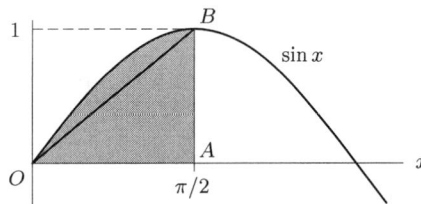

Compare the areas:
area OAB ¡ area under sine curve

$\frac{2}{\pi} \times$ area $OAB < \frac{2}{\pi} \times$ area under sine curve.

$\frac{2}{\pi} \times \frac{1}{2} \times 1 \times \frac{\pi}{2} < \frac{2}{\pi} \int_0^{\frac{\pi}{2}} \sin t \, dt$

$.5 < \frac{2}{\pi} \int_0^{\frac{\pi}{2}} \sin t \, dt$

(b) 0.65

4.

A car is moving along a straight road from A to B, starting from A at time $t = 0$. Below is the velocity (positive direction is from A to B) plotted against time.

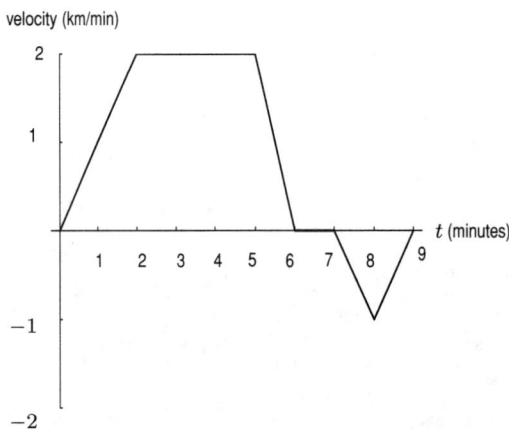

(a) How many kilometers away from A is the car at time $t = 2, 5, 6, 7,$ and 9?

(b) Carefully sketch a graph of the acceleration of the car against time. Label your axes.

ANSWER:

(a) Since distance is found by integrating velocity, we find the area under the curve:

$t = 2$, the distance from A is $\frac{1}{2}(2)(2) = 2$

$t = 5$, the distance from A is $2 + 3(2) = 8$

$t = 6$, the distance from A is $8 + \frac{1}{2}(1)(2) = 9$

$t = 7$, the distance from A is 9

$t = 9$, the distance from A is $9 - \frac{1}{2}(2)(1) = 8$

(b)

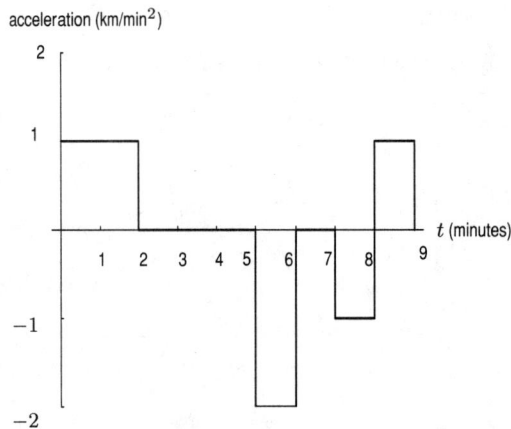

5. If an upper estimate of the area of a region bounded by the curve in Figure 5.3.143, the horizontal axis and the vertical lines $x = 3$ and $x = -3$ is 15, what is the upper estimate if the graph is shifted up one unit?

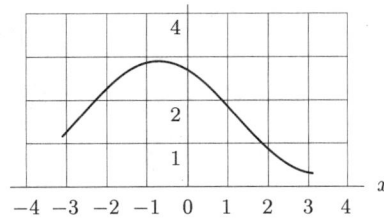

Figure 5.3.143

ANSWER:

The new area will be 6 units more, or 21 units.

6. Estimate the area of the region above the curve $y = \cos x$ and below $y = 1$ for $0 \le x \le \pi/2$.
ANSWER:

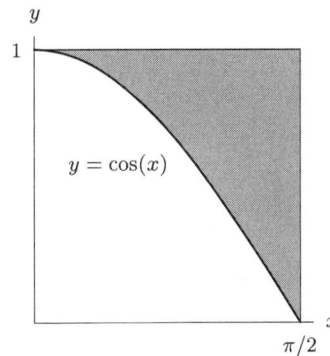

Figure 5.3.144

$$\text{Area=}\int_0^{\pi/2} (1 - \cos x)\,dx \approx 0.57$$

7. Estimate the area of the region between $y = \cos x$, $y = x$, $x = -\pi/2$, and $x = 0$.
ANSWER:

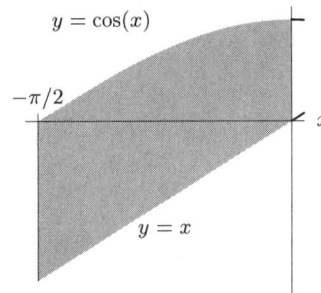

Figure 5.3.145

$$\text{Area=}\int_{-\pi/2}^0 \cos x\,dx + \int_{-\pi/2}^0 -x\,dx \approx 2.2337$$

8. Estimate the area of the region under the curve $y = -x^3 + 5$ and above the x-axis for $0 \leq x \leq \sqrt[3]{5}$.

ANSWER:

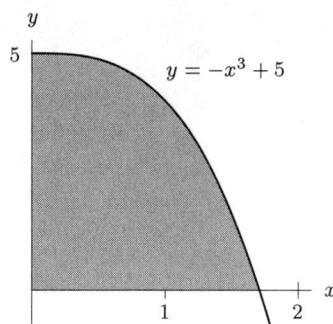

Figure 5.3.146

$$\text{Area} = \int_0^{\sqrt[3]{5}} (-x^3 + 5)\, dx \approx 6.41241$$

9. Estimate the area of the region under the curve $y = \sin(x/2)$ for $0 \leq x \leq 2$.

ANSWER:

Since $\sin(x/2) > 0$ for $0 \leq x \leq 2$, the area is given by

$$\text{Area} = \int_0^2 \sin(x/2)\, dx \approx 0.91939$$

10. Suppose $F(x) = 3\sin x + x + 5$. Find the total area bounded by $F(x)$, $x = 0$, $x = \pi$ and $y = 0$.

ANSWER:

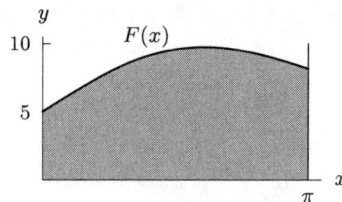

Figure 5.3.147

$$
\begin{aligned}
\text{Area} &= \int_0^\pi (3\sin x + x + 5)\, dx \\
&= \int_0^\pi 3\sin x\, dx + \int_0^\pi x\, dx + \int_0^\pi 5\, dx \\
&= -3\cos x\Big|_0^\pi + \frac{x^2}{2}\Big|_0^\pi + 5x\Big|_0^\pi \\
&= (-3(-1) + 3(1)) + \left(\frac{\pi^2}{2} - 0\right) + (5\pi - 0) \\
&= \frac{\pi^2}{2} + 5\pi
\end{aligned}
$$

Problems and Solutions   for Section 5.4

1. Equal numbers of two different species of ground squirrels are introduced into an area at time $t = 0$. They have the growth rates shown in Figure 5.4.148.

   (a) Which species has a larger population after 2 years? After 3 years?

   (b) Which species would you predict to have the largest population after 15 years?

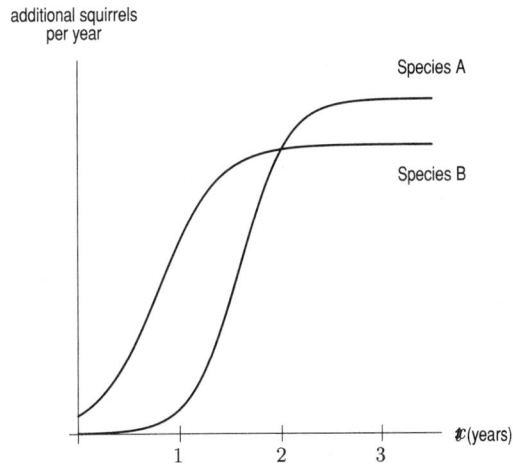

Figure 5.4.148

   ANSWER:

   (a) Until $t = 2$ years, the rate of growth of Species B is greater than the rate of growth for Species A. Thus the population of Species B is greater. After 3 years, the rate of growth of Species A has become greater than the rate of growth for Species B, so we need to look at the total area under each curve. In this case, the total area under the curve of Species B is still greater than the total area under the curve of Species A. Thus, the population of Species B is still greater.

   (b) It appears that the growth rate of Species A is leveling off at a higher rate than that of Species B. Therefore, after 15 years, we would expect the total area under the curve of Species A to become greater than the total area under the curve of Species A. Thus, the population of Species A will become greater.

2. A shop is open from 9am–7pm. The function $r(t)$, graphed below, gives the rate at which customers arrive (in people/hour) at time t. Suppose that the salespeople can serve customers at a rate of 80 people per hour.

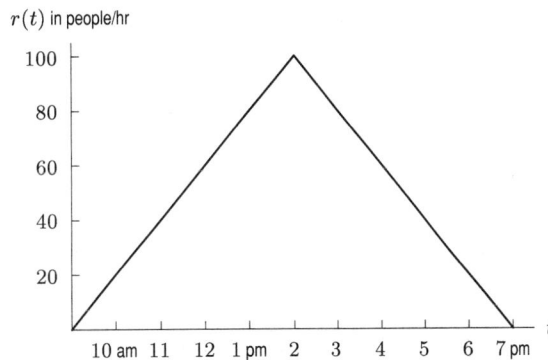

   (a) When do people have to start waiting in line before getting served? Explain clearly how you get your answer.

   (b) When is the line longest, and how many people are in the line then? Explain your answer.

   (c) When does the line vanish? Justify your answer.

ANSWER:

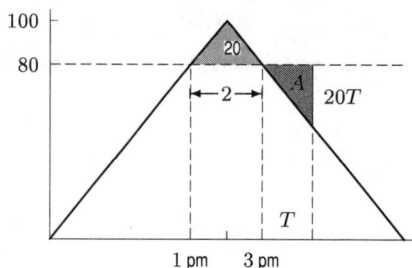

(a) The line starts forming when people arrive faster than they can be served, which happens when $r(t) > 80$. This happens first when $t \geq 1$ pm.

(b) The line builds up from 1pm to 3pm. After 3pm, the rate of arrivals falls <u>below</u> 80 and so the line starts to shrink again. The line is longest at 3pm, and the number in line is the shaded area above line at 80, i.e., length of line $= \frac{1}{2} \cdot 2 \cdot 20 = 20$ people.

(c) The line vanishes when an extra number served (over and above new arrivals) equals the 20 people in line before. This occurs when area is marked $A = 20$, slope of $r(t)$ (for $2 \leq t \leq 7$pm) is $\dfrac{-100}{5} = -20$. Thus, if T is the time beyond 3pm when the line vanishes, Area $= \dfrac{1}{2} \cdot T \cdot 20T = 10T^2 = 20$ so $T = \sqrt{2}$ or 1.41 hours $\approx$ 1 hour and 25 minutes, so around 4:25pm.

3. The Ethnic Food line at the Cougar Eat can serve customers at the rate of about 30 per hour. From 10 am until 4 pm one day, the rate R at which customers entered the line was about $R(t) = 45 - 5(t - 3)^2$ customers per hour at t hours past 10 am.

(a) About when did a waiting line begin to form?

(b) When was the waiting line the longest? About how long was it? How long did the last person in line at that time have to wait to be served?

(c) About how many customers were served between 10 am and 4 pm that day?

ANSWER:

(a) Let $S(t)$ be the rate of customers who can be served. We sketch $R(t)$ and $S(t)$ on the same axes, as shown in Figure 5.4.149.

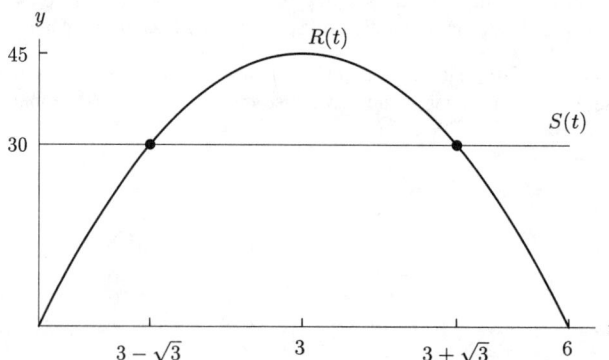

Figure 5.4.149

So the waiting line begins when $R(t) \geq S(t)$.

   Let $R(t) = S(t)$, i.e., $45 - 5(t - 3)^2 = 30$, and solve for t; we have $t = 3 \pm 1.732$ or $t = 1.268$ and $t = 4.732$. So the waiting line begins at 1.268 hours after 10 am, or at about 11:15 am.

(b) The waiting line is longest when $t = 4.732$ or at about 2:45 pm, and the number of customers in line is:

$$\int_{1.268}^{4.732} (15 - 5(t - 3)^2) \, dt = 37 \text{ customers.}$$

Since $37/30 \approx 1.23$, the last person in line at that time has to wait 1.23 hours to be served, which is until about 4 pm. (Note that there are more customers entering the line after 2:45, who will be served after 4 pm, but that their wait will be shorter.)

(c) The number is

$$\int_0^{1.268} (45 - 5(t - 3)^2)\, dt + \int_{1.268}^6 30\, dt,$$

or, equivalently,

$$180 - \int_0^{1.268} (30 - 45 + 5(t - 3)^2)^2\, dt = 180 - \int_0^{1.268} (5(t - 3)^2 - 15)\, dt = 180 - 17 = 163.$$

So the number of customers served between 10 am and 4 pm is 163.

4. After a foreign substance is introduced into the blood, the rate at which antibodies are made is given by

$$r(t) = \frac{t}{1 + t^2} \text{thousands of antibodies per minute}$$

where time t is measured in minutes and $0 \leq t \leq 4$. How many antibodies are in the blood after four minutes, if there were none at time $t = 0$?

ANSWER:

$$\int_{t=0}^4 \frac{t}{1 + t^2}\, dt = 1.4166 \text{ thousand antibodies} = 1417 \text{ antibodies.}$$

5. The rate of growth of the net worth of a company is given by $2000 - 12t^2$ dollars per year at time t years after its formation in 1990. How much will it increase in value between 1990 and 2000?

ANSWER:
$$\int_0^{10} 2000 - 12t^2\, dt = \$16{,}000$$

Problems and Solutions   for Section 5.5

1. The graph of f is shown in Figure 5.5.150. Find $F(3)$ if $F' = f$ and $F(0) = 0$.

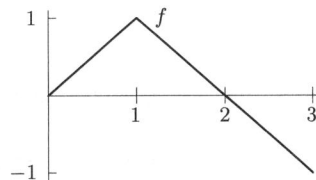

Figure 5.5.150

ANSWER:

$$F(3) - F(0) = \int_0^3 f\, dx, \text{ so we have } F(3) = \int_0^3 f\, dx + F(0) = 0.5.$$

For Problems 2– 3, circle the correct answer(s) or fill in the blanks. No reasons need be given.

2. If $r(t)$ represents the rate at which a country's debt is growing, then the increase in its debt between 1980 and 1990 is given by

(a) $\dfrac{r(1990) - r(1980)}{1990 - 1980}$

(b) $r(1990) - r(1980)$

(c) $\dfrac{1}{10} \displaystyle\int_{1980}^{1990} r(t)dt$

(d) $\displaystyle\int_{1980}^{1990} r(t)dt$

(e) $\displaystyle\frac{1}{10}\int_{1980}^{1990} r'(t)dt$

ANSWER:

(d) Fundamental Theorem

3. The graph of f'' is shown below.

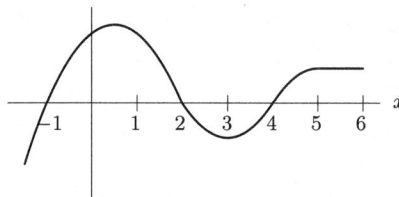

If f is increasing at $x = -1$, which of the following <u>must</u> be true? (Circle all that apply.)

(a) $f'(2) = f'(4)$
(b) $f'(4) > f'(-1)$
(c) $f'(4) > 0$
(d) $f(5) = f(6)$

ANSWER:

f increasing at $x = -1$ means $f'(-1) > 0$.

(a) false: $f'(4) - f'(2) = \displaystyle\int_2^4 f''(t)dt < 0$

(b) true: $f'(4) - f'(-1) = \displaystyle\int_{-1}^4 f''(t)dt > 0$

(c) true: $f'(4) > f'(-1) > 0$
(d) false: $f'(t) > 0$ for all $t > -1$ so $f(6) > f(5)$.

For Problems 4–6, decide whether each statement is true or false, and provide a short explanation or a counterexample.

4. $\displaystyle\int_{-1}^{1} \sqrt{1 - x^2}\, dx = \frac{\pi}{2}$.

ANSWER:

TRUE. The integral represents the area under the upper half of the circle radius 1 centered at the origin. The area is thus $\frac{\pi}{2}$.

5. If a function is concave UP, then the left-hand Riemann sums are always less than the right-hand Riemann sums with the same subdivisions, over the same interval.

ANSWER:

FALSE. On the example below, the function is concave up, yet the left-hand sum is clearly larger than the right-hand sum.

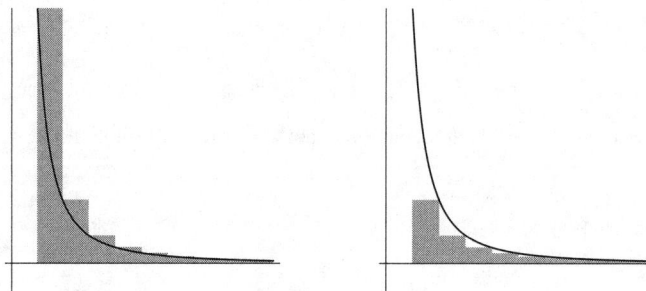

6. If $\int_a^b f(x)\,dx = 0$, then f must have at least one zero between a and b (assume $a \neq b$).

ANSWER:

TRUE, if we allow only continuous functions. FALSE, if we allow discontinuous functions. The function shown below has no roots, yet $\int_{-2}^4 f(x)\,dx = 0$.

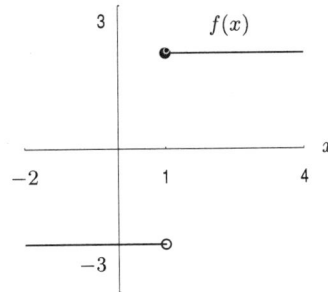

7. Complete the following sentences.

(a) If C is a cost function and $\displaystyle\int_{40}^{50} C'(q)\,dq = 55$, it means that _____.

(b) If $f'(t) = 2t$ is a production rate, measured in items/hour, then $\displaystyle\int_2^4 f(t)\,dt =$_____, which means that _____.

ANSWER:

(a) If C is a cost function and $\displaystyle\int_{40}^{50} C'(q)\,dq = 55$, it means that the cost of increasing production level from 40 to 50 is 55.

(b) $\displaystyle\int_2^4 f(t)\,dt = \int_2^4 2t\,dt = t^2\Big|_2^4 = 12$, which means that 12 items were produced from hour 2 to hour 4.

Problems and Solutions to Review Problems for Chapter 5 ━━━━━━━━━

1. Water is flowing into a container at an increasing rate, as shown in the following table. Give an upper and lower estimate for the total amount of water in the container after 30 minutes.

Table 5.5.34

time (minutes)	0	5	10	15	20	25	30
rate (gal/min)	5	7	9	12	15	19	24

ANSWER:

To find the upper estimate, take the largest value for each 5 minute period:

$$\text{Upper estimate} = 5\cdot 7 + 5\cdot 9 + 5\cdot 12 + 5\cdot 15 + 5\cdot 19 + 5\cdot 24 = 430 \text{ gal.}$$

To find the lower estimate, take the smallest value for each 5 minute period:

$$\text{Low er estimate} = 5\cdot 5\cdot 7 + 5\cdot 9 + 5\cdot 12 + 5\cdot 15 + 5\cdot 19 = 335 \text{ gal.}$$

2. Use Figure 5.5.151 to estimate $\displaystyle\int_0^{10} f(x)\,dx$.

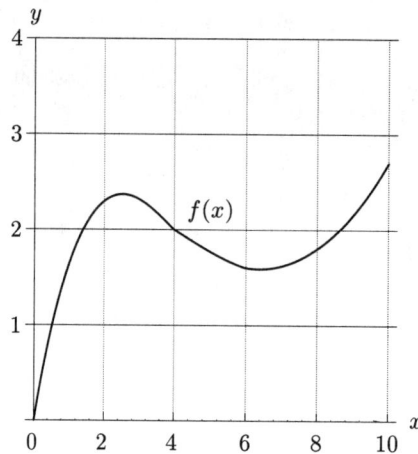

Figure 5.5.151

ANSWER:
Approximate the integral using left- and right-hand sums with $n = 5$ and $\Delta x = 2$.

$$\text{Left} - \text{hand sum} = 0 \cdot 2 + 2.3 \cdot 2 + 2 \cdot 2 + 1.6 \cdot 2 + 1.8 \cdot 2 = 15.4$$

$$\text{Right} - \text{hand sum} = 2.3 \cdot 2 + 2 \cdot 2 + 1.6 \cdot 2 + 1.8 \cdot 2 + 2.7 \cdot 2 = 20.8$$

Estimate the integral by taking the average:

$$\int_0^{10} f(x)\,dx \approx \frac{15.4 + 20.8}{2} = 18.1.$$

3. Use Figure 5.5.152 to find the values of

(a) $\displaystyle\int_a^b f(x)\,dx$ (b) $\displaystyle\int_a^c f(x)\,dx$ (c) $\displaystyle\int_a^d f(x)\,dx$ (d) $\displaystyle\int_a^d |f(x)|\,dx$

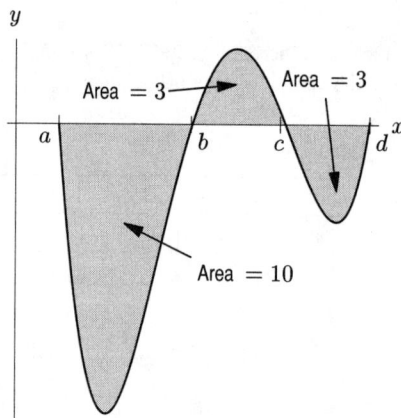

Figure 5.5.152

ANSWER:

(a) $\displaystyle\int_a^b f(x)\,dx = -10$

(b) $\displaystyle\int_a^c f(x)\,dx = \int_a^b f(x)\,dx + \int_b^c f(x)\,dx = -10 + 3 = -7$

(c) $\displaystyle\int_a^d f(x)\,dx = \int_a^b f(x)\,dx + \int_b^c f(x)\,dx + \int_c^d f(x)\,dx = -10 + 3 - 3 = -10$

(d) $\displaystyle\int_a^d |f(x)|\,dx = \int_a^b |f(x)|\,dx + \int_b^c |f(x)|\,dx + \int_c^d |f(x)|\,dx = 10 + 3 + 3 = 16$

4. The marginal cost function of producing a particular product is given by

$$C'(q) = 1000 - 20q$$

where q is quantity. If the fixed costs are $5000,

   (a) What is the total cost to produce 10 items?

   (b) If the items are sold for $600 each, what is the break even point?

   ANSWER:

   (a) Total cost = Fixed cost + Total variable cost, so

$$C(10) = 5000 + \int_0^{10} (1000 - 20x)\,dx$$
$$= \$16{,}000$$

   (b) The break even point is where $C=R$:

$$5000 + \int_0^q (1000 - 20x)\,dx = 600q.$$

   By trial and error, we determine that this happens at

$$q = 50 \text{ items.}$$

5. Data for a function G is given in Table 5.5.35. Estimate the following:

   (a) $G'(0.5)$ (b) $G'(0.0)$ (c) $G'(1.0)$ (d) $G''(0.5)$ (e) $\displaystyle\int_0^1 G(x)\,dx$

Table 5.5.35

x	0.0	0.1	0.2	0.3	0.4	0.5	0.6	0.7	0.8	0.9	1.0
$G(x)$	0.00	0.01	0.03	0.04	0.06	0.10	0.15	0.30	0.50	0.72	1.00

   ANSWER:

   (a) $G'(0.5) \approx \dfrac{0.15 - 0.10}{0.1} = 0.5.$

   (b) $G'(0.0) \approx \dfrac{0.01}{0.1} = 0.1.$

   (c) $G'(1.0) \approx \dfrac{1.00 - 0.72}{0.1} = 2.8.$

   (d) $G''(0.5) \approx \dfrac{1.5 - 0.5}{0.1} = 10$, since $G'(0.5) = 0.5$ (from part (a)), and $G'(0.6) = 1.5$.

   (e) $\displaystyle\int_0^1 G(x)\,dx \approx 0.1(0.01 + 0.03 + 0.04 + 0.06 + 0.1 + 0.15 + 0.3 + 0.5 + 0.72 + 1)$
   $= 2.91 \times 0.1 = 0.29.$

Breakfast at Cafeteria Charlotte

6. Below is the graph of the rate r in arrivals/minute at which students line up for breakfast. The first people arrive at 6:50 a.m. and the line opens at 7:00 a.m.

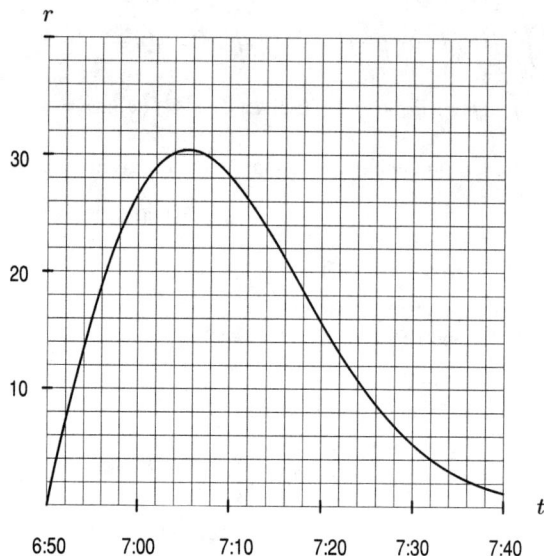

Suppose that once the line is open, checkers can check peoples' meal cards at a constant rate of 20 people per minute. Use the graph and this information to find an estimate for the following:

(a) The length of the line (i.e. the number of people) at 7:00 when the checkers begin.

(b) The length of the line at 7:10.

(c) The length of the line at 7:20.

(d) The rate at which the line is growing in length at 7:10.

(e) The length of time a person who arrives at 7:00 has to stand in line.

(f) The time at which the line disappears.

ANSWER:

(a) The length of the line at $7:00$ will simply be the number of people who arrived before $7:00$ a.m. This is just $\int_{6:50}^{7:00} r\, dt$. By counting squares, this turns out to be 150 students.

(b) This will simply be the [number of people who have arrived] - [number of people checked]

$$= \int_{6:50}^{7:10} r\, dt - 10(20)$$
$$= \int_{6:50}^{7:00} r\, dt + \int_{7:00}^{7:10} r\, dt - 200$$
$$\approx 150 + 280 - 200$$
$$= 230.$$

(c) Similarly, at $7:29$ we have the number of people in line

$$= \int_{6:50}^{7:20} r\, dt - 400$$
$$= 430 + \int_{7:10}^{7:20} r\, dt - 400$$
$$\approx 430 + 220 - 400$$
$$= 250.$$

(d) At $7:10$, the rate of arrivals is about 28 people per minute. The checking rate is 20 people per minute, so the line is growing at a rate of 8 people per minute.

(e) A person who arrives at $7:00$ has about 150 people waiting in front of her. At a checking rate of 20 people per minute, she will spend approximately 7.5 minutes in line.

(f) The total number of arrivals, from the graph, $\int_{6:50}^{7:40} r\, dt$, appears to be about 800. At a checking rate of 20 people per minute, this will take 40 minutes, beginning at $7:00$ a.m. So the line will disappear at $7:40$ a.m.

7. Let $f(v)$ be the amount of energy consumed by a flying bird, measured in joules per second, as a function of its velocity, measured in meters per second. The graph of f is shown in Figure 5.5.153. Note: flying at velocity 0 is called *hovering*.

(a) Explain why the graph is shaped the way it is, in terms of how birds fly.

(b) Let v_0 be the velocity that minimizes $f(v)$. Tell where v_0 is located.

(c) Let $a(v)$ be the amount of energy consumed by the same bird, measured in joules per meter. What is the relationship between $a(v)$ and $f(v)$?

(d) Let v_1 be the velocity that minimizes $a(v)$. Tell where v_1 is located, relative to v_0.

(e) Should the bird try to minimize $f(v)$ or $a(v)$? Why?

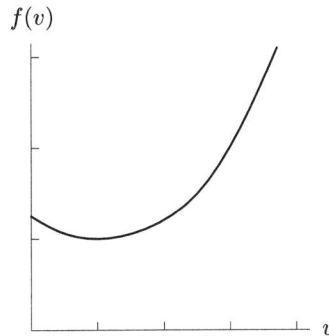

Figure 5.5.153

ANSWER:

(a) Without moving, a bird can not use the relative air flow to support itself, so it needs more energy for staying in the air, or hovering. At certain speeds with the best use of the relative air flow, the bird consumes less energy in terms of joules per second, compared with hovering, but to reach higher and higher speed, the bird needs more and more energy. That is why the graph is shaped this way.

(b) $v_0 \approx 10$ m/s. See Figure 5.5.154.

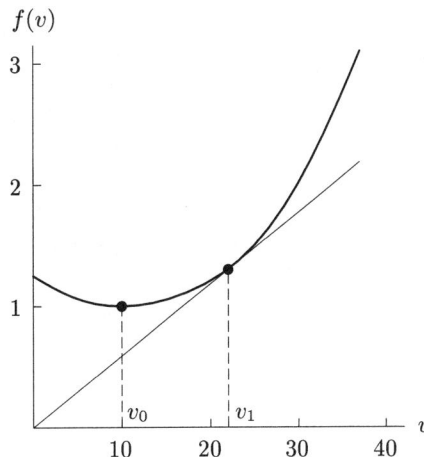

Figure 5.5.154

(c) Since $a(v)$ is measured in joules per meter, we have $a(v) = \dfrac{f(v)}{v}$, which is the slope of the line through $(0,0)$ and $(v, f(v))$. See Figure 5.5.154.

(d) From (c), we have $v_1 > v_0$.

(e) A bird should try to minimize $a(v)$ to cover more places to find food.

Problems and Solutions on Theorems about Definite Integrals

1. Suppose $f(t)$ is given by the graph to the right. Complete the table of values of the function $F(x) = \int_0^x f(t)\,dt$.

x	$F(x)$
0	
1	
2	
3	
4	
5	
6	

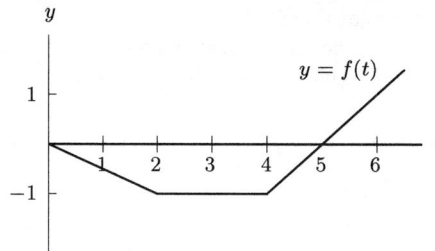

ANSWER:

x	$F(x)$
0	0
1	$-\frac{1}{4}$
2	-1
3	-2
4	-3
5	$-\frac{7}{2}$
6	-3

For Problems 2–3, assume that $\int_0^3 f(x)\,dx = 7$, $\int_0^3 g(x)\,dx = 5$, and $\int_3^5 f(x)\,dx = 15$.

2. Find the values of the integrals below.

   (a) $\int_0^3 (f(x) + g(x))\,dx$ **(b)** $\int_0^3 (2f(x) - 5g(x))\,dx$ **(c)** $\int_0^5 f(x)\,dx$

   ANSWER:

   (a) $\int_0^3 (f(x) + g(x))\,dx = \int_0^3 f(x)\,dx + \int_0^3 g(x)\,dx = 7 + 5 = 12.$

   (b) $\int_0^3 (2f(x) - 5g(x))\,dx = \int_0^3 2f(x)\,dx - \int_0^3 5g(x)\,dx = 2\int_0^3 f(x)\,dx - 5\int_0^3 g(x)\,dx = 2(7) - 5(5) = -11.$

   (c) $\int_0^5 f(x)\,dx = \int_0^3 f(x)\,dx + \int_3^5 f(x)\,dx = 7 + 15 = 22.$

3. **(a)** Sketch a possible graph of $f(x)$ on the interval $x = 0$ to $x = 5$.
   (b) Sketch a possible graph of $f(x)$ and $g(x)$ on the same axes, on the interval $x = 0$ to $x = 3$.
   ANSWER:

   (a) See Figure 5.5.155. Many other graphs are also possible.

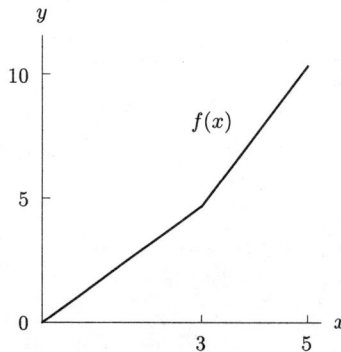

Figure 5.5.155

(b) See Figure 5.5.156. Many other graphs are also possible.

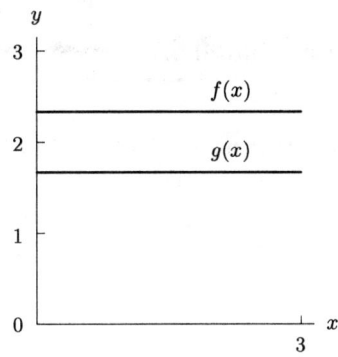

Figure 5.5.156

Chapter 6 Exam Questions

Problems and Solutions for Section 6.1

1. (a) Give a $N = 5$ term left Riemann sum approximation for the function in Figure 6.1.157.

$$\int_{100}^{350} h(x)\,dx \approx \underline{\hspace{1.5cm}} + \underline{\hspace{1.5cm}} + \underline{\hspace{1.5cm}} + \underline{\hspace{1.5cm}} + \underline{\hspace{1.5cm}}$$

$$= \underline{\hspace{1.5cm}}$$

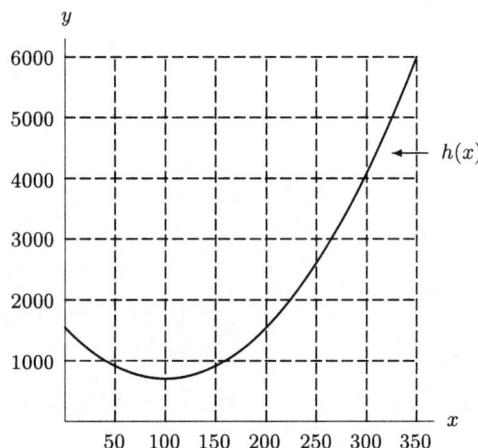

Figure 6.1.157

(b) Redraw Figure 6.1.157 and shade the area you actually computed in part (a).
(c) Is your approximate answer in part (a) too high or too low? Why?
(d) Give the $N = 5$ term right Riemann sum for the inverval in (a):

$$\int_{100}^{350} h(x)\,dx \approx \underline{\hspace{1.5cm}} + \underline{\hspace{1.5cm}} + \underline{\hspace{1.5cm}} + \underline{\hspace{1.5cm}} + \underline{\hspace{1.5cm}}$$

$$= \underline{\hspace{1.5cm}}$$

(e) Using your answers for parts (a) and (d), give your best estimate of $\int_{100}^{350} h(x)\,dx$.
(f) The average of $h(x)$ over the range $100 \leq x \leq 350$ is approximately \underline{\hspace{2.5cm}}.
(g) Draw the horizontal line on the graph that corresponds to the average of $h(x)$ over the range $100 \leq x \leq 350$.
ANSWER:
(a) $\int_{100}^{350} h(x)\,dx \approx 700 \times 50 + 900 \times 50 + 1600 \times 50 + 2600 \times 50 + 4000 \times 50 = 490{,}000.$
(b) See Figure 6.1.158.

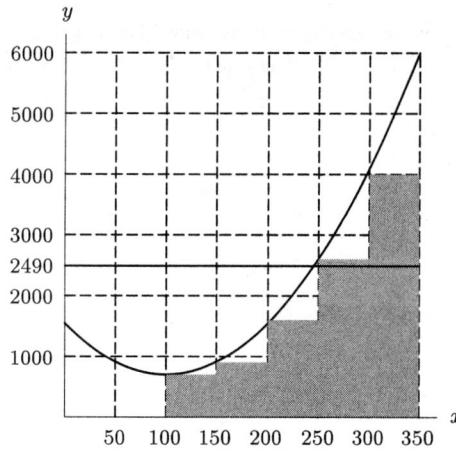

Figure 6.1.158

(c) Too low. The area computed in (a) is smaller than the area under $h(x)$.

(d) $\displaystyle\int_{100}^{350} h(x)\,dx \approx 900 \times 50 + 1600 \times 50 + 2600 \times 50 + 4000 \times 50 + 6000 \times 50 = 755{,}000.$

(e) We average our answers for parts (a) and (d) to get an estimate of $\dfrac{490{,}000 + 755{,}000}{2} = 622{,}500.$

(f) The average is approximately $\dfrac{490{,}000 + 755{,}000}{2 \times (350 - 100)} = 2490.$

(g) See (b).

2. Given the graph in Figure 6.1.159, arrange the following numbers in order from least to greatest. (a) $f'(1)$ (b) the average value of f on $[0, a]$ (c) $\displaystyle\int_0^1 f(x)dx$
(d) the average value of the rate of change of f over $[0, a]$ (e) $f'(0.2)$

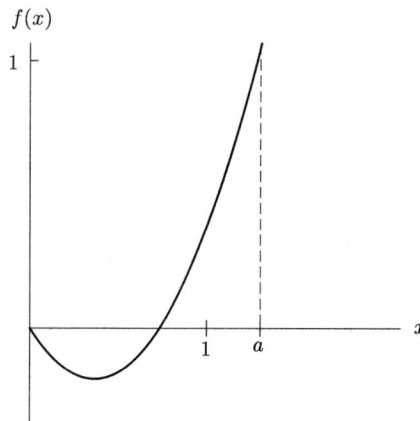

Figure 6.1.159

ANSWER:

$f'(1) \approx 2$; the average value of f on $[0, a]$ is about $1/10$; $-1 < \int_0^1 f(x)\,dx < 0$; the average value of the rate of change of f over $[0, a]$ is larger than $f'(1)$; and finally $f'(0.2) \approx -1$. So we have

$$(e) < (c) < (b) < (a) < (d).$$

3. A discount store specializing in fad items tries to maximize use of shelf space by maximizing the average weekly sales of each item it carries. It is found that for a particular item, the weekly sales $r(t)$ is given by the graph in Figure 6.1.160.

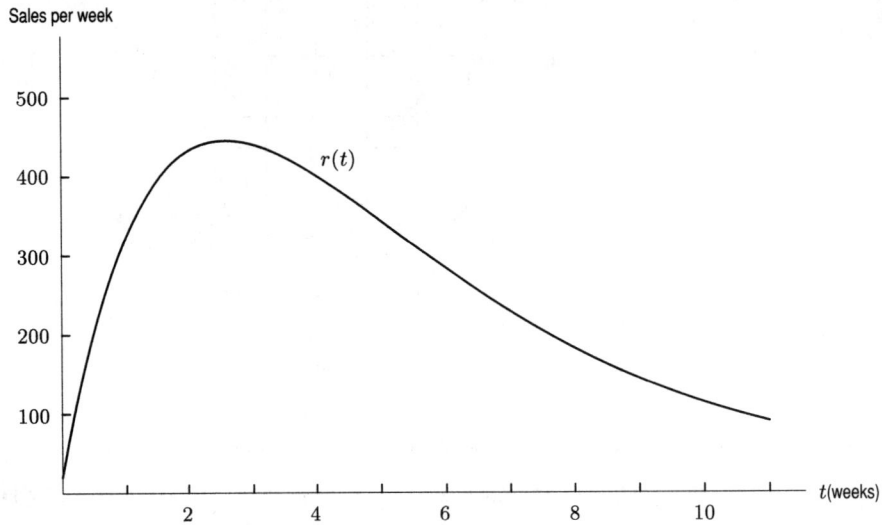

Figure 6.1.160

(a) Let $s(T)$ be the total sales of the item for the first T weeks.

   (i) On the graph of r above, shade a region representing $s(T)$ for a particular T.

   (ii) Sketch, on separate axes, a graph of s against T. Be careful about the scale.

(b) Write an equation expressing $s(T)$ in terms of $r(T)$.

(c) Let $a(T)$ be the average weekly sales of the item for the first T weeks.

   (i) Express $a(T)$ analytically in terms of $s(T)$ and/or $r(T)$.

   (ii) Explain how to read $a(T)$ from your graph of s in Problem a(ii).

(d) Our aim now is to find T so that $a(T)$ is a maximum, at which point it is time to clear the item from the shelves and make way for a new fad.

   (i) From your graph of s in Problem a(ii), estimate T so that $a(T)$ is maximum.

   (ii) Find an algebraic condition on $s(T)$ that makes $a(T)$ a maximum.

   (iii) From the graph of r above, estimate the time T at which $a(T)$ is maximum, and compare with your estimate in part (i).

   (iv) Find an algebraic condition on $r(T)$ that makes $a(T)$ a maximum.

   (v) Explain how the store's bookkeeper knows when it is time to clear the item from the shelves.

ANSWER:

(a) **(i)**

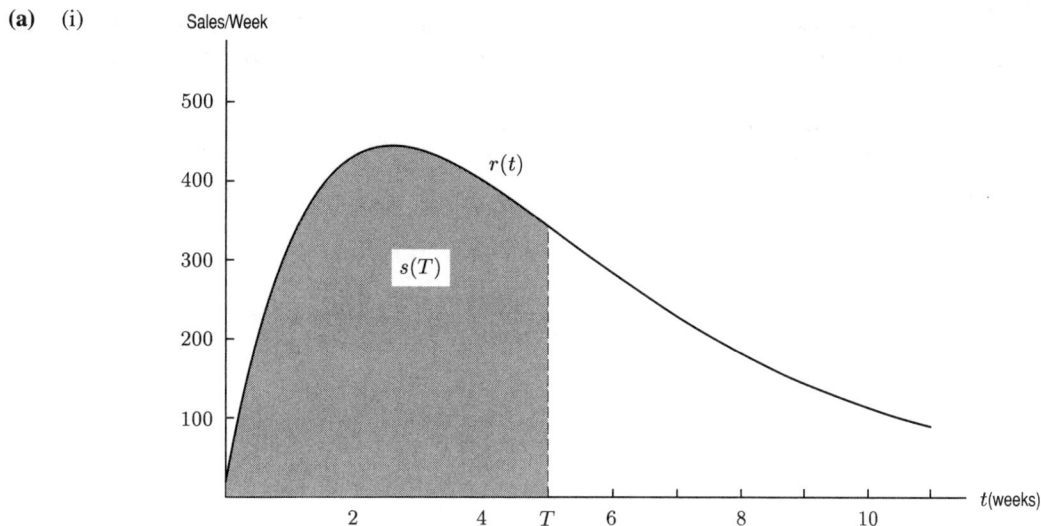

Figure 6.1.161

(ii)

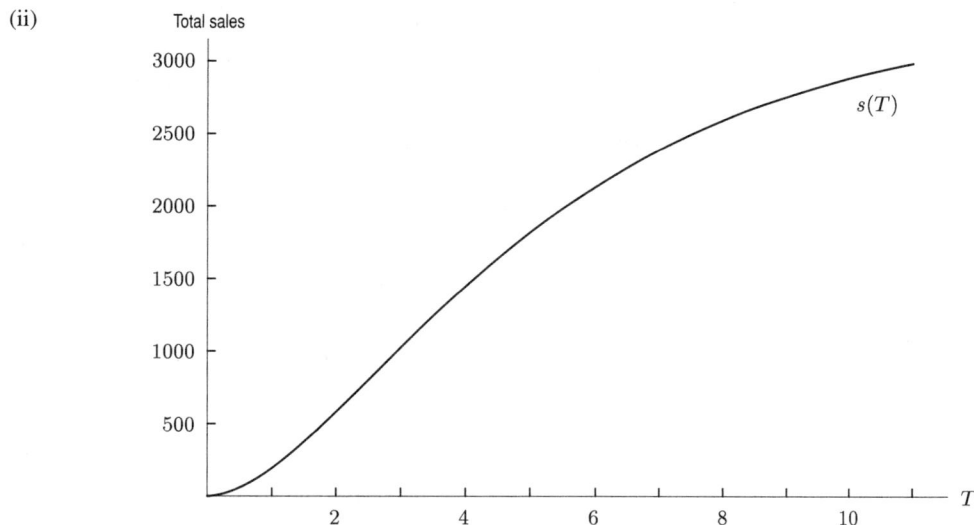

Figure 6.1.162

(b) $s(T) = \displaystyle\int_0^T r(t)\, dt.$

(c) **(i)** $a(T) = \dfrac{1}{T}\displaystyle\int_0^T r(t)\, dt = \dfrac{s(T)}{T}.$

   (ii) $a(T)$ is the value $s(T)$ over "run" T, i.e., it is the slope of the line through $(0,0)$ and $(T, s(T))$.

(d) **(i)** If we observe "height over run" in the graph of $s(T)$, $a(T$ is maximum at about $T = 5.5$.

   (ii) $s'(T) = \dfrac{s(T)}{T}.$

   (iii) Since $a(t)$ is the average weekly sales for the first T weeks, we need to find T such that the shaded area has highest average height. We have $T \approx 5.5$, which is the same as (i).

   (iv) $r(T) = s'(T) = \dfrac{s(T)}{T} = \dfrac{1}{T}\displaystyle\int_0^T r(t)\, dt$; or equivalently: $r(T)T = \displaystyle\int_0^T r(t)\, dt = s(T).$

   (v) When the total sales $s(T)$ for the first T weeks divided by T is equal to the current weekly sale $r(T)$.

4. Rashmi and Tia both go running from 7:00 am to 8:00 am. Both women increase their velocity throughout the hour, both beginning at a rate of 8 mi/hr. at 7:00 am and running at a rate of 12 mi/hr by 8:00 am. Rashmi's velocity increases at an increasing rate and Tia's velocity increases at a decreasing rate.

   (a) Who has run the greater distance in the hour? Explain your reasoning clearly and convincingly.

   (b) Who has the greatest average velocity, Rashmi or Tia, or do they have the same average velocity?

   ANSWER:

   (a) Consider the graphs of the two velocities:

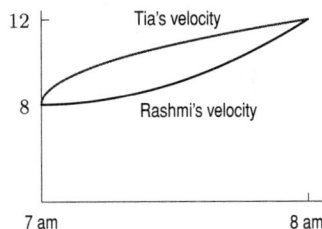

   Then, total distance covered $= \displaystyle\int_7^8 (\text{velocity})\,dt =$ (area under curve between 7 and 8).
   Area under Tia's curve is larger so she has covered more distance.

   (b) average velocity $=$ (total distance covered)/(time it took to cover it)
   Since both spend one hour running and Tia covered more ground, her average velocity must be greater!
   N.B. Average velocity is <u>not</u> defined to be $\dfrac{\text{change in velocity}}{\text{change in time}}$!! That would be average acceleration.

5. The size of an animal population varies according to the function $P(t) = 4000 + 500 \cos \frac{\pi}{6} t$, where t is in months from the beginning of the year.

   (a) What is the average rate at which the animals increase/decrease over the first three months of the year?

   (b) What is the average number of animals in the herd during the first three months of the year? Use an integral.

   ANSWER:

   (a)

$$P(0) = 4000 + 500 = 4500$$
$$P(3) = 4000 + 0.$$

   So the average rate of change is

$$\frac{4000 - 4500}{3} = -\frac{500}{3} \approx -167.$$

   Over the first 3 months the population is decreasing at an average of 167 per month.

   (b) $\dfrac{1}{3} \displaystyle\int_0^3 \left(4000 + 500 \cos \frac{\pi}{6} t\right) dt \approx 4318.$
   The average number of animals during the first 3 months is 4318.

6. Find the average value of the function over the given interval.

   (a) $h(x) = 2x + 2$ over $[1, 3]$
   (b) $f(x) = e^{2x}$ over $[0, 10]$

   ANSWER:

   (a) Average value $= \dfrac{1}{3-1} \displaystyle\int_1^3 (2x + 2)\,dx = \dfrac{1}{2}(12) = 6$

   (b) Average value $= \dfrac{1}{10-0} \displaystyle\int_0^{10} e^{2x}\,dx \approx \dfrac{1}{10}(2.4 \times 10^8) = 2.4 \times 10^7$

7. The average value of $y = h(x)$ equals a for $0 \le x \le 3$, and equals b for $3 \le x \le 9$. What is the average value of $h(x)$ for $0 \le x \le 9$?

   ANSWER:
   We know that the average value of $h(x) = a$ for $0 \le x \le 3$, so $\dfrac{1}{3-0} \displaystyle\int_0^3 h(x)\,dx = a$, and thus $\displaystyle\int_0^3 h(x) = 3a.$

Similarly, we know

$$\frac{1}{9-3} \int_3^9 h(x)\,dx = b, \text{ so } \int_3^9 h(x) = 6b.$$

The average value for $0 \le x \le 9$ is given by

$$\text{Average value} = \frac{1}{9-0} \int_0^9 h(x)\,dx$$

$$= \frac{1}{9}\left(\int_0^3 h(x)\,dx + \int_3^9 h(x)\,dx \right)$$

$$= \frac{1}{9}(3a + 6b) = \frac{a + 2b}{3}$$

Problems and Solutions   for Section 6.2

For Problems 1–2, refer to Figure 6.2.163, which shows the demand and supply curves for a product.

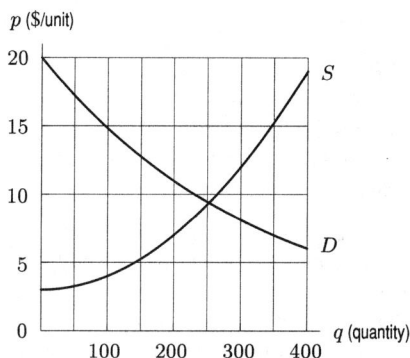

Figure 6.2.163

1. **(a)** Estimate the equilibrium price and equilibrium quantity.
   (b) Estimate the consumer surplus. Draw it on a graph similar to Figure 6.2.163.
   (c) Estimate the producer surplus. Draw it on a graph similar to Figure 6.2.163.
   (d) Estimate the total gains from trade.
   ANSWER:

   (a) From the graph, we estimate that $p^* \approx 9.5$ and $q^* \approx 250$.
   (b) Consumer surplus covers an area in the graph of about 4.5 boxes, and each box has area $(5)(50) = 250$, so the consumer surplus is about $(4.5)(250) = \$1125$. See Figure 6.2.164.

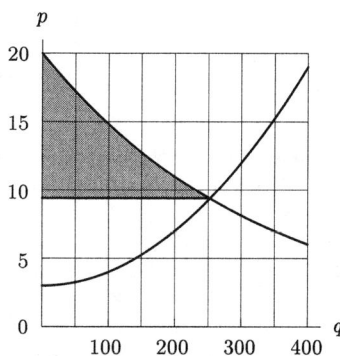

Figure 6.2.164

(c) Producer surplus covers an area in the graph of about 4 boxes, so the producer surplus is about $(4)(250) = \$1000$. See Figure 6.2.165.

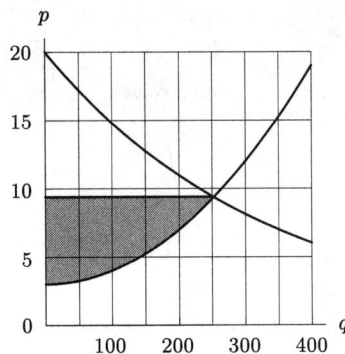

Figure 6.2.165

(d) Total gains from trade = consumer surplus + producer surplus $\approx 1125 + 1000 = \$2125$.

2. Suppose a price of \$12 is artificially imposed.

(a) At this price, what quantity will consumers buy?

(b) Estimate the consumer surplus, the producer surplus, and the total gains from trade at this price. Compare your answers to your answers in Problem 1, and discuss the effect of price controls on the consumer surplus, producer surplus, and total gains from trade in this case.

ANSWER:

(a) When $p = 12$, $q \approx 170$ (from the graph), so at a price of \$12, consumers will buy about 170 units.

(b) At a price of \$12, consumer surplus is the area shown in Figure 6.2.166. This area is about equal to that of 2.5 boxes on the graph, and each box has an area of $(50)(5) = 250$, so the consumer surplus is approximately $(2.5)(250) = \$625$. This is substantially less than the consumer surplus of the equilibrium price.

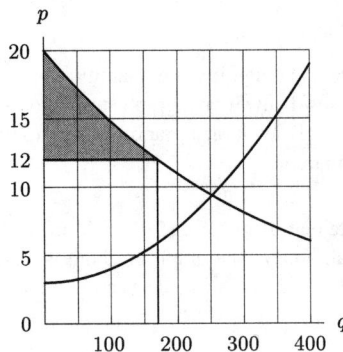

Figure 6.2.166

Producer surplus is the area shown in Figure 6.2.167. This area is about 5.5 boxes, so producer surplus is approximately $(5.5)(250) = \$1375$. This is larger than the producer surplus at the equilibrium price.

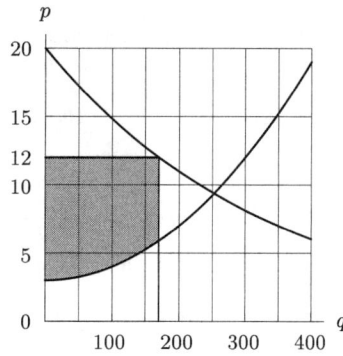

Figure 6.2.167

At $p = 12$, total gains from trade $= 625 + 1375 = \$2000$, which is less than the total gain from trade at the equilibrium price. Total gains from trade will always go down if a price is artificially imposed.

3. Supply and demand curves for a product are given by the equations

$$\text{Demand}: p = 80 - 7.15q$$

$$\text{Supply}: p = 0.2q^2 + 10$$

where p is price in $ and q is quantity.

(a) Find the equilibrium price.

(b) Compute the consumer and producer surplus.

ANSWER:

(a) The equilibrium price is where demand equals supply:

$$0.2q^2 + 10 = 80 - 7.15q$$

$$0.2q^2 + 7.15q - 70 = 0$$

This gives $q = 8$, and at this quantity $p = 80 - 7.15(8) = \$22.80$.

(b) Consumer surplus = the area under the demand curve and above $q = 22.8$, which is

$$\int_0^8 (80 - 7.15q - 22.8)\, dq = \$228.80.$$

Producer surplus = the area under $q = 22.80$ and above the supply curve, which is

$$= \int_0^8 (22.8 - (0.2q^2 + 10))\, dq = \$68.27.$$

4. Supply and demand data are given in the following tables.

(a) Which table shows supply and which shows demand?

(b) Estimate equilibrium price and quantity.

(c) Estimate consumer and producer surplus.

Table 6.2.36

q(quantity)	0	10	20	30	40	50
p(dollars per unit)	180	139	108	85	68	55

Table 6.2.37

q(quantity)	0	10	20	30	40	50
p(dollars per unit)	50	73	108	148	190	241

ANSWER:

(a) Table 6.2.36 shows demand because as price increases, quantity decreases. Table 6.2.37 shows supply because as price increases, quantity increases.

(b) The equilibrium point is where the tables are the same: at $p = \$108$ and $q = 20$.

(c) To estimate consumer surplus take upper and lower estimate of the area:

$$\text{Upper estimate} = (180 - 108)10 + (139 - 108)10 = 1030$$

$$\text{Lo w er estimate} = (139\text{-} 108)10 + (108 - 108)10 = 310$$

$$\text{Average} = \frac{1030 + 310}{2} = \$670.$$

To estimate producer surplus take upper and lower estimate of the area:

$$\text{Upper estimate} = (108 - 50)10 + (108 - 73)10 = 930$$

$$\text{Low er estimate} = (108\text{-} 73)10 + (108 - 108)10 = 350$$

$$\text{Av erage} = \frac{930 + 350}{2} = \$640.$$

5. Suppose the government fixes the price of a commodity at an artificially high price, $p+$, above the equilibrium price, p. Sketch two graphs with supply and demand curves. Label p and a possible point $p +$. On one graph, shade in the total gains from trade at price p. On the other graph, shade in the total gains from trade at point $p +$. Which is greater?

ANSWER:

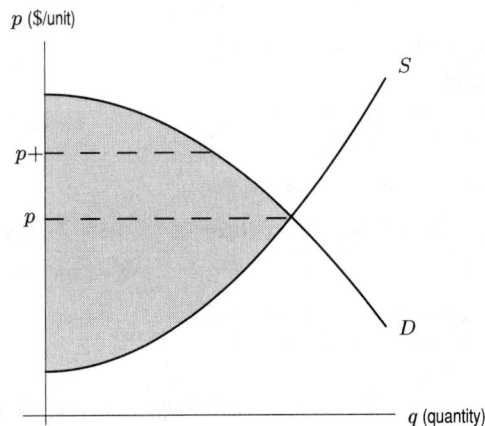

Figure 6.2.168

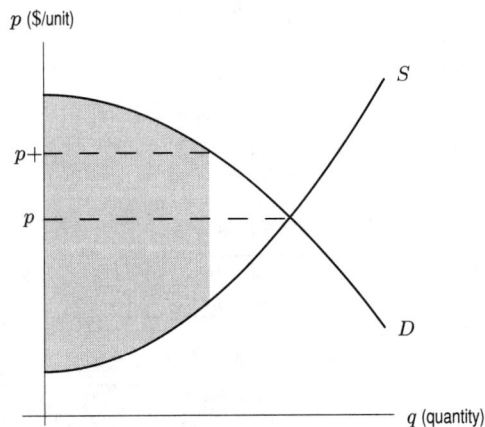

Figure 6.2.169

The total gains from trade at point p are greater.

Problems and Solutions   for Section 6.3

1. Find the present value and future value of an income stream of $1000 a year, for a period of 5 years, if the interest rate is 8%.

ANSWER:

We use the formulas for present and future value, given that the rate of deposits, $S(t)$, is 1000, and that the interest rate, r is 0.08:

$$\text{Present value} = \int_0^M S(t)e^{-rt}\, dt = \int_0^5 1000e^{-0.08t}\, dt = 4121.$$

$$\text{Future value} = \text{Present value} \cdot e^{rM} = 4121e^{(0.08)(5)} = 6147.81.$$

So the present value is $4121 and the future value is $6147.81.

Problems and Solutions   for Section 6.4

1. A person's annual salary starts out at $30,000. Find a formula for the person's salary t years later, and find the salary after 5 years and after 30 years, in each case below:

(a) Each year, the person receives a $2000 raise.

(b) Each year, the person recieves a 5% raise.

ANSWER:

(a) Let the salary be represented by S (in dollars). Then we have the formula $S = 30{,}000 + 2000t$. Substituting 5 and 30 for t in this formula, we find that after 5 years, $S = \$40{,}000$ and after 30 years, $S = \$90{,}000$.

(b) In this case the formula is $S = 30{,}000(1.05)^t$. Substituting again, we find that after 5 years, $S = \$38{,}288$, and after 30 years, $S = \$129{,}658$.

2. The initial size of a bacteria population is 10,000. Fill in the table assuming the bacteria increases by

(a) 500 bacteria per hour

(b) 4% per hour

Table 6.4.38

time (hours)	0	1	2	2	4	5
population						

ANSWER:

(a)

Table 6.4.39

time (hours)	0	1	2	2	4	5
population	10,000	10,500	11,000	11,500	12,000	12,500

(b)

Table 6.4.40

time (hours)	0	1	2	2	4	5
population	10,000	10,400	10,816	11,248	11,698	12,166

3. The number of US households with cable television was 42,237,140 in 1986 and was 64,654,160 in 1996.[4] Find the (absolute) average rate of change and the relative (percent) rate of change in the number of US households with cable television during this time period. Give units with the answers, and interpret them in terms of number of households with cable television.

---

[4]*The World Almanac and Book of Facts 1998* (Mahwah, NJ: K-111 Reference Corporation).

ANSWER:

The average rate of change is $\dfrac{64{,}654{,}160 - 42{,}237{,}140}{1996 - 1986} = 2{,}241{,}702$ households per year. So the number of US households with cable television went up at an average rate of 2,241,702 households every year between 1986 and 1996.

The relative rate of change is $\dfrac{2{,}241{,}702}{42{,}237{,}140} = 0.053 = 5.3\%$ per year. So the number of US households with cable televisions went up at an average rate of 5.3% per year between 1986 and 1996.

4. A graph of the relative growth rate of a population is given in Figure 6.4.170. By approximately what percentage does the population change over the 10 year period?

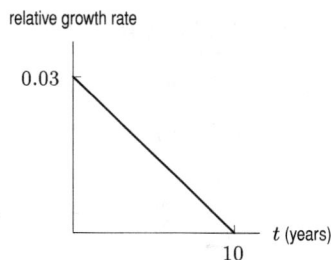

Figure 6.4.170

ANSWER:

The change in $\ln P(t)$ is the integral of the relative growth rate. The integral is the area under the graph in Figure 6.4.170, which is $\dfrac{1}{2}(10)(0.03) = 0.15$. We have $\ln P(10) - \ln P(0) = 0.15$, so

$$\ln\left(\frac{P(10)}{P(0)}\right) = 0.15$$
$$\frac{P(10)}{P(0)} = e^{0.15} = 1.162.$$

We see that $P(10) = 1.162 \cdot P(0)$, so the population has grown by 16.2% during this 10 year period.

5. The following figure shows the relative rate of change of a population. By approximately what factor does the population change over the 20 year period?

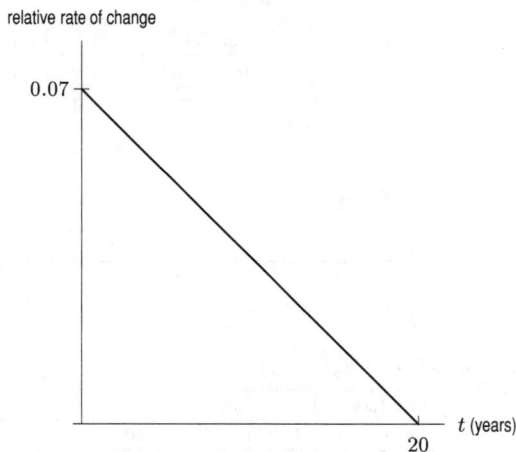

Figure 6.4.171

ANSWER:
We know that

$$\int_0^{20} \frac{P'(t)}{P(t)}\, dt = \text{the area under the graph} = \frac{1}{2}(20)(0.07) = 0.7.$$

Thus,

$$\ln\left(\frac{P(20)}{P(0)}\right) = \int_0^{20} \frac{P'(t)}{P(t)}\, dt = 0.7$$

$$\frac{P(20)}{P(0)} = e^{0.7} \approx 2.01$$

$$P(20) = 2.01 P(0)$$

i.e. the population changed by a factor of 2.01.

6. The following graph shows the rate of change of a population. By approximately what percentage does the population change over the 50 year period?

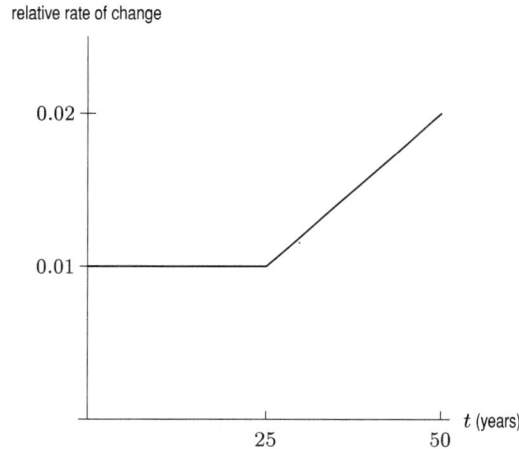

relative rate of change

Figure 6.4.172

ANSWER:

We first find $\int_0^{50} \frac{P'(t)}{P(t)}\, dt$. We get this by computing the area under the graph in the figure:

$$50(0.01) + \frac{1}{2}(25)(0.01) = 0.625.$$

We now use the equation

$$\ln\left(\frac{P(50)}{P(0)}\right) = \int_0^{50} \frac{P'(t)}{P(t)}\, dt = 0.625$$

Thus,

$$\frac{P(50)}{P(0)} = e^{0.625} \approx 1.87$$

$$P(50) = 1.87 P(0).$$

Thus the population changes by approximately 87%.

7. The following graph shows the rate of change of a population. By approximately what percentage does the population change over the 50 year period?

relative rate of change

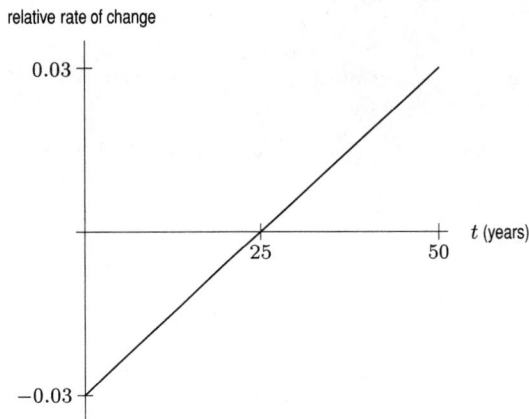

Figure 6.4.173

ANSWER:

We first find $\int_0^{50} \dfrac{P'(t)}{P(t)} \, dt$. We get this by computing the area under the graph in the figure:

$$-\frac{1}{2}(0.03)(25) + \frac{1}{2}(0.03)(25) = 0.$$

We now use the equation

$$\ln\left(\frac{P(50)}{P(0)}\right) = \int_0^{50} \frac{P'(t)}{P(t)} \, dt = 0.$$

Thus,

$$\frac{P(50)}{P(0)} = e^0 = 1$$

$$P(50) = P(0).$$

Thus the population did not change.

Problems and Solutions to Review Problems for Chapter 6 ────────

1. The population of a city t years after 2000 is predicted to be $P = 2.3e^{0.02t}$ million.

   (a) What was the population in the year 2000?
   (b) What is the population predicted to be in the year 2020?
   (c) What is the predicted average population between 2000 and 2020?

   ANSWER:

   (a) $P(0) = 2.3e^{0.02(0)} = 2.3$ million.
   (b) $P(20) = 2.3e^{0.02(20)} = 3.4$ million.
   (c) The average population between 2000 and 2020 is given by

   $$\text{Av erage } \text{alue} = \frac{1}{20 - 0} \int_0^{20} 2.3e^{0.02t} \, dt$$

   $$= \frac{2.3}{20} \int_0^{20} e^{0.02t} \, dt$$

   $$= 2.8 \text{ million.}$$

2. Using Figure 6.4.174, list the following numbers from least to greatest:

   (a) The average value of f on $0 \le x \le 5$

   (b) $\int_0^5 f(x) \, dx$

(c) $\displaystyle\int_0^5 |f(x)|\,dx$

(d) $f'(4)$

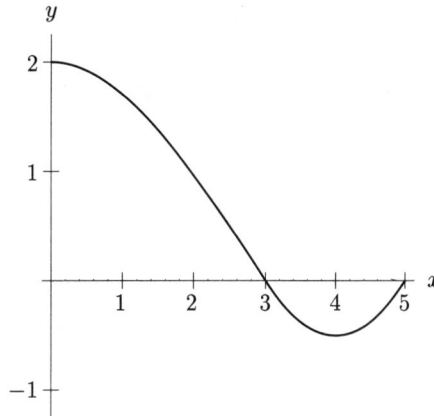

Figure 6.4.174

ANSWER:

The average value of f on $0 \le x \le 5$ is given by

$$\text{Average value} = \frac{1}{5-0}\int_0^5 f(x)\,dx = \frac{1}{5}\int_0^5 f(x)\,dx.$$

Since $\displaystyle\int_0^5 f(x)\,dx > 0$, we get (a)<(b). Also, since $f(x) \le |f(x)|$, we have (b)≤(c). Finally, $f'(4) = 0$, so we get (d)≤(a)≤(b)≤(c).

3. Supply and demand curves for a product are given in Figure 6.4.175.

   (a) Estimate equilibrium price and quantity.
   (b) Estimate consumer and producer surplus.

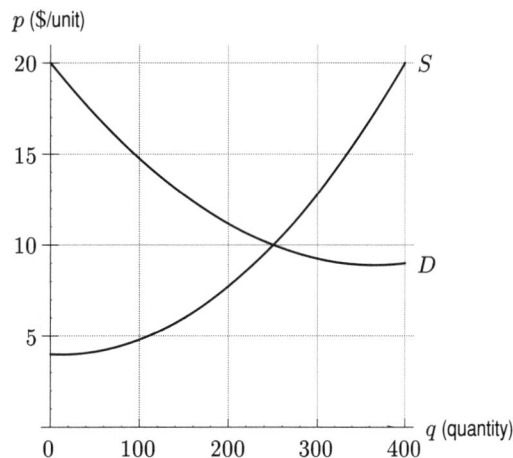

Figure 6.4.175

ANSWER:

(a) The equilibrium price and quantity occur where the supply and demand curves cross: at $p = \$10, q = 250$.

(b) The consumer surplus is the area between the demand curve and the horizontal line at $p = 10$, which is approximately 2 complete squares or \$1000. The producer surplus is the area between the horizontal line at $p = 10$ and the supply curve, which is also approximately 2 complete squares or \$1000.

4. Suppose you want to have $20,000$ in 5 years time in a bank account earning 3% interest, compounded continuously.

(a) If you make one lump sum deposit now, how much should you deposit?

(b) If you would rather deposit money continuously throughout this 5 year period, at what rate should you deposit it?

ANSWER:

(a) Using the equation for future value with future value = 20,000, $r = 0.03$, and $t = 5$, we have:

$$20,000 = Pe^{0.03(5)}$$

$$P = \frac{20,000}{e^{0.03(5)}} = \$17,214.$$

(b) Suppose you deposit money at a continuous rate of $\$S$ per year. Then, using $P = 17,214$ from part (a), we have

$$17,214 = \int_0^5 Se^{-0.03t}\, dt$$
$$= 4.643S$$

Thus $S = 17,214/4.643 = \$3708.$

Chapter 7 Exam Questions

Problems and Solutions   for Section 7.1

1. Find an antiderivative:

   (a) $f(x) = 3x^2 + 2$

   (b) $h(t) = \dfrac{1}{t}$

   ANSWER:

   (a) $x^3 + 2x + C$

   (b) $\ln |t| + C$

2. Find an antiderivative $G(z)$ with $G'(z) = g(z)$ and $G(0) = 2$.

$$g(z) = z - \sqrt{z}$$

   ANSWER:

   $G(z) = \frac{1}{2}z^2 - \frac{2}{3}z^{3/2} + C$. Since $G(0) = 2$, $C = 2$, and thus $G(z) = \frac{1}{2}z^2 - \frac{2}{3}z^{3/2} + 2$.

3. Find an antiderivative $F(x)$ of $f(x) = \sin x$ such that $F(0) = 2$.

   ANSWER:

   We know that the derivative of $-\cos x$ is $\sin x$, so $F(x)$ will take the form $F(x) = -\cos x + C$. We know that $F(0) = 2$, so we have $2 = -\cos 0 + C$, or $2 = -1 + C$; so $C = 3$. Thus we have $F(x) = -\cos x + 3$.

4. Evaluate the following integrals.

   (a) $\displaystyle\int (x^2 + 6x - 5)\, dx$ (b) $\displaystyle\int \left(\frac{1}{x} + \frac{1}{x^2}\right) dx$ (c) $\displaystyle\int 6\sqrt{x}\, dx$

   ANSWER:

   (a) $\displaystyle\int (x^2 + 6x - 5)\, dx = \frac{x^3}{3} + 3x^2 - 5x + C.$

   (b) $\displaystyle\int \left(\frac{1}{x} + \frac{1}{x^2}\right) dx = \ln x - \frac{1}{x} + C.$

   (c) $\displaystyle\int 6\sqrt{x}\, dx = 6\left(\frac{2}{3}x^{3/2}\right) + C = 4x^{3/2} + C.$

5. Find the indefinite integral:

   (a) $\displaystyle\int (p^2 + 3p + 4)\, dp$

   (b) $\displaystyle\int \cos\theta\, d\theta$

   (c) $\displaystyle\int e^{3t}\, dt$

   ANSWER:

   (a) $\frac{1}{3}p^3 + \frac{3}{2}p^2 + 4p + C$

   (b) $\sin\theta + C$

   (c) $\frac{1}{3}e^{3t} + C$

6. Find the indefinite integrals in each case below. Simplify your answer.

   (a) $\displaystyle\int (8x^3 + 9x^2 - 6x + 5)\, dx$ (b) $\displaystyle\int (\cos t + 5\sin t)\, dt$ (c) $\displaystyle\int \frac{6}{\sqrt{x}}\, dx$

   (d) $\displaystyle\int (3e^x + 5)\, dx$ (e) $\displaystyle\int (t^2 + \frac{1}{t^2})\, dt$

   ANSWER:

   (a) $\displaystyle\int (8x^3 + 9x^2 - 6x + 5)\, dx = 2x^4 + 3x^3 - 3x^2 + 5x + C.$

   (b) $\displaystyle\int (\cos t + 5\sin t)\, dt = \sin t - 5\cos t + C.$

(c) $\int \dfrac{6}{\sqrt{x}}\, dx = \int \dfrac{6}{x^{-1/2}}\, dx = 6\dfrac{x^{1/2}}{1/2} = 12\sqrt{x} + C.$

(d) $\int (3e^x + 5)\, dx = 3e^x + 5x + C.$

(e) $\int \left(t^2 + \dfrac{1}{t^2}\right) dt = \dfrac{t^3}{3} + \dfrac{t^{-1}}{-1} + C = \dfrac{t^3}{3} - \dfrac{1}{t} + C.$

Problems and Solutions for Section 7.2

In Problems 1– 6, find the integral using integration by substitution. Check by differentiation.

1. $\int (x + 7)^4\, dx$

ANSWER:

Let $w = x + 7$. Then $dw = w'(x)\, dx = dx$. Rewriting, we have

$$\int (x+7)^4\, dx = \int w^4\, dw = \frac{1}{5}w^5 + C = \frac{1}{5}(x+7)^5 + C$$

2. $\int t^2 \sin\left(t^3\right) dt$

ANSWER:

Let $w = t^3$. Then $dw = w'(t)\, dt = 3t^2\, dt$, so $\frac{1}{3}\, dw = t^2\, dt$. Rewriting, we have

$$\int t^2 \sin\left(t^3\right) dt = \int \frac{1}{3}\sin w\, dw = -\frac{1}{3}\cos w + C = -\frac{1}{3}\cos\left(t^3\right) + C$$

3. $\int \dfrac{y}{y^2 + 2}\, dy$

ANSWER:

Let $w = y^2 + 2$. Then $dw = w'(y)\, dy = 2y\, dy$, so $\frac{1}{2}\, dw = y\, dy$. Rewriting, we have

$$\int \frac{y}{y^2+2}\, dy = \int \frac{\frac{1}{2}\, dw}{w} = \frac{1}{2}\ln|w| + C = \frac{1}{2}\ln|y^2 + 2| + C$$

4. $\int \cos\theta(\sin\theta + 3)^2\, d\theta$

ANSWER:

Let $w = \sin\theta + 3$. Then $dw = w'(\theta)\, d\theta = \cos\theta\, d\theta$. Rewriting, we have

$$\int \cos\theta(\sin\theta + 3)^2\, d\theta = \int w^2\, dw = \frac{1}{3}w^3 + C = \frac{1}{3}(\sin\theta + 3)^3 + C$$

5. $\int xe^{2x^2}\, dx$

ANSWER:

Let $w = 2x^2$. Then $dw = w'(x)\, dx = 4x\, dx$, so $\frac{1}{4}\, dw = x\, dx$. Rewriting, we have

$$\int xe^{2x^2}\, dx = \int \frac{1}{4}e^w\, dw = \frac{1}{4}e^w + C = \frac{1}{4}e^{2x^2} + C$$

6. $\int z^2 \sqrt{z^3 + 7}\, dz$

ANSWER:

Let $w = z^3 + 7$. Then $dw = w'(z)\, dz = 3z^2\, dz$, so $\frac{1}{3}\, dw = z^2\, dz$. Rewriting, we have

$$\int z^2 \sqrt{z^3 + 7}\, dz = \int \frac{1}{3}\sqrt{w}\, dw = \frac{1}{3}\cdot\frac{2}{3}w^{3/2} + C = \frac{2}{9}\left(z^3 + 7\right)^{3/2} + C$$

7. Find the integrals. Check by differentiating.

(a)

$$\int 6y(y^2 + 6)^3 \, dy$$

(b)

$$\int \frac{4x}{\sqrt{9 - x^2}} \, dx$$

ANSWER:

(a) Let $u = y^2 + 6$
$du = 2y \, dy$

$$\int 6y(y^2 + 6)^3 \, dy = 3 \int (y^2 + 6)^3 (2y \, dy) = 3 \int u^3 \, du$$

$$= \frac{3u^4}{4} + C = \frac{3}{4}(y^2 + 6)^4 + C$$

(b) Let $u = 9 - x^2$
$du = -2x \, dx$

$$\int \frac{4x}{\sqrt{9 - x^2}} \, dx = -2 \int \frac{-2x \, dx}{\sqrt{9 - x^2}} = -2 \int \frac{du}{\sqrt{u}} = -2 \int u^{-1/2} \, du$$

$$= -4u^{1/2} + C = -4(9 - x^2)^{1/2} + C$$

Problems and Solutions for Section 7.3

1. Use antiderivatives and the Fundamental Theorem of Calculus to evaluate the definite integral $\int_1^3 9x^2 \, dx$. Show your work.

ANSWER:

$$\int_1^3 9x^2 \, dx = 3x^3 \Big|_1^3 = 81 - 3 = 78.$$

2. Use antiderivatives and the Fundamental Theorem of Calculus to find the average value of $f(x) = x^2 + 1$ on the interval $x = 0$ to $x = 3$. Show your work.

ANSWER:

We use the formula $\dfrac{1}{b - a} \int_a^b f(x) \, dx$ for the average value of a function:

$$\text{Average value} = \frac{1}{3 - 0} \int_0^3 (x^2 + 1) \, dx$$

$$= \frac{1}{3} \left[\left(\frac{x^3}{3} + x \right) \Big|_0^3 \right]$$

$$= \frac{1}{3}((9 + 3) - (0 + 0))$$

$$= 4.$$

3. Evaluate the definite integral $\int_0^3 (x^2 + 4x + 1) \, dx$ exactly using the Fundamental Theorem of Calculus. Show your work.

ANSWER:

$$\int_0^3 (x^2 + 4x + 1) \, dx = \left(\frac{x^3}{3} + 2x^2 + x \right) \Big|_0^3 = (9 + 18 + 3) - 0 = 30.$$

4. Use the Fundamental Theorem of Calculus to evaluate $\int_{-2}^2 (3x^2 - 4x + 5) \, dx$.

ANSWER:

We have $\displaystyle\int_{-2}^{2} (3x^2 - 4x + 5)\, dx = x^3 - 2x^2 + 5x \Big|_{-2}^{2}$

$$= \left((2^3) - 2(2)^2 + 5(2)\right) - \left((-2)^3 - 2(-2)^2 + 5(-2)\right)$$
$$= 8 - 8 + 10 + 8 + 8 + 10 = 36$$

5. Use the Fundamental Theorem to evaluate the definite integral exactly.

(a) $\displaystyle\int_{1}^{3} \frac{1}{x^3}\, dx$

(b) $\displaystyle\int_{-\pi/4}^{\pi/4} \sin\theta\, d\theta$

ANSWER:

(a)

$$\int_{1}^{3} \frac{1}{x^3}\, dx = \int_{1}^{3} x^{-3}\, dx = -\frac{1}{2}x^{-2}\Big|_{1}^{3} = -\frac{1}{2}\left(3^{-2} - 1^{-2}\right) = -\frac{1}{2}\left(\frac{1}{9} - 1\right) = \frac{4}{9}$$

(b)

$$\int_{-\pi/4}^{\pi/4} \sin\theta\, d\theta = -\cos\theta\Big|_{-\pi/4}^{\pi/4} = -\left(\cos\left(\frac{\pi}{4}\right) - \cos\left(-\frac{\pi}{4}\right)\right) = -\left(\frac{\sqrt{2}}{2} - \frac{\sqrt{2}}{2}\right) = 0$$

6. Use integration by substitution and the Fundamental Theorem to evaluate

$$\int_{0}^{4} \frac{1}{x+3}\, dx.$$

ANSWER:

Let $w = x + 3$. Then $dw = dx$. Also, when $x = 0$, $w = 3$, and when $x = 4$, $w = 7$. Thus we have

$$\int_{0}^{4} \frac{1}{x+3}\, dx = \int_{3}^{7} \frac{1}{w}\, dw = \ln|w|\Big|_{3}^{7} = \ln 7 - \ln 3 = \ln\left(\frac{7}{3}\right).$$

7. Propellant is leaking from the pressurized fuel tanks of the space shuttle at a rate of $r(t) = 15e^{-0.1t}$ psi per second at time t in seconds.

(a) At what rate, in psi per second, is pressure reducing at 15 seconds?

(b) How many psi have leaked during the first minute?

ANSWER:

(a) At $t = 15$, $r(t) = 15e^{-0.1(15)} \approx 3.35$ psi/sec.

(b) To find the amount of propellant leaked during the first minute, we integrate the rate from $t = 0$ to $t = 60$.

$$\int_{0}^{60} 15e^{-0.1t}\, dt = -\frac{15}{0.1}e^{-0.1t}\Big|_{0}^{60} = 149.628\text{psi}.$$

Problems and Solutions for Section 7.4

1. Suppose $F'(x) = 3^x$ and $F(0) = 6$. Find $F(b)$ for $b = 0$, 0.5, 1, 1.5, 2.

ANSWER:

Apply the Fundamental Theorem with $a = 0$ to get

$$F(b) - F(0) = \int_{0}^{b} F'(t)\, dt = \int_{0}^{b} 3^x\, dx.$$

Since $F(0) = 6$, $F(b) = 6 + \int_0^b 3^x \, dx$.

Using a calculator to estimate $\int_0^b 3^x \, dx$ for each value of b, we get

$$F(0) = 6$$
$$F(0.5) = 6 + 0.67 = 6.67$$
$$F(1) = 6 + 1.82 = 7.82$$
$$F(1.5) = 6 + 3.82 = 9.82$$
$$F(2) = 6 + 7.28 = 13.28$$

2. Figure 7.4.176 shows the graph of $f(x)$. If $F' = f$ and $F(0) = 3$, find $F(b)$ for $b = 1, 2, 3, 4$.

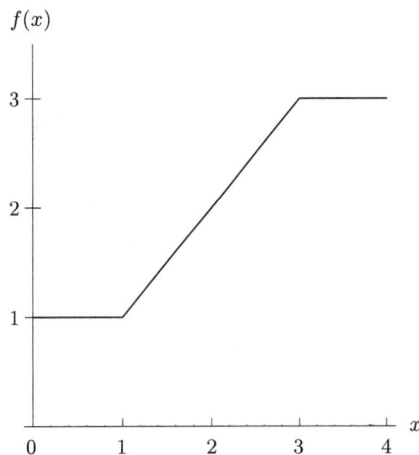

Figure 7.4.176

ANSWER:

From the Fundamental Theorem, $\int_0^b f(t) \, dt = F(b) - F(0)$. Thus $F(b) = 3 + \int_0^b f(t) \, dt$. Using the figure to estimate $\int_0^b f(t) \, dt$, we have

$$F(1) = 3 + 1 = 4$$
$$F(2) = 3 + 2.5 = 5.5$$
$$F(3) = 3 + 5 = 8$$
$$F(4) = 3 + 8 = 11$$

3. Figure 7.4.177 shows a graph of $f'(x)$. Find the intervals where f is increasing and decreasing and the x-coordinates of any local maxima and minima of f. Then sketch a possible graph for f.

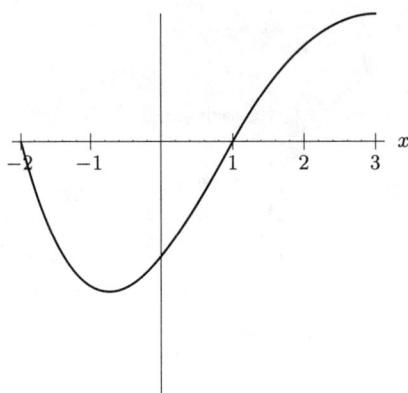

Figure 7.4.177

ANSWER:

$f'(x)$ is positive for $1 < x < 3$ and negative for $-2 < x < 1$. Thus f is increasing for $1 < x < 3$ and decreasing for $-2 < x < 1$. Since $f'(x) = 0$ for $x = -2, 1$, these are either local maximums or local minimums. In this case, $x = -2$ is a local maximum, and $x = 1$ is a local minimum.

One possible graph:

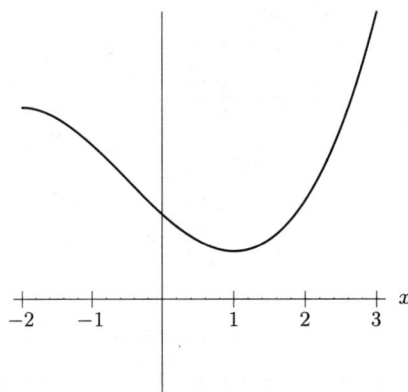

Figure 7.4.178

4. Using Figure 7.4.179, sketch a graph of $g(x)$ satisfying $g(0) = 1$. Label each critical point with its coordinates.

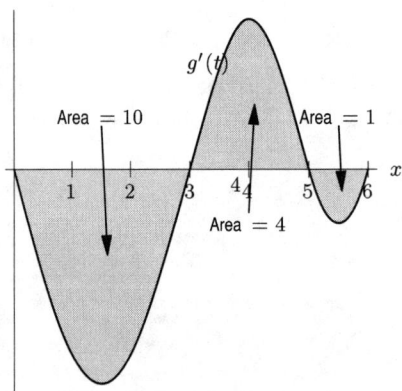

Figure 7.4.179

ANSWER:

$g'(t)$is positive for $3 < t < 5$, so g is increasing there. $g'(t)$is negative for $0 < t < 3$ and $5 < t < 6$, so g is decreasing there. $g'(t) = 0$ for $t = 0, 3, 5, 6$, so those are the critical points. The local maximums are at $t = 0$ and 5, and the local minimums are at $t = 3$ and 6.

From the Fundamental Theorem, $\int_0^b g'(t)\,dt = g(b) - g(0)$, so $g(b) = 1 + \int_0^b g'(t)\,dt$. Using the areas given in the figure, we get

$$g(3) = 1 - 10 = -9$$
$$g(5) = 1 - 10 + 4 = -5$$
$$g(6) = 1 - 10 + 4 - 1 = -6.$$

Thus we have the following graph:

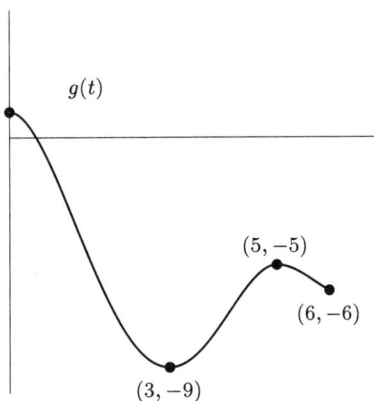

Figure 7.4.180

5. (a) The graph of $g'(x)$ is given in Figure 7.4.181. Sketch the graph of $g(x)$ on the axes in Figure 7.4.182. Assume $g(0) = 2000$.

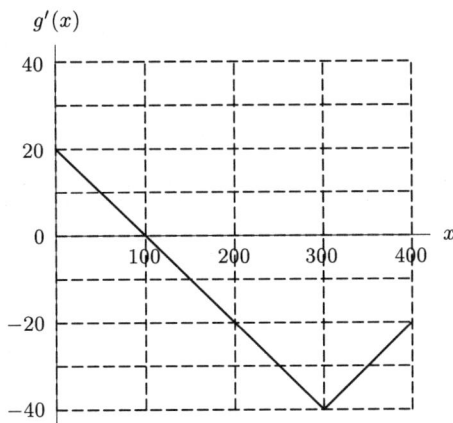

Figure 7.4.181

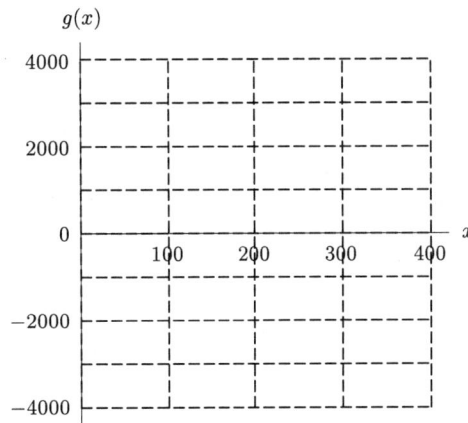

Figure 7.4.182

(b) Complete the following tables of values.

Table 7.4.41

x	0	100	200	300	400
$g(x)$	2000				

Table 7.4.42

x	100	200	300	400
$g'(x)$				

(c) Determine whether each of the following is positive or negative:

(i) $g(50)$	(ii) $g(150)$	(iii) $g(350)$
(iv) $g'(50)$	(v) $g'(150)$	(vi) $g'(350)$
(vii) $g''(50)$	(viii) $g''(150)$	(ix) $g''(350)$

(d) What is happening on the graph of $g(x)$ at
    (i) $x = 100$?
    (ii) $x = 300$?
    ANSWER:

(a) See Figure 7.4.183.

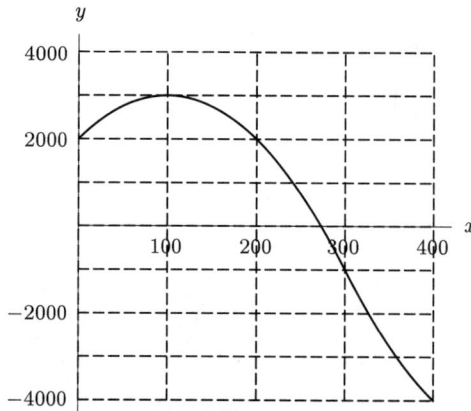

Figure 7.4.183

(b) See Tables 7.4.43 and 7.4.44.

Table 7.4.43

x	0	100	200	300	400
$g(x)$	2000	3000	2000	−1000	−3800

Table 7.4.44

x	100	200	300	400
$g'(x)$	0	−20	−40	−20

(c)

(i) $g(50) > 0$	(ii) $g(150) > 0$	(iii) $g(350) < 0$
(iv) $g'(50) > 0$	(v) $g'(150) < 0$	(vi) $g'(350) < 0$
(vii) $g''(50) < 0$	(viii) $g''(150) < 0$	(ix) $g''(350) > 0$

(d) (i) At $x = 100$, $g(x)$ is a local maximum.
    (ii) At $x = 300$, $g(x)$ is a inflection point.

6. A graph of the derivative $f'(x)$ is given in Figure 7.4.184. Sketch two possible graphs of the function $f(x)$, one with $f(0) = 0$ and one with $f(0) = 2$.

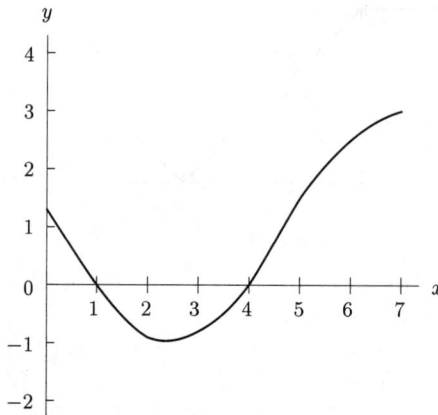

Figure 7.4.184

ANSWER:

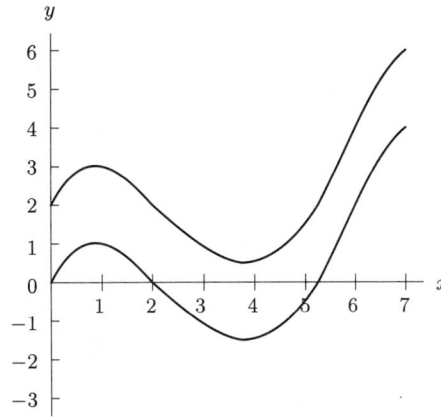

Figure 7.4.185

7. The graph below represents the *rate of change* of a function f with respect to x; i.e., it is a graph of f'.

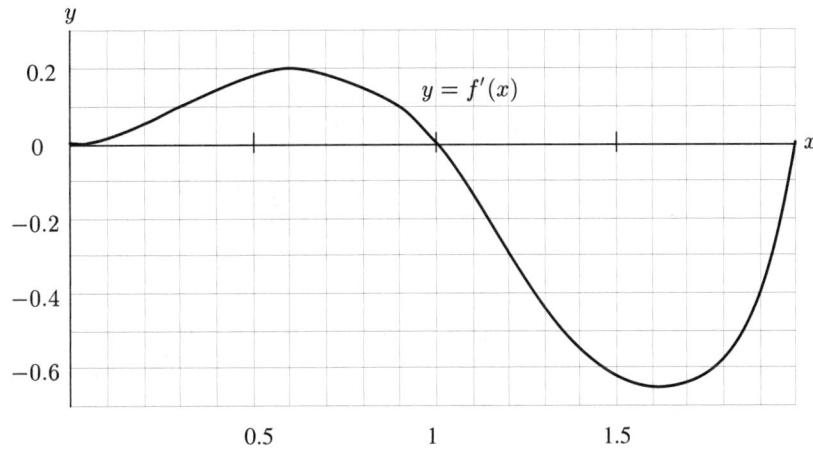

You are told that $f(0) = 0$. On what intervals is f increasing? On what intervals is it decreasing? On what intervals is the graph of f concave up? Concave down? Is there any value $x = a$ other than $x = 0$ in the interval $0 \leq x \leq 2$ where $f(a) = 0$? If not, explain why not, and if so, give the approximate value of a.

ANSWER:

f is increasing where f' is positive, namely from 0 to 1, and is decreasing where f' is negative, between 1 and 2. f is concave up where f' is increasing, namely on the intervals $[0, 0.6]$ and $[1.6, 2]$ and concave down on the interval $[0.6, 1.6]$. Finding an a, $0 \leq a \leq 2$, such that $f(a) = 0$ is equivalent to finding an a such that

$$\int_0^a f'(x)\, dx = 0$$

We see that at $x = 1.4$, $\int_0^{1.4} f'(x)\, dx$ is approximately zero, since the area above the x-axis between 0 and 1 cancels the area below the x-axis between 1 and 1.4, so $a \approx 1.4$.

8.

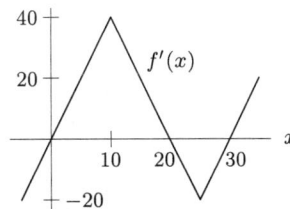

Figure 7.4.186

If Figure 7.4.186 shows the graph of $f'(x)$, sketch $f(x)$ if $f(0) = 100$.
ANSWER:

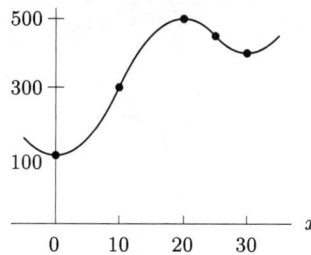

Figure 7.4.187

9. Given the values of the derivative $f'(x)$ in the table and that $f(0) = 40$, estimate $f(x)$ for $x = 2, 4, 6$.

Table 7.4.45

x	0	2	4	6
$f'(x)$	3	15	27	39

ANSWER:
By the Fundamental Theorem of Calculus, we know that

$$f(2) - f(0) = \int_0^2 f'(x)\,dx.$$

Using a left-hand sum, we estimate $\int_0^2 f'(x)\,dx \approx (3)(2) = 6$.

Using a right-hand sum, we estimate $\int_0^2 f'(x)\,dx \approx (15)(2) = 30$.

Averaging, we have $\int_0^2 f'(x)\,dx \approx \dfrac{6+30}{2} = 18$.

We know $f(0) = 40$, so

$$f(2) = f(0) + \int_0^2 f'(x)\,dx \approx 40 + 18 = 58.$$

Similarly, we estimate $\int_2^4 f'(x)\,dx \approx \dfrac{(15)(2) + (27)(2)}{2} = 42$

So $f(4) = f(2) + \int_2^4 f'(x)\,dx \approx 58 + 42 = 100$

Similarly, $\int_4^6 f'(x)\,dx \approx \dfrac{(27)(2) + (39)(2)}{2} = 66$

So $f(6) = f(4) + \int_4^6 f'(x)\,dx \approx 100 + 66 = 166$.

The values are shown in the table:

Table 7.4.46

x	0	2	4	6
$f(x)$	40	58	100	166

10.

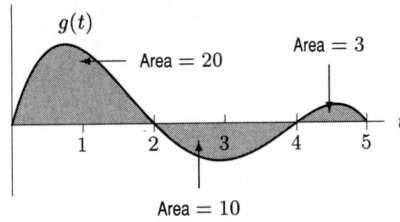

Figure 7.4.188

Using Figure 7.4.188, sketch 2 graphs of antiderivatives $G_1(t)$ and $G_2(t)$ of $g(t)$ satisfying $G_1(0) = 10$ and $G_2(0) = -5$. Label each critical point with its coordinates.

ANSWER:

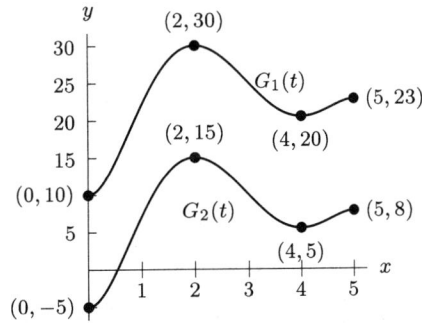

Figure 7.4.189

11. The rate of change in concentration of a certain medication in a person's body, $H'(t)$, in micrograms per milliliter per minute, is -1 for the first 2 seconds. Then it increases at a constant rate for 2 seconds, reaching 1 at $t = 4$. Then it remains constant for 1 second. Sketch $H'(t)$ and $H(t)$, assuming $H(0) = 6$.

ANSWER:

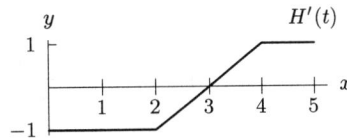

Figure 7.4.190

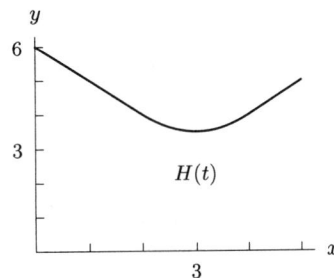

Figure 7.4.191

12. Figure 7.4.192 shows the graph of f. If $F' = f$ and $F(0) = 0$, sketch $F(t)$.

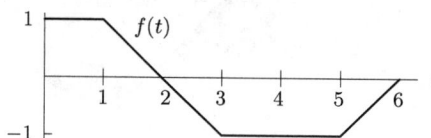

Figure 7.4.192

ANSWER:

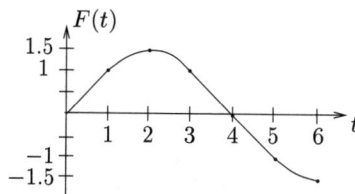

Figure 7.4.193

Problems and Solutions to Review Problems for Chapter 7

1. Find the indefinite integral:

$$\int (x^2 + 3)\, dx$$

ANSWER:
$$\int (x^2 + 3)\, dx = \frac{1}{3}x^3 + 3x + C$$

2. Find the indefinite integral:

$$\int (3e^x - 2\sin x + \cos x)\, dx$$

ANSWER:
$$\int (3e^x - 2\sin x + \cos x)\, dx = 3e^x + 2\cos x + \sin x + C$$

3. Find the indefinite integral using integration by substitution. $\int \dfrac{2e^x}{2 + e^x}\, dx$

ANSWER:
Let $w = 2 + e^x$. Then $dw = w'(x)\, dx = e^x\, dx$. Rewriting, we have

$$\int \frac{2e^x}{2 + e^x}\, dx = \int \frac{2}{w}\, dw = 2\ln|w| + C = 2\ln|2 + e^x| + C$$

4. Find the indefinite integral using integration by substitution. $\int \dfrac{\sin\sqrt{x}}{\sqrt{x}}\, dx$

ANSWER:
Let $w = \sqrt{x}$. Then $dw = w'(x)\, dx = \frac{1}{2\sqrt{x}}\, dx$, so $2\, dw = \frac{dx}{\sqrt{x}}$. Rewriting, we have

$$\int \frac{\sin\sqrt{x}}{\sqrt{x}}\, dx = \int 2\sin w\, dw = -2\cos w + C = -2\cos\sqrt{x} + C$$

5. Use the Fundamental Theorem to evaluate the definite integral:

$$\int_0^3 \frac{1}{\sqrt{x}} \, dx$$

ANSWER:

$$\int_0^3 \frac{1}{\sqrt{x}} \, dx = \left. (2\sqrt{x}) \right|_0^3 = 2\sqrt{3}$$

6. Use the Fundamental Theorem to evaluate the definite integral:

$$\int_0^{\pi/2} (\sin \theta + \cos \theta) \, dx$$

ANSWER:

$$\int_0^{\pi/2} (\sin \theta + \cos \theta) \, dx = \left. (-\cos \theta + \sin \theta) \right|_0^{\pi/2} = 2$$

7. Use the Fundamental Theorem to find the average value of $f(x) = e^{0.2x}$ on the interval between $x = 0$ and $x = 4$. Illustrate on a graph of $f(x)$.

ANSWER:

$$\text{Average value} = \frac{1}{b-a} \int_a^b f(x) \, dx = \frac{1}{4} \int_0^4 e^{0.2x} \, dx = \frac{1}{4} \left. \left(\frac{1}{0.2} e^{0.2x} \right) \right|_0^4 \approx 1.53$$

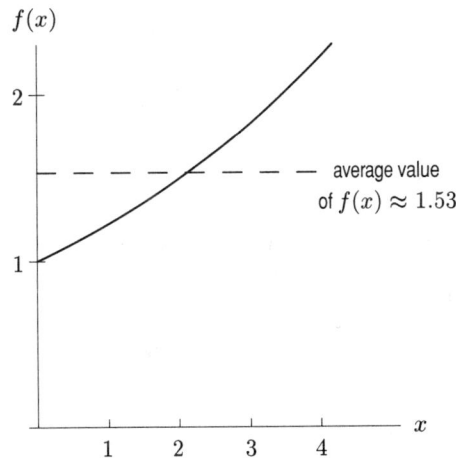

Figure 7.4.194

8. Use the Fundamental Theorem to determine the value of b if the area under the graph of $f(x) = 3x^2 + 1$ between $x = 0$ and $x = b$ is 30. Assume $b > 0$.

ANSWER:

The area is given by

$$\int_0^b (3x^2 + 1) \, dx = \left. (x^3 + x) \right|_0^b = b^3 + b = 30$$

Thus, $b = 3$.

9. Figure 7.4.195 shows the derivative $F'(x)$. If $F(0) = 3$, find the value of $F(4)$, $F(6)$, and $F(10)$.

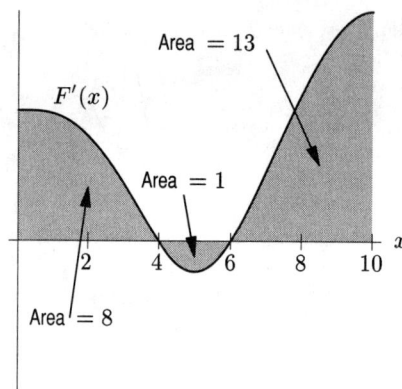

Figure 7.4.195

ANSWER:

By the Fundamental Theorem,

$$\int_0^b F'(x)\,dx = F(b) - F(a),$$

so

$$F(b) = F(0) + \int_0^b F'(x)\,dx.$$

Thus, using the areas from the figure,

$$F(4) = F(0) + \int_0^4 F'(x)\,dx = 3 + 8 = 11$$

$$F(6) = F(0) + \int_0^6 F'(x)\,dx = 3 + 8 - 1 = 10$$

$$F(10) = F(0) + \int_0^{10} F'(x)\,dx = 3 + 8 - 1 + 13 = 23$$

10. A mountain stream runs into a catch basin at the mouth of a canyon. Water can flow out of the catch basin at the maximum rate of 300 cubic feet per second. The rate of flow of the stream is normally much less than that, but for a 10-day period during the spring runoff the rate of flow of the stream is approximately

$$r(n) = 16n(10 - n) + 44$$

cubic feet per second at day n, $0 \le n \le 10$.

(a) On which day (for what value of n) is the rate of flow of the stream at a maximum? What is the peak flow of the stream?

(b) On which day (for what value of n) does water begin to back up in the catch basin?

(c) On which day (for what value of n) is the water level in the catch basin at its highest?

(d) How much water must the catch basin hold in order to prevent flooding?

ANSWER:

(a) Let $(r(n))' = 0$; then we have $n = 5$. So on the 5th day the rate of the flow is maximum and the peak flow of the stream is $r(5) = 400 + 44 = 444$ cubic feet per second.

(b) Let $r(n) = 300$ and solve for n; then we have $n = 2$ and $n = 8$. So at day 2, water begins to back up.

(c) At day 8 the water level is at its highest.

(d)

$$\int_2^8 r(n) - 300\,dn = \int_2^8 16n(10 - n) + 44 - 300\,dn$$

$$= \int_2^8 (160n - 16n^2 - 256)\,dn$$

$$= 16 \int_2^8 (10n - n^2 - 16)\, dn = 16\left(5n^2 - \frac{n^3}{3} - 16n\right)\bigg|_2^8$$
$$= 576.$$

So in order to prevent flooding the catch basin must hold at least 576 cubic feet of water.

Chapter 8 Exam Questions

Problems and Solutions   for Section 8.1

1. A density function for the daily calorie intake of a certain species is given in Figure 8.1.196.

   (a) Find the value of c.
   (b) What percent have daily calorie intake less than 20?
   (c) What percent have daily calorie intake between 40 and 50?

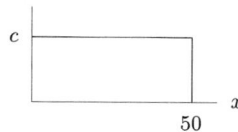

Figure 8.1.196

   ANSWER:

   (a) The area under the curve is $50c$. Since the area under a density function must be 1, we have $50c = 1$ and so $c = \dfrac{1}{50} = 0.02$.
   (b) To find the percent with daily calorie intake under 20, we find the area under the curve between 0 and 20, which is $20(0.02) = 0.4$. So 40% have daily calorie intake less than 20.
   (c) To find the percent with daily calorie intake between 40 and 50, we find the area under the curve between 40 and 50, which is $10(0.02) = 0.2$. So 20% have daily calorie intake between 40 and 50.

2. Scores on a certain state-wide test range from 0 to 100. Sketch a possible graph of the density function in each case below. It is not necessary to put a scale on the vertical axis.

   (a) Scores are equally distributed between 0 and 100.
   (b) There are many high scores and many low scores, with very few scores in the middle range.
   (c) The density function looks like a bell-shaped curve centered at 70. (There are many scores around 70 and fewer and fewer scores as you move away from 70.)

   ANSWER:

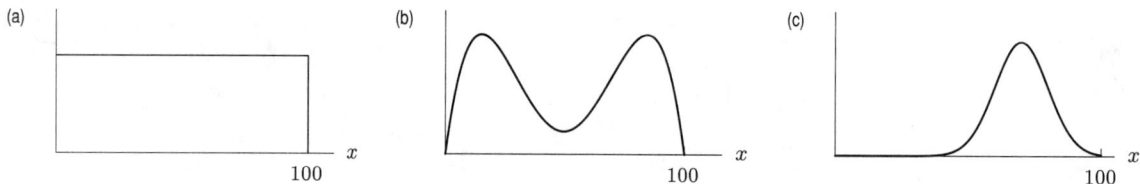

Figure 8.1.197

3. A density function for the lifetime of a certain type of frog is shown in Figure 8.1.198.

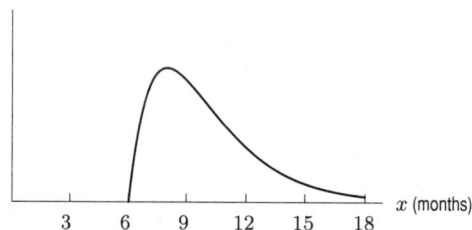

Figure 8.1.198

   (a) What is the most likely lifetime for a frog of this type?

(b) Is the frog's lifetime more likely to be between 6 and 7 months or between 9 and 10 months?

ANSWER:

(a) The most likely lifetime is the highest point on the graph, which is at about 8 months.

(b) Since there is a greater area under the graph between 6 and 7 months than between 9 and 10 months, the frog's lifetime is more likely to be between 6 and 7 months.

4. The distribution of heights, x, in meters, of a group of shrubs is represented by the density function $p(x)$ (no shrubs are higher then 1.5 meters). Calculate the percentage of shrubs which are:

(a) Less than 0.5 meters high.

(b) Between 0.5 and 1 meter high.

(c) More than 1 meter high.

What is the sum of the percentages?

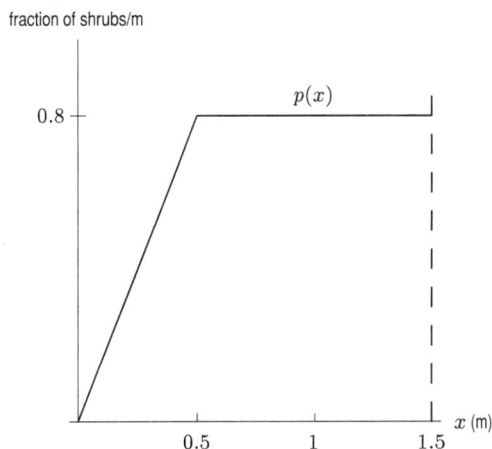

Figure 8.1.199

ANSWER:

Use the areas from the figure to compute the appropriate integrals:

(a)
$$\int_0^{0.5} p(x) = \frac{1}{2}(0.5)(0.8) = 0.2 = 20\%$$

(b)
$$\int_{0.5}^{1} p(x) = (0.5)(0.8) = 0.4 = 40\%$$

(c)
$$\int_1^{1.5} p(x) = (0.5)(0.8) = 0.4 = 40\%$$

$20\% + 40\% + 40\% = 100\%$.

5. Suppose that $p(x)$ is the density function for the ages of U.S. women. What does the statement $p(40) = 0.03$ tell us?

ANSWER:

It means that for some small interval Δt around 40, the fraction of U.S. women with ages in this interval is approximately $p(40)\Delta t = 0.03\Delta t$.

Problems and Solutions for Section 8.2

1. Suppose $p(x)$ is the density function for a certain distribution and $P(x)$ is the cumulative distribution function for the same distribution. Explain how to find the percent of the distribution between $x = 3$ and $x = 5$ using each one of these two formulas.

ANSWER:

The percent of the distribution between $x = 3$ and $x = 5$ is found by finding the area under the graph of $p(x)$ from $x = 3$ to $x = 5$, which amounts to finding the value of the integral $\int_3^5 p(x)\,dx$. Since $P(x)$ is the antiderivative of $p(x)$, this amounts to finding the value of $P(5) - P(3)$.

2. A density function is given in Figure 8.2.200. Sketch the corresponding cumulative distribution function.

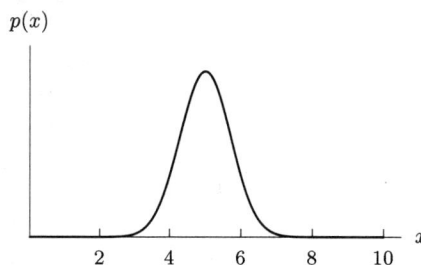

Figure 8.2.200

ANSWER:

This density function is positive between about 3 and about 7, so the cumulative distribution function will be increasing between $x = 3$ and $x = 7$. It is rising most rapidly at $x = 5$. See Figure 8.2.201 for a possible sketch.

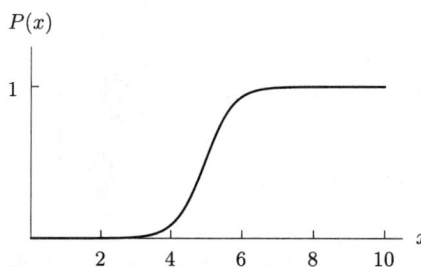

Figure 8.2.201

3. The cumulative distribution function for the time to complete a step on an assembly line is given in Table 8.2.47.

   (a) What fraction of the steps take less than or equal to 10 minutes to complete?
   (b) What fraction take from 12 and 16 minutes to complete?
   (c) What fraction take more than 20 minutes?

Table 8.2.47

t (min)	6	8	10	12	14	16	18	20
$P(t)$	0	0.03	0.15	0.21	0.54	0.78	0.86	0.97

ANSWER:

   (a) By the definition of the cumulative distribution function and from the table, we can see that 0.15, or 15%, of the steps take less than or equal to 10 minutes to complete.
   (b) From the table, we see that 21% of the steps take less than 12 minutes to complete, and that 78% take less than 16 minutes to complete, so that the fraction of steps that take from 12 and 16 minutes to complete must be $78\% - 21\% = 57\%$.
   (c) From the table, we see that 97% of the steps take less than or equal to 20 minutes to complete, so that $100\% - 97\% = 3\%$ of the steps must take more than 20 minutes to complete.

4. Sketch possible cumulative distribution graphs for the total annual sales of the following items. Put the date on the horizontal axis with January at zero and fraction of total annual sales on the vertical axis.

(a) Hot chocolate

(b) Sunscreen

(c) Milk

ANSWER:

(a) Sales are high in the winter, practically stop in the summer, then go up again as winter returns. Thus the fraction of total sales goes up at first, then levels off, then goes up again.

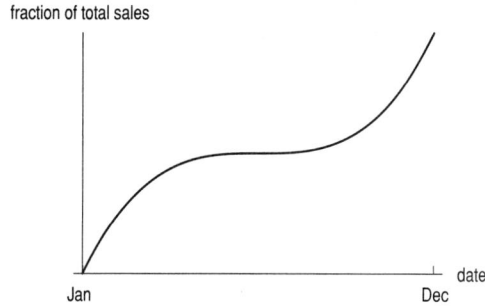

Figure 8.2.202

(b) Sales are about zero in the winter, go way up in the summer, then back to zero again as winter returns. Thus the fraction of total sales starts out near zero for the first few months, then goes up quickly through the summer, then levels off again.

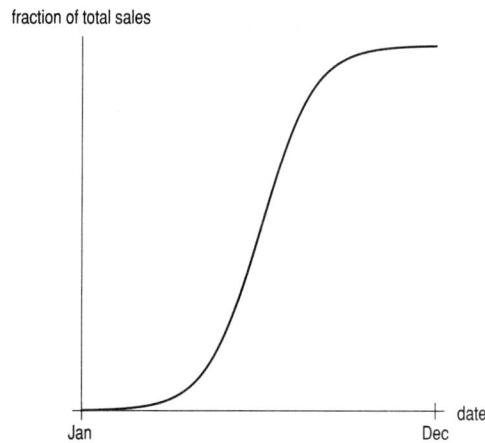

Figure 8.2.203

(c) Sales are about constant throughout the year, so the fraction of total sales goes up at a constant rate.

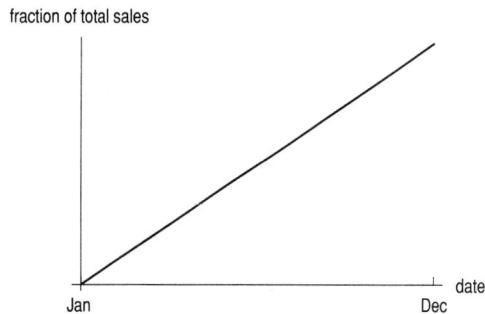

Figure 8.2.204

5. The density function for the height of trees in a forest is given by $p(x) = 0.003x^2$, where x is height in meters and the tallest tree is 10 meters.

   (a) Find the cumulative distribution function, $P(x)$, of the density function.
   (b) Find the probability that a tree is between 5 and 7 meters tall.
   (c) Find the probability that a tree is over 9 meters tall.

   ANSWER:

   (a) The cumulative distribution function is given by

$$P(x) = \int_0^x 0.003t^2 \, dt = 0.001t^3 \Big|_0^x = 0.001x^3.$$

   (b) The probability that a tree is between 5 and 7 meters tall is given by

$$\int_5^7 0.003x^2 \, dx \approx 0.218.$$

   (c) The probability that a tree is over 9 meters tall is given by

$$\int_9^{10} 0.003x^2 \, dx \approx 0.271.$$

Problems and Solutions for Section 8.3

1. The probability of waiting no more than m minutes for a taxi on a certain street corner is

$$prob(t \leq m) = \int_0^m \frac{1}{3} e^{-t/3} dt.$$

   (a) Find the probability of waiting no more than 3 minutes.
   (b) Find the probability of waiting more than 10 minutes.
   (c) Find the median waiting time.
   (d) Find the mean waiting time.

   ANSWER:

   (a) $\text{Prob}(t \leq 3) = \displaystyle\int_0^3 \frac{1}{3} e^{-\frac{t}{3}} dt = 0.632.$

   (b) $\text{Prob}(t > 10) = 1 - \text{Prob}(t \leq 10) = 1 - \displaystyle\int_0^{10} \frac{1}{3} e^{-\frac{t}{3}} dt = 1 + e^{-\frac{t}{3}} \Big|_0^{10} = \frac{1}{e^{\frac{10}{3}}} = 0.036.$

   (c) Let $\displaystyle\int_0^T \frac{1}{3} e^{-\frac{t}{3}} dt = 0.5$ and solve for T. We have the median $T = -3 \ln 0.5 = 2.0794.$
   (d) The mean is

$$\int_0^\infty \frac{t}{3} e^{-\frac{t}{3}} dt = \int_0^\infty -t \, d\left(e^{-\frac{t}{3}}\right)$$

$$= -te^{-\frac{t}{3}} \Big|_0^\infty + \int_0^\infty e^{-\frac{t}{3}} dt$$

$$= -te^{-\frac{t}{3}} \Big|_0^\infty - 3e^{-\frac{t}{3}} \Big|_0^\infty$$

$$= 3.$$

     So the mean waiting time is 3 minutes.

2. The density function for the time to complete a certain task is approximately equal to $p(t) = 0.18e^{-0.18t}$, where t is time in minutes and $0 \le t \le 50$. Find the median and mean time to complete the task and interpret each answer.

ANSWER:

The median is the time m such that $\displaystyle\int_0^m 0.18e^{-0.18t}\, dt = 0.5$. Using trial and error, we see that $m = 3.85$. This means that half the time, the task is completed in less than or equal to 3.85 minutes.

The mean is equal to $\displaystyle\int_0^{50} t(0.18e^{-0.18t})\, dt = 5.55$. This means that if all the different times it could take to complete a task were averaged, the result would be 5.55 minutes.

3. The final exam scores for a calculus course were approximately normally distributed with mean $\mu = 73$ and standard deviation $\sigma = 9$. The maximum possible score was 100.

   (a) What is the probability that a randomly selected student received an A grade (90% or greater)?

   (b) What percentage of students failed the exam (scores of 60 or lower)?

   ANSWER:

   (a) The probability that a randomly selected student received an A grade is given by

$$\int_{90}^{100} \frac{1}{9\sqrt{2\pi}} e^{\frac{-(x-73)^2}{2\cdot 9^2}}\, dx \approx 0.028.$$

   (b) The percentage of students who failed the exam is given by

$$\int_{0}^{60} \frac{1}{9\sqrt{2\pi}} e^{\frac{-(x-73)^2}{2\cdot 9^2}}\, dx \approx 0.074 \text{ or } 7.4\%.$$

4. The lifespan of a bug is approximately normally distributed with mean $\mu = 9$ days and standard deviation $\sigma = 2.5$ days. Assume a maximum possible lifespan of 3 weeks.

   (a) What is the probability of a randomly selected bug living less than a week?

   (b) What is the probability of a randomly selected bug living more than 2 weeks?

   ANSWER:

   (a) The probability of a randomly selected bug living less than a week is given by

$$\int_{0}^{7} \frac{1}{2.5\sqrt{2\pi}} e^{\frac{-(x-9)^2}{2\cdot 2.5^2}}\, dx \approx 0.21.$$

   (b) The probability of a randomly selected bug living more than 2 weeks is given by

$$\int_{14}^{21} \frac{1}{2.5\sqrt{2\pi}} e^{\frac{-(x-9)^2}{2\cdot 2.5^2}}\, dx \approx 0.02$$

Problems and Solutions to Review Problems for Chapter 8 ━━━━

1. Using the following figure, calculate the value of c if p is a density function.

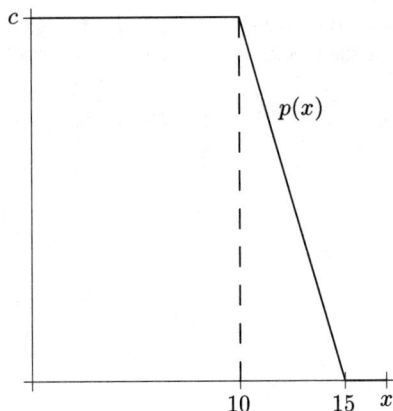

Figure 8.3.205

ANSWER:

We know that $\int_0^{15} p(x)\,dx = 1$. Also, $\int_0^{15} p(x)\,dx$ = the area under the graph of $p(x)$ between 0 and 15. This is a rectangle plus a triangle, so the area $=10c + \frac{1}{2}(5)c = 12.5c$. Thus, $12.5c = 1$, so $c = 1/12.5 = 0.08$.

2. Sketch a possible density function representing the heights of the following groups of people. Use the same horizontal scale on each graph. It is not necessary to put a scale on the vertical axis.

   (a) The group consists of all people in a community.
   (b) The group consists of only the adults in a community.
   (c) The group consists of only the children in a community.

   ANSWER:

   (a)

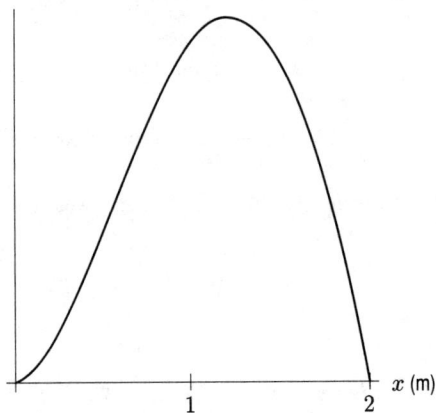

Figure 8.3.206

(b)

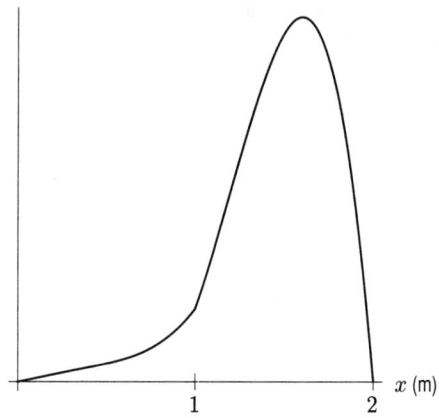

Figure 8.3.207

(c)

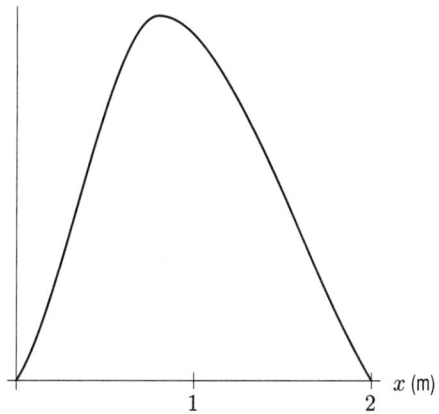

Figure 8.3.208

3. Suppose scores from a standardized test measure from 0 to 100. Pick the graph that best represents the probability density function and the cumulative distribution function if

(a) Scores were equally distributed between 0 and 100.
(b) Most scores were in the middle, with few extremely high or low scores.
(c) There were many high scores, a moderate number of low scores, and few low scores.

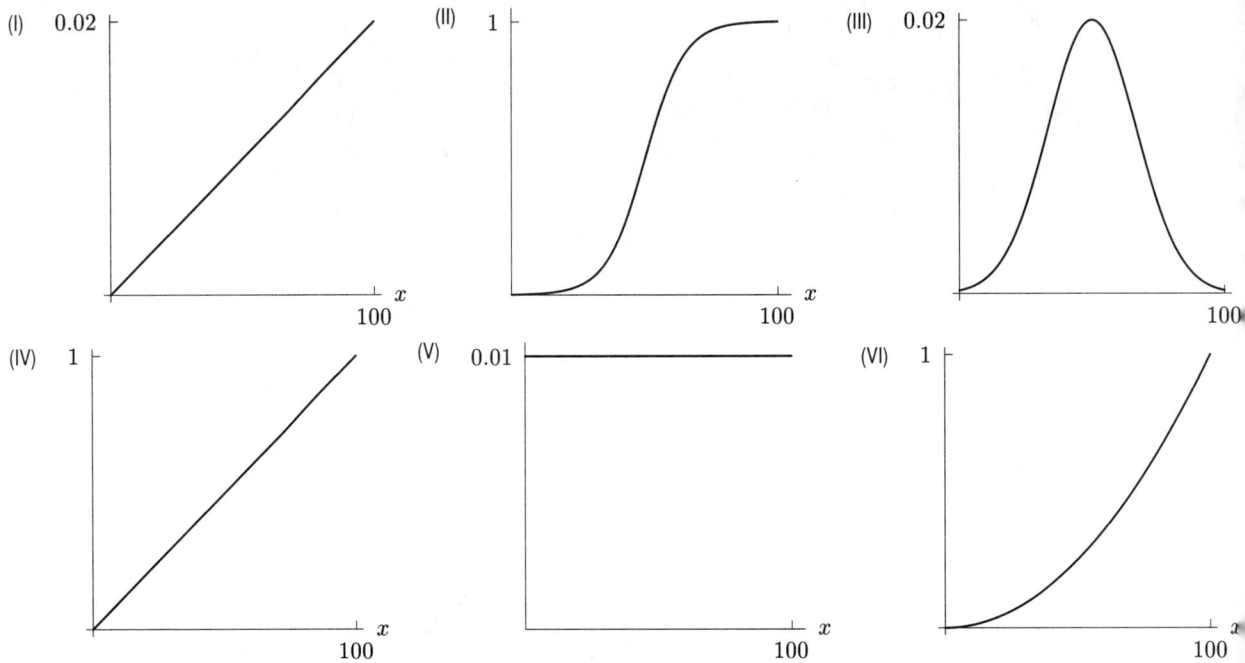

Figure 8.3.209

ANSWER:

(a) Density function: (V)
     Distribution function: (IV)
(b) Density function: (III)
     Distribution function: (II)
(c) Density function: (I)
     Distribution function: (VI)

4. Let $p(t) = 0.2e^{-0.2t}$ be the density function for call-back time by an answering service with t = time in minutes and $0 \le t \le 30$. Calculate and interpret the median and mean.

ANSWER:

The median is given by solving the following for T :

$$\int_0^T 0.2e^{-0.2t}\,dt = 0.5$$

$$-e^{-0.2t}\Big|_0^T = 0.5$$

$$1 - e^{-0.2T} = 0.5$$

$$e^{-0.2T} = 0.5$$

$$T = \frac{\ln 0.5}{-0.2} \approx 3.45 \text{ minutes.}$$

Thus, about half of the calls were returned in 3.45 minutes or less.

The mean is given by the equation

$$\int_0^{30} t0.2e^{-0.2t}\,dt \approx 4.91 \text{ minutes.}$$

Thus, the average call-back time was 4.91 minutes.

5. The speeds of cars on a freeway are approximately normally distributed with a mean $\mu = 78$ mph and standard deviation $\sigma = 5$ mph. Assume a maximum speed of 100 mph.

(a) If speeding tickets are given to cars traveling faster than 80 mph, what is the probability that a randomly selected car is going fast enough to get a ticket?

(b) What fraction of cars are going between 75 and 80 mph?

   ANSWER:

(a) The probability that a randomly selected car is going fast enough to get a ticket is given by

$$\int_{80}^{100} \frac{1}{5\sqrt{2\pi}} e^{\frac{-(x-78)^2}{2\cdot 5^2}} \, dx \approx 0.34.$$

(b) The fraction of cars going between 75 and 80 mph is given by

$$\int_{75}^{80} \frac{1}{5\sqrt{2\pi}} e^{\frac{-(x-78)^2}{2\cdot 5^2}} \, dx \approx 0.38 = \frac{19}{50}.$$

Chapter 9 Exam Questions

Problems and Solutions for Section 9.1

1. The profit, P, from producing a product is expressed as a function of the cost, C, of producing the product and the revenue, R, from selling the product. Do you expect P to be an increasing or decreasing function of C? Of R? Explain.

    ANSWER:

    P is a decreasing function of C because as cost increases, profits will decrease. P is an increasing function of R because as revenue increases, profits will increase.

2. The cost, C, of renting a car is \$30 per day and \$0.20 per mile. Write a cost function $C = f(d, m)$, for the total cost of renting a car where d is the number of days and m is the number of miles. Graph f for $d =$1, 2, and 3. Comment on the appearance of the graph.

    ANSWER:

    $C = 30d + 0.20m.$

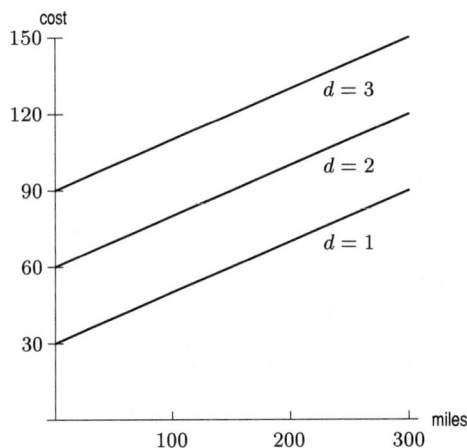

Figure 9.1.210

The costs all rise at the same rate, but have different starting points.

3. The demand for cola on campus is D liters per day. This demand is a function of the price, c, of cola (in dollars/liter) and the price, m, of milk (dollars/liter), and is given by $D(c, m) = 200c - \dfrac{500}{m}$.

    (a) Use the same axes to represent graphically the demand for cola as a function of the price of cola per liter if the price for milk is \$2, \$2.50, and \$3 respectively.

    (b) How do the prices of cola and milk influence the demand for cola?

    ANSWER:

    (a)

    $$D(c, 2) = 200c - 250;$$
    $$D(c, 2.5) = 200c - 200;$$
    $$D(c, 3) = 200c - 167.$$

    So we have the graph in Figure 9.1.211.

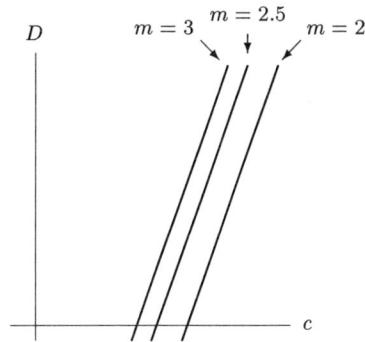

Figure 9.1.211

(b) For a fixed price of milk, increasing the price of cola will increase the demand. For a fixed price of cola, increasing the price of milk will also increase the demand.

4. A company sells two products. The fixed costs for the company are $3000, the variable costs for the first product are $4 per unit, and the variable costs for the second product are $10 per unit.

   (a) If the company produces q_1 units of the first product and q_2 units of the second product, write a formula for the total cost, C, as a function of q_1 and q_2.
   (b) If $C = f(q_1, q_2)$, find $f(500, 400)$ and interpret the result.
   (c) Is C an increasing or decreasing function of q_1? Of q_2?

   ANSWER:

   (a) $C = 3000 + 4q_1 + 10q_2$.
   (b) $f(500, 400) = 3000 + 4(500) + 10(400) = \$9000$. This means that if the company produces 500 units of the first product and 400 units of the second product, the total cost will be $9000.
   (c) C is an increasing function of both q_1 and q_2, since if the company produces more of either product, the total cost will go up.

5. Make a table of values for the cost function given in Problem 4, using $q_1 = 0$, 100, 200 and 300, and $q_2 = 0$, 100, 200 and 300.
   ANSWER:

Table 9.1.48

		q_1		
	0	100	200	300
0	3000	3400	3800	4200
100	4000	4400	4800	5200
200	5000	5400	5800	6200
300	6000	6400	6800	7200

(q_2 labels the rows)

Problems and Solutions for Section 9.2

1. Sketch a contour diagram of $f(x, y) = 2x - y + 1$. Include at least four labeled contours.
   ANSWER:
   If $f(x, y) = 0$, we have $0 = 2x - y + 1$, or $y = 2x + 1$, a line with y-intercept 1 and slope 2. Similarly, if $f(x, y) = 1$, we have $y = 2x$; if $f(x, y) = 2$, we have $y = 2x - 1$; and so on. It can be shown that for $f(x, y) = a$, we have the line $y = 2x + 1 - a$, which has slope 2 and y-intercept $1 - a$. So the contour diagram looks like a set of parallel lines, all with slope 2, as shown in Figure 9.2.212.

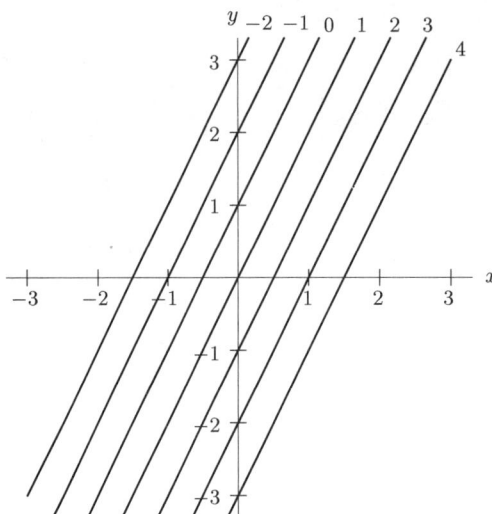

Figure 9.2.212

2. The heat index tells you how hot it feels as a result of the combination of temperature and humidity. Figure 9.2.213 gives a contour diagram for the heat index as a function of temperature and humidity.

   (a) If the humidity is 30%, what temperature feels like 90°F?
   (b) Estimate the heat index if the temperature is 90°F and the humidity is 50%.
   (c) Heat exhaustion is likely to occur when the heat index is 105° or higher. If the temperature is 100°, at about what humidity level does heat exhaustion become likely?
   (d) Fix the humidity level at 50% and consider the heat index as a function of temperature. As the temperature rises, does the heat index rise more rapidly at milder temperatures (70° to 90°) or at very hot temperatures (90° to 110°)? Justify your answer using the contour diagram.
   (e) Is the heat index an increasing or decreasing function of temperature? Is the heat index an increasing or decreasing function of humidity?

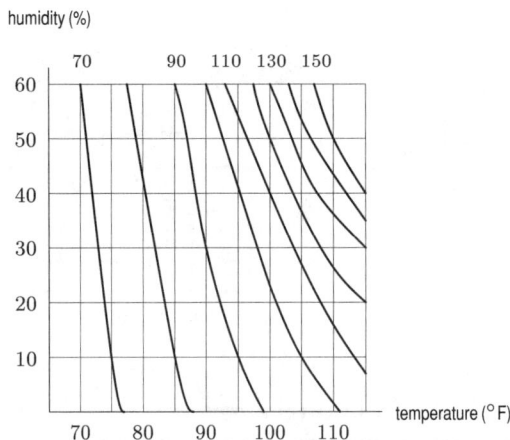

Figure 9.2.213

   ANSWER:
   (a) Since the contour line marked 90 goes through the point $(90, 30)$, 90°F feels like 90°F at 30% humidity.
   (b) The point $(90, 50)$ lies approximately on the contour line 95, so the heat index for a temperature of 90°F and a humidity of 50% is about 95.
   (c) The contour line 105 goes approximately through the point $(100, 32)$, so if the temperature is 100°F, exhaustion becomes likely at about 32% humidity.

(d) Looking along the line $y = 50$ on the contour diagram, we find that the points of intersection of the contour lines with $y = 50$ occur more frequently (are spaced togther more closely) as x increases. This tells us that the heat index rises more rapidly at very hot temperatures than at milder temperatures.

(e) Since the heat index increases for an increase in either humidity or temperature, the heat index is an increasing function of both.

3. Sketch a possible contour diagram (or topographical map) for terrain which consists of one hill, which is very steep on one side and is a gentle rise on the other.

ANSWER:

See Figure 9.2.214 for one possible diagram.

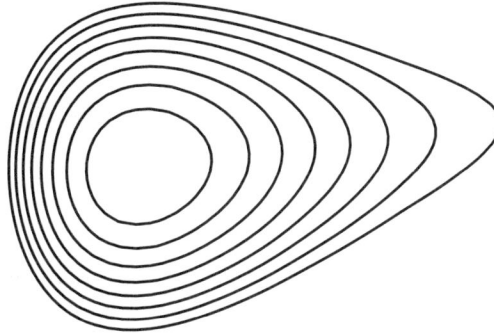

Figure 9.2.214

4. Draw a contour diagram for the function $C = 30d + 0.20m$. Include contours for $C = 30, 60, 90,$ and 120.

ANSWER:

If $C = 30$, we have $30d + .02m = 30$, which is a line with d-intercept 1 and m-intercept 150.

If $C = 60$, we have $30d + .02m = 60$, which is a line with d-intercept 2 and m-intercept 300.

If $C = 90$, we have $30d + .02m = 90$, which is a line with d-intercept 3 and m-intercept 450.

If $C = 120$, we have $30d + .02m = 120$, which is a line with d-intercept 4 and m-intercept 600.

Thus, we get a set of parallel lines, all with slopes of -150.

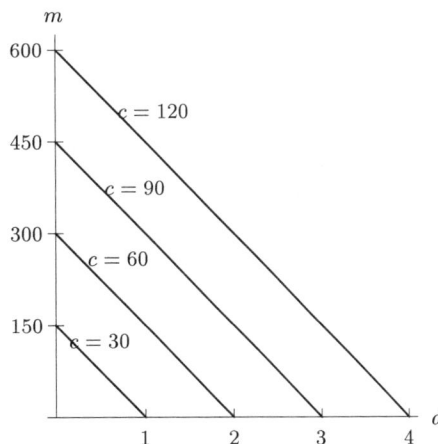

Figure 9.2.215

5. Plants need varying amounts of water and sunlight to grow. The following contour diagrams each show the growth of a different plant as a function of the amount of water and sunlight.

(a) Which diagram would represent a plant that does best in the desert (little water, lots of sunshine)?

(b) Which diagram would represent a plant that does best in a tropical rain forest floor (lots of water, little sunshine)?

(c) Which diagram would represent a plant that does best in a temperate climate (moderate water, moderate sunshine)?

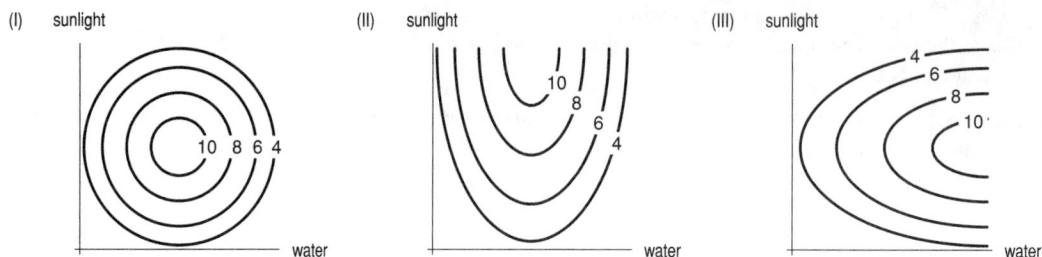

(I) sunlight   (II) sunlight   (III) sunlight

10 8 6 4   10 8 6 4   4 6 8 10

water   water   water

Figure 9.2.216

ANSWER:

(a) II
(b) III
(c) I

Problems and Solutions   for Section 9.3

1. The number of tons, N, of a product produced at a factory in a day is a function of the number of workers, w, and the number of hours worked, h. We have $N = f(w, h)$. Interpret each of the following equations in terms of production at this factory.
   (a) $f(5, 8) = 300$ **(b)** $f_w(5, 8) = 45$ **(c)** $f_h(5, 8) = 28$
   ANSWER:

   (a) This equation means that if the factory has 5 workers who work for 8 hours, it will produce 300 tons of the product.
   (b) This equation tells us that when the factory has 5 workers who work for 8 hours, the instantaneous rate of increase of production relative to number of workers is 45. In other words, adding one additional worker will cause the amount produced to go up by about 45 tons.
   (c) This equation tells us that when the factory has 5 workers who work for 8 hours, the instantaneous rate of increase of production relative to number of hours worked is 28. In other words, making all the workers work for one more hour will cause the amount produced to go up by about 28 tons.

2. A table of values for $f(x, y)$ is given in Table 9.3.49.

   (a) Is f_x positive or negative? Is f_y positive or negative? Explain.
   (b) Estimate $f_x(6, 10)$ and $f_y(6, 10)$.

Table 9.3.49

		x					
		0	2	4	6	8	10
	0	500	510	525	560	590	640
	5	440	450	470	500	540	610
y	10	410	420	445	480	520	575
	15	390	405	430	460	490	525
	20	375	385	410	435	475	500

ANSWER:

(a) Since f increases as x increases, f_x must be positive. Since f decreases as y increases, f_y is negative.

(b) We can estimate $f_x(6, 10)$ by looking at $\dfrac{\Delta f}{\Delta x}$ for the smallest value of Δ that we have. This gives us $f_x(6, 10) \approx$
$\dfrac{520 - 480}{8 - 6} = \dfrac{40}{2} = 20$. Similarly, we have $f_y(6, 10) \approx \dfrac{\Delta f}{\Delta y} = \dfrac{460 - 480}{15 - 6} = \dfrac{-20}{5} = -4$.

3. The following figure is a contour diagram for $z = f(x, y)$. Is $f_x(x, y)$ positive or negative? Is $f_y(x, y)$ positive or negative? Estimate $f(1, 4), f_x(1, 4)$, and $f_y(1, 4)$.

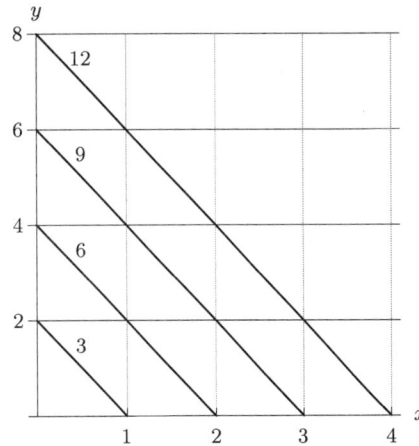

Figure 9.3.217

ANSWER:

$f_x(x, y)$ is positive because as x increases, the z-contour increases. $f_y(x, y)$ is positive because as y increases, the z-contour increases. $f(1, 4)$ is on the contour $z = 9$, so $f(1, 4) = 9$.

To find $f_x(1, 4)$, look at the point (2,4) on the contour $z = 12$. z increases by 3 when x increases by 1. Thus,

$$f_x(1, 4) \approx \frac{\Delta z}{\Delta x} = \frac{3}{1} = 3.$$

To find $f_y(1, 4)$, look at the point (1,6) on the contour $z = 12$. z increases by 3 when y increases by 2. Thus,

$$f_y(1, 4) \approx \frac{\Delta z}{\Delta y} = \frac{3}{2} = 1.5.$$

4. The amount, $A = f(w, h)$, of your paycheck is a function of your hourly wage, w, and the number, h, of hours you work.

(a) Would you expect $f_w(w, h)$ to be positive or negative? Why?

(b) Explain the meaning of the statement $f_h(10, 34) = 10$ in terms of your paycheck.

ANSWER:

(a) $f_w(w, h)$ would be positive, because as w increases, your salary would increase.

(b) This means that an increase of 1 hour worked from 34 hours results in a wage increase of $10.

5. An airline's revenue, $R = f(x, y)$, is a function of the number of full price tickets, x, and the number of discount tickets, y, sold. When 300 full price and 600 discount tickets are sold, $R = \$225,000$. Use the partial derivatives $f_x(300, 600) = 350$ and $f_y(300, 600) = 200$ to estimate revenue when:

(a) $x = 303$ and $y = 600$.

(b) $x = 300$ and $y = 605$.

(c) $x = 305$ and $y = 604$.

ANSWER:

(a) $f_x(300, 600) = 350$ means that at $x = 300$, R increases by 350 for every additional full price ticket sold. Thus, $R \approx 225,000 + 3(350) = \$226,050$.

(b) $f_y(300, 600) = 200$ means that at $y = 600$, R increases by 200 for every additional discount ticket sold. Thus, $R \approx 225,000 + 5(200) = \$226,000$.

(c) $\Delta R \approx$ change due to change in x + change due to change in y

$$= 5(350) + 4(200).$$

Thus $R \approx 225,000 + 5(350) + 4(200) = \$227,550$.

Problems and Solutions for Section 9.4

1. Find f_x and f_y in each of the following cases.
   (a) $f(x, y) = x^2 y^3$ (b) $f(x, y) = x^2 + 5xy + y^2$
   ANSWER:

   (a) $f_x = 2xy^3$ and $f_y = 3x^2 y^2$.
   (b) $f_x = 2x + 5y$ and $f_y = 5x + 2y$.

2. Find $\dfrac{\partial V}{\partial r}$ and $\dfrac{\partial V}{\partial h}$ if $V = \dfrac{4}{3}\pi r^2 h$.
   ANSWER:
   $\dfrac{\partial V}{\partial r} = \dfrac{4}{3}\pi(2r)h = \dfrac{8}{3}\pi rh$ and $\dfrac{\partial V}{\partial h} = \dfrac{4}{3}\pi r^2$.

3. Suppose the Cobb-Douglas production function for a company is given by $P = 50L^{0.70}K^{0.40}$, where P is production in tons, L is the number of workers, and K is the capital investment, in thousands of dollars.

   (a) What is the production if the company has a capital investment of 25 thousand dollars and the company employs 15 workers?
   (b) Find $\dfrac{\partial P}{\partial L}$ and $\dfrac{\partial P}{\partial K}$ at the values given in part (a), and interpret them in terms of production at this company.
   ANSWER:

   (a) Using the function given, we have $P = 50(15)^{0.70}(25)^{0.40} = 1206.2$ tons produced.

   (b) We have $\dfrac{\partial P}{\partial L} = 50(0.70L^{-0.30})K^{0.40} = 35\dfrac{K^{0.40}}{L^{0.30}}$. Substituting $L = 15$ and $K = 25$, we get $\dfrac{\partial P}{\partial L} = 35\dfrac{(25)^{0.40}}{(15)^{0.30}} =$
   56.3 tons per worker. This means that with a capital investment of 25 thousand dollars and 15 workers at the company, if the amount of workers at the company increased by 1, the amount produced would increase by about 56.3 tons.
   We also have $\dfrac{\partial P}{\partial K} = 50L^{0.70}(0.40K^{-0.60}) = 20\dfrac{L^{0.70}}{K^{0.60}}$. Substituting, we get $\dfrac{\partial P}{\partial K} = 20\dfrac{(15)^{0.70}}{(25)^{0.60}} = 19.3$ tons per thousand dollars. This means that with a capital investment of 25 thousand dollars and 15 workers at the company, if an additional thousand dollars was spent on capital, the amount produced would increase by about 19.3 tons.

4. The volume of a cylinder is given by $V = f(r, h) = \pi r^2 h$, where r is the radius of the base and h is the height, both in centimeters. Find $f(5, 8)$, $f_r(5, 8)$, and $f_h(5, 8)$. Interpret each in terms of volume.
   ANSWER:
   $f(5, 8) = \pi(5^2)(8) = 200\pi$ cm^3 is the volume of a cylinder with radius 5 cm and height 8 cm.
   $f_r(r, h) = 2\pi rh$, so $f_r(5, 8) = 2\pi(5)(8) = 80\pi$ cm^3. Thus, at a radius of 5 cm and height of 8 cm, each 1 cm increase in the radius increases the volume by approximately 80π cm^3.
   $f_h(r, h) = \pi r^2$, so $f_h(5, 8) = 25\pi$ cm^3. Thus, at a radius of 5 cm and height of 8 cm, each 1 cm increase in the height increases the volume by approximately 25π cm^3.

5. For $f(x, y) = \dfrac{x^2}{y}$, calculate all four partial second-order derivatives and confirm that the mixed partials are equal.
   ANSWER:

$$f_x(x, y) = \frac{2x}{y} \quad f_y(x, y) = \frac{-x^2}{y^2}$$

$$f_{xx}(x, y) = \frac{2}{y} \quad f_{xy}(x, y) = \frac{-2x}{y^2} = f_{yx}(x, y) \quad f_{yy}(x, y) = \frac{2x^2}{y^3}$$

Problems and Solutions for Section 9.5

1. A contour diagram of $f(x, y)$ is given in Figure 9.5.218. List the obvious critical points on this region. Which points are at a local maximum? Which points are at a local minimum? What are the global maximum and global minimum on this region?

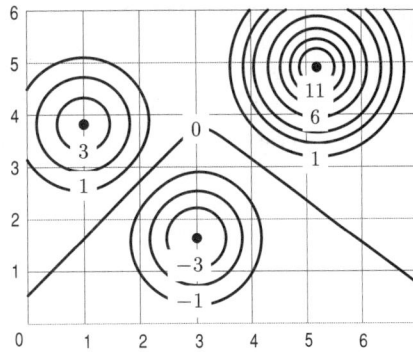

Figure 9.5.218

ANSWER:

Obvious critical points are at the points $(1, 3.8)$, $(3, 1.6)$, and $(5.2, 4.9)$. (Another possible critical point is at $(3, 3.8)$.) The function has local maxima at the points $(1, 3.8)$ and $(5.2, 4.9)$, and a local minimum at $(3, 1.6)$. The global maximum is about 12 and the global minimum is about -4.

2. Find all the critical points of $f(x, y) = x^3 + y^2 - 2x^2 + 3y - 8$ and determine whether each is a local maximum, local minimum, or neither.

ANSWER:

To find the critical points, set both partial derivatives equal to zero:

$$f_x(x, y) = 3x^2 - 4x = 0$$

$$f_y(x, y) = 2y + 3 = 0$$

Solving this system gives the critical points (0,-3/2), (4/3,-3/2). To check whether each is a local maximum, local minimum, or neither, compute the second-order partial derivatives:

$$f_{xx}(x, y) = 6x - 4 \qquad f_{xy}(x, y) = f_{yx}(x, y) = 0 \qquad f_{yy}(x, y) = 2$$

At (0,-3/2),

$$D = f_{xx}f_{yy} - f_{xy}^2 = -8 < 0.$$

Thus, (0,-3/2) is neither a local maximum nor a local minimum.

At (4/3,-3/2),

$$D = f_{xx}f_{yy} - f_{xy}^2 = 8 > 0.$$

Thus, (4/3,-3/2) is a local minimum.

3. Two products are manufactured in quantities q_1 and q_2 and sold at prices \$8 and \$12, respectively. The cost of producing them is given by

$$C = q_1^2 + 2q_2^2 + 8.$$

Find the maximum profit that can be made.

ANSWER:

Profit = revenue - cost, so we have

$$P = 8q_1 + 12q_2 - (q_1^2 + 2q_2^2 + 8)$$

$$= 8q_1 + 12q_2 - q_1^2 - 2q_2^2 - 8.$$

Finding the partial derivatives and setting them equal to zero,

$$P_{q_1} = 8 - 2q_1 = 0, \text{ so } q_1 = 4$$

$$P_{q_2} = 12 - 4q_2 = 0, \text{ so } q_2 = 3$$

Using the second derivative test,

$$P_{q_1 q_1} = -2 \quad P_{q_1 q_2} = P_{q_2 q_1} = 0 \quad P_{q_2 q_2} = -4$$

Thus, $D = (-2)(-4) - 0 = 8 > 0$. So P has a local maximum at (4,3). $P(4, 3) = \$26$.

4. Find all critical points of $f(x, y) = x^2 + 3xy - 12y$. Determine whether each critical point is a local maximum, a local minimum, or neither.

ANSWER:

First we find the partial derivatives and set them equal to 0. We have $f_x = 2x + 3y = 0$ and $f_y = 3x - 12 = 0$. The second equation tells us that x must be 4, and substituting $x = 4$ into the first equation gives us $(4, -\frac{8}{3})$ as the only critical point of f. To determine what type of critical point this is, we look at the discriminant D, which is equal to $f_{xx}f_{yy} - (f_{xy})^2 = (2)(0) - 3^2$, which is less than 0. So this critical point is a saddle point. (It is also possible to use a table of values instead of the discriminant to determine this fact.)

5. Suppose $f(x, y) = x^2 + Ax + y^2 + By + C$. What values of A, B and C give $f(x, y)$ a local minimum value of 12 at the point $(2, 3)$?

ANSWER:

At a local minimum, $f_x = 0$, so $2x + A = 0$ when $x = 2$. Thus $4 + A = 0$ and so $A = -4$. Also, at a local minimum, $f_y = 0$, so $2y + B = 0$ when $y = 3$. Thus $6 + B = 0$ and so $B = -6$. So $f(x, y) = x^2 - 4x + y^2 - 6y + C$. To find C, we use the fact that $f(2, 3) = 12$; substituting gives us the equation $12 = 2^2 - 4(2) + 3^2 - 6(3) + C$, and solving this equation yields $C = 25$.

Problems and Solutions   for Section 9.6

1. Use Lagrange multipliers to find the maximum or minimum values of $f(x, y) = 3xy$, subject to the constraint $4x + y = 50$.

ANSWER:

Use $g(x) = 4x + y = 50$. Differentiating gives

$$f_x(x, y) = 3y \quad f_y(x, y) = 3x$$

$$g_x(x, y) = 4 \quad g_y(x, y) = 1$$

leading to the equations

$$3y = \lambda(4)$$
$$3x = \lambda(1)$$
$$4x + y = 50.$$

Solving these gives $\lambda = 18.75$, $x = 6.25$, and $y = 25$. For these values, $f(x, y) = 468.75$.

2. Use Lagrange multipliers to find the maximum or minimum values of $f(x, y) = x^2 + y^2$, subject to the constraint $2x - y = 5$.

ANSWER:

Use $g(x) = 2x - y = 5$. Differentiating gives

$$f_x(x, y) = 2x \quad f_y(x, y) = 2y$$

$$g_x(x, y) = 2 \quad g_y(x, y) = -1$$

leading to the equations

$$2x = \lambda(2)$$
$$2y = \lambda(-1)$$
$$2x - y = 5.$$

Solving these gives $\lambda = 2$, $x = 2$, and $y = -1$. For these values, $f(x, y) = 5$.

3. A company operates two plants that make the same product. If the two plants produce quantities x and y of the product, the cost functions are $C_1 = 8.5 + 0.03x^2$ and $C_2 = 5.2 + 0.04y^2$. The total market demand $q = x + y$ for the product is related to the selling price p by $p = 60 - 0.04q$. How much should each plant produce to maximize the company's profit?

ANSWER:

The profit is

$$\pi = q \cdot p - (C_1 + C_2).$$

Thinking q and p are functions of x and y, we have

$$\pi_x = p - 0.04q - 0.006x$$
$$\pi_y = p - 0.04q - 0.08y.$$

Let $\pi_x = \pi_y = 0$ and solve for x and y; we have $x = 652$, $y = 49$. So the profit will be maximum when $x = 652$ and $y = 49$.

4. The production function for a company is $P = 300x^{0.75}y^{0.25}$, where P is the amount produced given x units of labor and y units of equipment. Each unit of labor costs $1000 and each unit of equipment costs $300.

   (a) Assume the goal of the company is to maximize production given a fixed budget of $50,000. What is the function to be optimized? What is the constraint equation? What is the meaning of the Lagrange multiplier λ in this situation?

   (b) Now assume that the goal of the company is to minimize costs at a fixed production goal of 8000 units produced. What is the function to be optimized? What is the constraint equation? What is the meaning of the Lagrange multiplier λ in this situation?

   ANSWER:

   (a) In this case, the company wishes to maximize $f(x, y) = 300x^{0.75}y^{0.25}$ given the constraint equation $1000x + 300y = 50,000$. The Lagrange multiplier represents the additional units that can be produced if the budget is increased by $1.

   (b) In this case, the company wishes to minimize $f(x, y) = 1000x + 300y$ given the constraint equation $300x^{0.75}y^{0.25} = 8000$. The Lagrange multiplier represents the amount it would cost the company to produce one more unit.

5. We wish to find the maximum value of $f(x, y) = 8xy$ subject to the constraint equation $2x + y = 20$.

   (a) First, we'll try trial and error. If $x = 1$, what is y? What is $f(x, y)$? If $x = 4$, what is y? What is $f(x, y)$? If $x = 7$, what is y? What is $f(x, y)$?

   (b) Now, use the method of Lagrange multipliers to find the values of x and y that make $f(x, y)$ maximum subject to the given constraint. What is the maximum value of $f(x, y)$?

   ANSWER:

   (a) In each of these cases we find y using the constraint equation $2x + y = 20$, and the value of $f(x, y)$ using $f(x, y) = 8xy$. See Table 9.6.50.

Table 9.6.50

x	y	$f(x, y)$
1	18	144
4	12	384
7	6	336

   (b) The three equations are $8y = \lambda \cdot 2$, $8x = \lambda \cdot 1$, and $2x + y = 20$. From the first two equations, we have $8x = 4y$, so $y = 2x$. Substituting this result into the third equation, we have $2x + (2x) = 20$, so $x = 5$. Therefore the maximum value of f occurs at $x = 5$ and $y = 10$ (using the constraint equation to find y), and the maximum value is $8(5)(10) = 400$.

6. A mirror is formed by joining a half-circle to a rectangle, as in Figure 9.6.219. If the total circumference of the mirror is 200 cm, what should the radius of the half circle be for the area of the mirror to be a maximum?

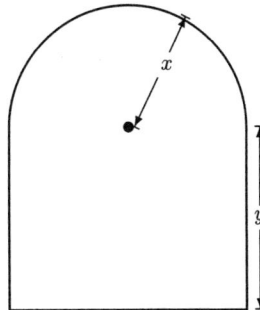

Figure 9.6.219

ANSWER:

We have the total area of the mirror:

$$A(x, y) = 2xy + \frac{\pi x^2}{2},$$

and the constraint:

$$200 = 2y + 2x + \pi x.$$

Solve $\begin{cases} 2y + \pi x = \lambda(2 + \pi) \\ 2x = 2\lambda \\ 200 = 2y + 2x + \pi x \end{cases}$

We have $x = y = 200/(4 + \pi)$. Therefore the radius should be $200/(4 + \pi) \approx 28$cm.

Problems and Solutions to Review Problems for Chapter 9 ━━━━━━━━━━

1. A television salesman earns a fixed salary of $8.50 per hour plus a $20 commission for each television he sells.

   (a) If h is the number of hours worked and s is the number of televisions sold, write a formula for $E(h, s)$, his total earnings.
   (b) Find and interpret $E(40, 31)$.
   (c) Is E an increasing or decreasing function of h? Of s?

   ANSWER:

   (a) Total earnings = hourly wages + commissions, so we get

   $$E = 8.5h + 20s.$$

   (b) $E(40, 31) = 8.5(40) + 20(31) = \$960$. This is his earnings for working 40 hours and selling 31 televisions.
   (c) E is an increasing function of both h and s because either working more hours or selling more televisions will increase earnings.

2. Sketch a possible contour diagram showing the population density of a region in a city where the center is

   (a) A garbage dump.
   (b) The city center.

   Justify your answer.

   ANSWER:

   One possible diagram:

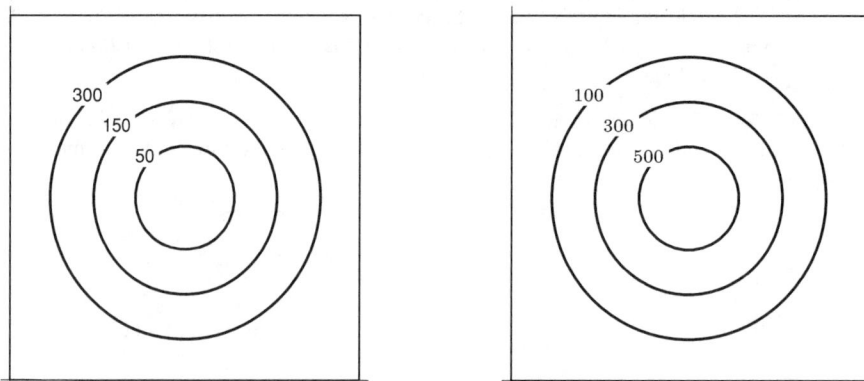

Figure 9.6.220

   (a) Not many people would want to live near a garbage dump, so the population is low in the center and increases farther out.
   (b) The population would likely be highest in the city center, but decrease as you moved farther out into the suburbs.

3. The monthly car payment for a new car is a function of three variables

$$P = f(A, r, t)$$

where A is the amount borrowed in dollars, r is the annual interest rate, and t is time in months before the car is paid off.

(a) Interpret, in financial terms, the meaning of the statement $f(20{,}000, 0.09, 48) = 498$.

(b) Would you expect each of the three partial derivatives to be increasing of decreasing? Why?

ANSWER:

(a) $f(20{,}000, 0.09, 48) = 498$ means that the monthly payment for a car costing $20,000 financed for 48 months at an annual rate of 9% would be $498.

(b) $\dfrac{\partial P}{\partial A}$ would be positive, because as the amount financed increases, the payments increase.

$\dfrac{\partial P}{\partial r}$ would be positive, because as the interest rate increases, the payments increase.

$\dfrac{\partial P}{\partial t}$ would be negative, because as the number of months the car is finances increases, the payments decrease.

4. If $f(x, y) = x^2 y + xy^2 + 3x$, find $f_x(x, y)$ and $f_y(x, y)$.

ANSWER:

$$f_x(x, y) = 2xy + y^2 + 3$$
$$f_y(x, y) = x^2 + 2xy$$

5. If $P = 5e^{-rt}$, find all 4 second order partial derivatives and confirm that the mixed partials are equal.

ANSWER:

$$\frac{\partial P}{\partial r} = -5te^{-rt} \qquad\qquad\qquad \frac{\partial P}{\partial t} = -5re^{-rt}$$

$$\frac{\partial^2 P}{\partial r^2} = 5t^2 e^{-rt} \qquad\qquad\qquad \frac{\partial^2 P}{\partial t^2} = 5r^2 e^{-rt}$$

$$\frac{\partial^2 P}{\partial r \partial t} = (-5r)(-te^{-rt}) + (-5)(e^{-rt}) = 5(rt - 1)e^{-rt}$$

$$\frac{\partial^2 P}{\partial t \partial r} = (-5t)(-re^{-rt}) + (-5)(e^{-rt}) = 5(rt - 1)e^{-rt} = \frac{\partial^2 P}{\partial r \partial t}$$

6. Find all the critical points of $f(x, y) = x^2 + 4xy + y^2 + 2$ and determine whether each is a local maximum, local minimum, or neither.

ANSWER:

To find the critical points, set both partial derivatives equal to zero:

$$f_x(x, y) = 2x + 4y = 0$$

$$f_y(x, y) = 4x + 2y = 0$$

Solving this system gives $x = 0, y = 0$, so f has only one critical point: (0,0). To check whether it is a local maximum, local minimum, or neither, compute the second-order partial derivatives:

$$f_{xx}(x, y) = 2 \qquad\quad f_{xy}(x, y) = f_{yx}(x, y) = 4 \qquad\quad f_{yy}(x, y) = 2$$

$$D = f_{xx}f_{yy} - f_{xy}^2 = 2 \cdot 2 - 4^2 = -12 < 0.$$

Thus, (0,0) is neither a local maximum nor a local minimum.

7. A company manufactures x units of one item and y units of another. The total cost, C, in dollars, of producing these two items is given by the function

$$C = 3x^2 + xy + 4y^2 + 1000.$$

(a) Use Lagrange multipliers to find the minimum cost subject to the constraint that at least 100 items (total) must be produced.

(b) Give the value of the Lagrange multiplier, λ, and interpret it in terms of cost.

ANSWER:

(a) Using $f(x, y) = C$ and $g(x, y) = x + y = 1000$, we get the following partial derivatives:

$$f_x = 6x + y \quad f_y = x + 8y$$
$$g_x = 1 \quad g_y = 1$$

This leads to the system of equations

$$6x + y = \lambda$$
$$x + 8y = \lambda$$
$$x + y = 100$$

Which gives the solutions $x = 58.3$ and $y = 41.7$. Rounding to whole numbers of items, $x = 58$ and $y = 42$, which gives $C = \$20,584$.

(b) From the equations in part (a), $\lambda = 391.7$. This means that when the production quota of 100 is raised by 1 unit, cost will increase by about $391.70.

Problems and Solutions on Deriving the Formula for a Regression Line ⎯⎯⎯⎯

1. Find the least squares line for the five points $(1, 3)$, $(4, 5)$, $(2, 4)$, $(2, 3)$, and $(0, 2)$ by following the steps given below.

(a) Find each of the following four sums:

(i) $\quad SX = \displaystyle\sum_{i=1}^{4} x_i$

(ii) $\quad SX = \displaystyle\sum_{i=1}^{4} y_i$

(iii) $\quad SXX = \displaystyle\sum_{i=1}^{4} x_i^2$

(iv) $\quad SXY = \displaystyle\sum_{i=1}^{4} x_i y_i$

(b) Substitute the values for these sums into the following equations. Recall that n is the number of data points, so that $n = 5$ in this case.

(i) $\quad b = \dfrac{(SXX)(SY) - (SX)(SXY)}{n(SXX) - (SX)^2}$

(ii) $\quad m = \dfrac{n(SXY) - (SX)(SY)}{n(SXX) - (SX)^2}$

(c) Substitute the values for b and m from part (b) into the equation for the least squares line $y = b + mx$. Graph the five points and the least squares line together.

ANSWER:

(a)   (i) $SX = 1 + 4 + 2 + 2 + 0 = 9$.

(ii) $SY = 3 + 5 + 4 + 3 + 2 = 17$.

(iii) $SXX = 1^2 + 4^2 + 2^2 + 2^2 + 0^2 = 1 + 16 + 4 + 4 + 0 = 25$.

(iv) $SXY = 1 \cdot 3 + 4 \cdot 5 + 2 \cdot 4 + 2 \cdot 3 + 0 \cdot 2 = 3 + 20 + 8 + 6 + 0 = 37$.

(b)   (i) $b = \dfrac{(25)(17) - (9)(37)}{(5)(25) - (9)^2} = \dfrac{425 - 333}{125 - 81} = \dfrac{92}{44} = 2.09$.

(ii) $m = \dfrac{(5)(37) - (9)(17)}{(5)(25) - (9)^2} = \dfrac{185 - 153}{44} = \dfrac{32}{44} = 0.73$.

(c) We get $y = 2.09 + 0.73x$. See Figure 9.6.221.

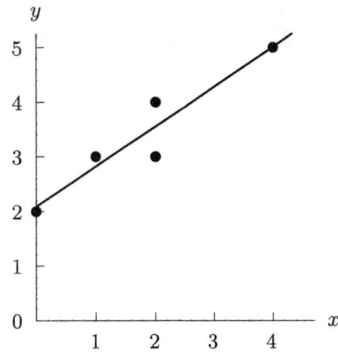

Figure 9.6.221

Chapter 10 Exam Questions

Problems and Solutions   for Section 10.1 ━━━━━━━━━━━━━━━━━

1. Match the graphs in Figure 10.1.222 with the following descriptions:

   (a) The speed of a car merging onto the freeway.
   (b) The temperature of a glass of soda that has ice added to it and then is left to sit in the sun.
   (c) The temperature in a large city over the course of a year.
   (d) The amount of a natural resource remaining in the earth, if it is being consumed at an increasing rate.

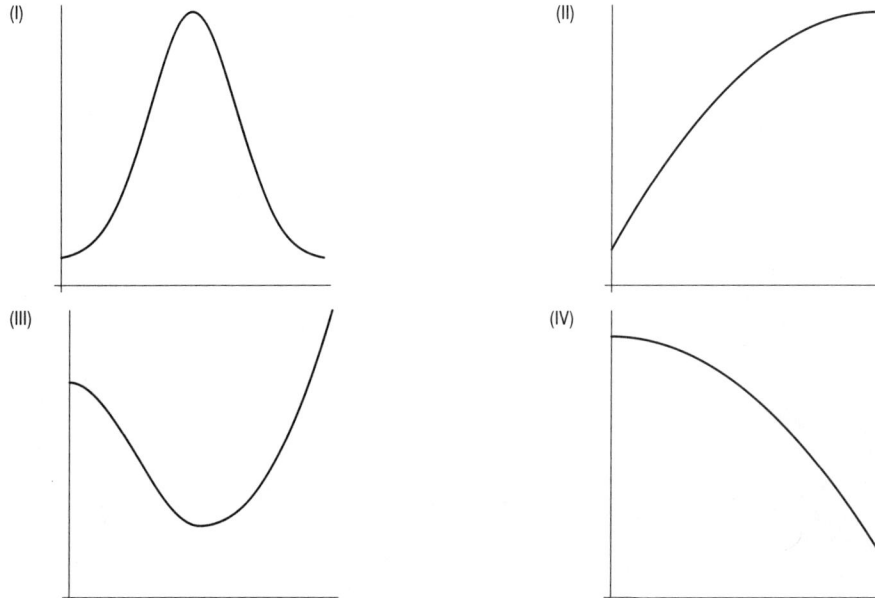

Figure 10.1.222

ANSWER:

(a) II
(b) III
(c) I
(d) IV

2. A population of rodents grows at a rate proportional to the size of the population. Write a differential equation for the size of the population, P, as a function of time, t. Is the constant of proportionality positive or negative? Explain your reasoning.
   ANSWER:
   The rate of change of P is proportional to P, so

   $$\frac{dP}{dt} = kP.$$

   Since the population is growing, dP/dt is positive. Thus k is also positive.

3. A bank account initially containing $1000 earns interest at a continuous rate of 3% per year. Deposits are made into the account at a constant rate of $500 per year. Write a differential equation for the balance, B, in the account as a function of time, t, where t is time in years.
   ANSWER:
   The balance, B, is increasing at a proportional rate of 3% per year, plus the additional deposits. Thus,

   $$\frac{dB}{dt} = 0.03B + 500.$$

4. A drug is administered intravenously to a patient at a rate of 10 mg per day. About 40% of the drug in the patient's body is metabolized and leaves the body each day. Write a differential equation for the amount of the drug, D, in the body as a function of time, t, in years.

ANSWER:

The amount of the drug, D, is increasing at a constant rate of 10 mg per day and decreasing at a rate of $0.40D$ per day. Thus,

$$\frac{dD}{dt} = 10 - 0.40D.$$

5. A quantity Q satisfies the differential equation

$$\frac{dQ}{dt} = 10 - 3Q.$$

(a) Is Q increasing or decreasing at $Q = 0$? At $Q = 10$?
(b) For what value of Q is the rate of change equal to zero?

ANSWER:

(a) At $Q = 0$, $dQ/dt = 15 > 0$, so Q is increasing. At $Q = 10$, $dQ/dt = -15 < 0$, so Q is decreasing.
(b) $dQ/dt = 0$ when $15 - 3Q = 0$, or when $Q = 5$.

Problems and Solutions   for Section 10.2

1. In Kenya, the population P for the recent past has obeyed the growth model $\dfrac{dP}{dt} = ktP$, with t the number of years since 1990.

(a) Show that $P = Ae^{kt^2/2}$ is a solution of this differential equation.
(b) Express P as a function of t, given that the population of Kenya in 1990 was 24.64 million and in 1992 was 26.16 million.
(c) Estimate the population of Kenya today.

ANSWER:

(a) $\dfrac{d}{dt}\left(Ae^{kt^2/2}\right) = Ae^{kt^2/2} \cdot (kt) = ktP.$

(b) We solve for A and k:

$$24.64 = A$$
$$26.16 = Ae^{2k}.$$

We have $A = 24.64$ and $k = 0.03$. So $P = 24.64e^{0.015t^2}$.

(c) For 1998, let $t = 8$; then $P(8) = 24.64e^{0.015 \times 64} = 64.35$ million people. The population of Kenya according to this model is about 64.35 million in 1998.

2. (a) What conditions are put on $y = Ce^{kt}$ if this function is a solution to the differential equation $\dfrac{dy}{dt} = 2y$? Show your work to justify your answer.
(b) What additional conditions are put on if $y = 25$ when $t = 0$?

ANSWER:

(a) Since $\dfrac{dy}{dt} = 2y$, we have that $\dfrac{d(Ce^{kt})}{dt} = Ce^{kt} \cdot k = 2Ce^{kt}$. Thus k must equal 2.
(b) If $y = 25$ when $t = 0$, then we have $25 = Ce^{2(0)}$, or $C = 25$. So we know that the function must be $y = 25e^{2t}$.

3. Match the following solutions to the following differential equations. A differential equation may have no solutions or more than one solution.

(a) $y = 3x$ (b) $y = e^{3x}$ (c) $y = x$ (d) $y = 3e^x$ (e) $y = x^3$

(i) $\dfrac{dy}{dx} = y$ (ii) $\dfrac{dy}{dx} = 3x$ (iii) $\dfrac{dy}{dx} = 3y$ (iv) $\dfrac{dy}{dx} = \dfrac{y}{x}$ (v) $\dfrac{dy}{dx} = 3\dfrac{y}{x}$

ANSWER:

We can "work backwards" by finding $\dfrac{dy}{dx}$ for each solution. For (a), we have $\dfrac{dy}{dx} = 3 = \dfrac{y}{x}$; for (b), we have $\dfrac{dy}{dx} = 3e^{3x} = 3y$; for (c), $\dfrac{dy}{dx} = 1 = \dfrac{y}{x}$ again; for (d), $\dfrac{dy}{dx} = 3e^x = y$; and for (e), $\dfrac{dy}{dx} = 3x^2 = 3\dfrac{x^3}{x} = 3\dfrac{y}{x}$. So we match the solutions to the equations as follows: (a) and (c) with (iv), (b) with (iii), (d) with (i), (e) with (v), and none with (ii).

4. Find the values of k (if there are any) for which $y = x^2 + k$ is a solution to the differential equation $2y - x\dfrac{dy}{dx} = 10$.

ANSWER:

We substitute $x^2 + k$ for y in the equation and solve for k:

$$2(x^2 + k) - x\frac{d(x^2 + k)}{dx} = 10$$
$$2(x^2 + k) - x(2x) = 10$$
$$2x^2 + 2k - 2x^2 = 10$$
$$2k = 10$$
$$k = 5.$$

5. Determine whether each of the following is a solution to the differential equation
$x\dfrac{dy}{dx} - 3y = 0$. Show your work to justify your answer.

(a) $y = x^2$ **(b)** $y = x^3$

ANSWER:

(a) First we differentiate $y = x^2$ with respect to x, to get $\dfrac{dy}{dx} = 2x$. We can now write the differential equation $x\dfrac{dy}{dx} - 3y = 0$ in terms of x and check whether the solution makes sense. Substituting x^2 for y and $2x$ for $\dfrac{dy}{dx}$, we get

$$x(2x) - 3(x^2) = 0$$
$$2x^2 - 3x^2 = 0,$$

which makes no sense. So $y = x^2$ is not a solution to the differential equation.

(b) This is done in the same manner as part (a). Differentiating with respect to x gives $\dfrac{dy}{dx} = 3x^2$, and substituting gives

$$x(3x^2) - 3(x^3) = 0$$
$$3x^3 - 3x^3 = 0,$$

which is correct. So $y = x^3$ is a solution to the differential equation.

6. (a) Find the general solution of the differential equation $\dfrac{dy}{dx} = -4x + 3$.
  (b) Find the solutions satisfying $y(1) = 5$ and $y(1) = 0$.

ANSWER:

(a) $y = \displaystyle\int (-4x + 3)\, dx$, so the solution is $y = -2x^2 + 3x + C$.

(b) If $y(1) = 5$, we have $-2(1)^2 + 3(1) + C = 5$ and so $C = 4$. Thus we have the solution $y = -2x^2 + 3x + 4$.

If $y(1) = 0$, we have $-2(1)^2 + 3(1) + C = 0$ and so $C = -1$. Thus we have the solution $y = -2x^2 + 3x - 1$.

7. Find the solutions of the initial value problems:

(a) $\dfrac{dK}{dt} = 3 - \cos 2t$ when $K(0) = -10$.

(b) $\dfrac{dP}{dt} = -10e^{-t}$ when $P(0) = 20$.

ANSWER:

(a) $K = \displaystyle\int (3 - \cos 2t)\, dt = 3t - \dfrac{\sin 2t}{2} + C$. If $K(0) = -10$, then $3(0) - \dfrac{\sin(2 \cdot 0)}{2} + C = -10$ and $C = 10$.

Thus $K = 3t - \dfrac{\sin 2t}{2} - 10$.

(b) $P = \displaystyle\int -10e^{-t}\, dt = 10e^{-t} + C$. If $P(0) = 20$, then $10e^0 + C = 20$ and $C = 10$. Thus $P = 10e^{-t} + 10$.

8. Show that $y = 3\cos 3t$ satisfies $\dfrac{d^2y}{dt^2} + 9y = 0$.

ANSWER:

If $y = 3\cos 3t$, then $\dfrac{dy}{dt} = -9\sin 3t$, and $\dfrac{d^2y}{dt^2} = -27\cos 3t$.

So $\dfrac{d^2y}{dt^2} + 9y = -27\cos 3t + 9(3\cos 3t) = 0$.

Problems and Solutions for Section 10.3

1. Sketch the slope field for the differential equation $\dfrac{dy}{dx} = \dfrac{x}{y}$, for $-2 \le x \le 2$ and $-2 \le y \le 2$.

ANSWER:

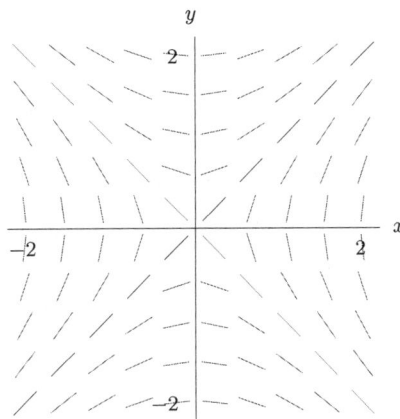

Figure 10.3.223

2. Consider the slope field for $dy/dx = y - x$. Complete the table by filling in the slope of the line segment at each point.

Table 10.3.51

point	(0,0)	(0,1)	(1,1)	(2,1)	(1,2)	(-1,-1)	(-2,-1)	(-1,-2)
slope								

ANSWER:

Compute the slope by substituting the coordinates at each point into the equation $dy/dx = y - x$.

Table 10.3.52

point	(0,0)	(0,1)	(1,1)	(2,1)	(1,2)	(-1,-1)	(-2,-1)	(-1,-2)
slope	0	1	0	-1	1	0	1	-1

3. (a) Which of the following differential equations goes with the slope field shown in Figure 10.3.224?

    (i) $\dfrac{dy}{dx} = y$ (ii) $\dfrac{dy}{dx} = \dfrac{x}{y}$ (iii) $\dfrac{dy}{dx} = x^2 + y^2$ (iv) $\dfrac{dy}{dx} = x - y$

    (b) Sketch the solutions through the points $(0, 2)$ and $(-2, 2)$.

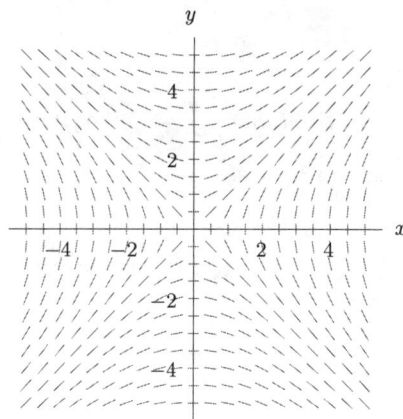

Figure 10.3.224

ANSWER:

(a) There are many ways to see which answer is correct. One quick way is to note that the value of $\dfrac{dy}{dx}$ in the slope field shown is 0 at all points along the y-axis, i.e. at all points where $x = 0$. Substituting $x = 0$ for each equation given, we find that answer (b) is the only case which yields $\dfrac{dy}{dx} = 0$. Testing other points confirms that $\dfrac{x}{y}$ is the correct solution.

(b)

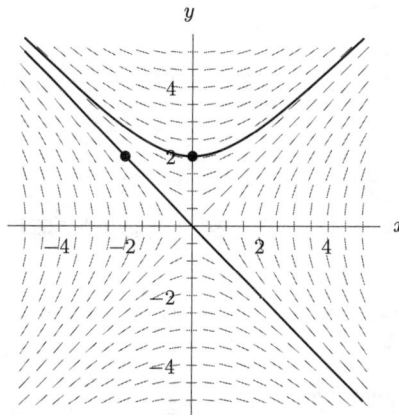

Figure 10.3.225

4. Draw the slope field for the differential equation $\dfrac{dy}{dx} = x - y$ at the nine points $(0, 0)$, $(0, 1)$, $(0, 2)$, $(1, 0)$, $(1, 1)$, $(1, 2)$, $(2, 0)$, $(2, 1)$, and $(2, 2)$.

ANSWER:

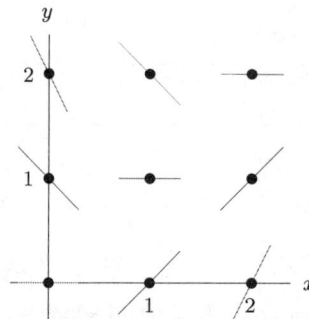

Figure 10.3.226

5. If a slope field for $\dfrac{dy}{dx}$ has constant slopes where x is constant, what do you know about $\dfrac{dy}{dx}$?

ANSWER:

We know that $\dfrac{dy}{dx}$ depends on x only.

Problems and Solutions   for Section 10.4

1. Solve the differential equation $\dfrac{dy}{dx} = -8y$, subject to the initial condition that $y = 100$ when $x = 0$.

ANSWER:

The solution has the form $y = Ce^{kx}$, and in this case we get $k = -8, C = 100$. Thus, $y = 100e^{-8x}$.

2. Money in a bank account grows continuously at an annual rate of 4%. Suppose \$10,000 is put into an account at time $t = 0$.

   (a) Write a differential equation for B, the balance of the account after t years.
   (b) Solve the differential equation.

   ANSWER:

   (a) We know that the change in the balance is equal to 0.04 times the balance. Thus,

$$\frac{dB}{dt} = 0.04B.$$

   (b) The solution has the form $B = Ce^{0.04t}$, and since $B = 10,000$ at time $t = 0$, we get $B = 10,000e^{0.04t}$.

3. An anti-inflammatory drug has a half-life in the human body of about 8 hours.

   (a) Use the half-life to find the constant k in the differential equation

$$\frac{dQ}{dt} = -kQ,$$

   where Q is the quantity of the drug in the body t hours after the drug is administered.
   (b) At what time will 40% of the original dose remain in the body?

   ANSWER:

   (a) Since the half-life is 8 hours, we know that the quantity, Q, remaining when $t = 8$ is equal to $0.5Q_0$, where Q_0 is the original amount. We substitute into the solution to the differential equation, $Q = Q_0 e^{-kt}$, and solve for k :

$$Q = Q_0 e^{-kt}$$
$$0.5Q_0 = Q_0 e^{-k(8)}$$
$$0.5 = e^{-8k}$$
$$\ln 0.5 = -8k$$
$$k = \frac{\ln 0.5}{-8} \approx 0.0866.$$

   (b) Substituting into the same equation, this time solving for t :

$$0.4Q_0 = Q_0 e^{-0.0866t}$$
$$0.4 = e^{-0.0866t}$$
$$\ln 0.4 = -0.0866t$$
$$t = \frac{\ln 0.4}{-0.0866} \approx 10.58.$$

   Thus, 40% of the drug will remain in the body at $t = 10.58$, or about 10 1/2 hours.

4. The amount of medicine present in the blood of a patient decreases due to metabolization according to the exponential decay model. One hour after a dose was given, there were 3.7 ng/cm^3 present, and an hour later there were 2.5 ng/cm^3. When will there be less than 0.5 ng/cm^3 present, assuming no more medicine is taken?

   ANSWER:

We solve for A and k:

$$Ae^{-k} = 3.7$$
$$Ae^{-2k} = 2.5.$$

We have $k = \ln(3.7/2.5) = 0.392$, and $A = 3.7e^{0.392} = 5.476$. So the amount of medicine in the blood t hours later is

$$M = 5.476e^{-0.392t}.$$

We solve the following for t:

$$0.5 = 5.476e^{-0.392t}.$$

We have

$$t = \frac{\ln 10.952}{0.392} = 6.1 \text{ hours.}$$

There will be less than 0.5ng/cm^3 present after 6.1 hours.

5. If no more pollutants are dumped in a lake, the amount of pollution in the lake will decrease at a rate proportional to the amount of pollution present.

   (a) Write a differential equation for Q, the amount of pollution present, as a function of time t.
   (b) Solve the differential equation. If there are 400 units of pollution present initially and 184 units left after 8 years, write a formula for the quantity present, Q, as a funcion of the number of years, t.
   (c) How much pollution will remain 10 years after the initial time?

   ANSWER:

   (a) $\dfrac{dQ}{dt} = kQ$ for some negative constant k.

   (b) $\dfrac{dQ}{dt} = kQ$ has $Q = Ce^{kt}$ as a general solution, for some constant k. Substituting 400 for C (the value when $t = 0$ in the known case), 184 for Q and 8 for t, we can solve the equation for k:

   $$184 = 400e^{k \cdot t}$$
   $$\ln\left(\frac{184}{400}\right) = \ln(e^{8k}) = 8k$$
   $$k = \frac{\ln\left(\frac{184}{400}\right)}{8} \approx -0.097.$$

   So the formula is $Q = 400e^{-0.097t}$.

   (c) To find the amount of pollution 10 years after the initial time, we substitute 10 for t in the formula, to get $Q = 400e^{-0.097 \cdot 10} \approx 151.6$. So there are about 151.5 units of pollution present after 10 years.

6. Give the solution to the differential equation $\dfrac{dP}{dt} = 0.15P$, given that $P = 10$ when $t = 0$.

   ANSWER:

   We know that the general solution is $P = Ce^{0.15t}$, and since we know that $P = 10$ fot $t = 0$, we know that $C = 10$; so the solution is $P = 10e^{0.15t}$.

Problems and Solutions   for Section 10.5

1. A yam gains or loses heat according to Newton's law. The kitchen has temperature 70° and the oven is preheated to 350°. The yam, initially at room temperature, is placed in the oven, and after one hour has temperature 175°. The yam is then removed from the oven and left sitting on the table for one half hour. What is its temperature at that point?

   ANSWER:

   Let $T = A + Ce^{kt}$. Initially we have $70 = 350 + C$, i.e. $C = -280$. Now we can solve the following for k:

   $$175 = 350 - 280e^{k}.$$

   We get $k = -0.47$. After the yam is removed from the oven, we have

   $$T = 70 + Be^{-0.47t},$$

and $175 = 70 + b$ implies that $B = 105$. So

$$T = 70 + 105e^{-0.47t}.$$

Now

$$T(0.5) = 70 + 105e^{-0.47 \times 0.5} = 153°.$$

So the temperature is $153°$.

2. A certain bank account earns interest at the rate of 5% compounded continuously. Money is being withdrawn from the account in a continuous stream at the constant rate of $100,000 per year.

   (a) If B represents the account balance, write a differential equation modeling how B changes in time.
   (b) Solve the differential equation in (a), given an initial balance of B_0.
   (c) What size should B_0 be in order for the account never to be depleted?

   ANSWER:

   (a) We have $\dfrac{dB}{dt} = 0.05B - 100,000$.
   (b) The general solution is

   $$B = Ae^{0.05t} + 2,000,000,$$

   where A is any constant. For an initial balance B_0 we have $B = B_0 e^{0.05t} - 2,000,000(e^{0.05t} - 1)$, or

   $$B = (B_0 - 2,000,000)e^{0.05t} + 2,000,000.$$

   (c) For the account never to be depleted, B_0 should be equal to or bigger than $2,000,000.

3. Find the general solution of each of the following differential equations.

   (a) $\dfrac{dy}{dx} = \sin x + e^x$ (b) $\dfrac{dy}{dx} = 100 - y$

   ANSWER:

   (a) $y = e^x - \cos x + c$
   (b) $y = 100 - ce^{-x}$

4. According to Newton, the rate at which the temperature of water in a swimming pool changes is directly proportional to the difference between the temperature L outside and the temperature T of the water in the pool.

   (a) Complete the equation $\dfrac{dT}{dt} = k($ $)$, where k is a positive constant.
   (b) Suppose the temperature outside stays at a constant $26°$C for two hours, and that during this time the temperature of the water increases from $20°$C to $23°$C. Give a complete solution of the differential equation in part (a).
   (c) Give an equilibrium solution to the equation in part (b). Is this equilibrium stable or unstable?

   ANSWER:

   (a) $dT/dt = k(L - T)$.
   (b) The general solution is

   $$T(t) = L - ce^{-kt}.$$

   Now apply $L = 26$, $T(0) = 20$ and $T(2) = 23$. We have $c = 6$ and $k = \ln 2/2 \approx 0.35$, so

   $$T(t) = 26 - 6e^{-0.35t}.$$

   (c) Let $dT/dt = k(L - T) = 0$. We have $T = L = 26$, which is a stable equilibrium.

5. A lake with a constant volume V, in km^3, contains a quantity of Q km^3 pollutant. Clean water enters the lake and causes a total outflow of r km^3 per year. The rate at which the pollutant decreases at any time t equals the product of the pollutant Q per volume V and the rate r at which the water flows out of the lake.

   (a) Give the differential equation which describes the rate at which the pollutant decreases.
   (b) Find the general solution of this differential equation.
   (c) If $V = 12 \times 10^3$ km^3 and $r = 65$ km^3 per year, how long will it take for the pollutant to decrease to half of its original quantity?

   ANSWER:

   (a) $\dfrac{dQ}{dt} = -\dfrac{Q}{V} \cdot r = -\dfrac{r}{V} \cdot Q$
   (b) $Q(t) = Ce^{-\frac{r}{V}t}$ is the general solution where C is any constant.

(c) Solve for t: $\frac{1}{2}C = Ce^{-\frac{65}{12,000}t}$. We have $-\ln 2 = -\frac{65}{12,000}t$, or

$$t = \frac{12,000\ln 2}{65} \approx 128.$$

So it will take about 128 years for the pollutant to decrease to half of its original quantity.

6. Find the solution of the differential equation $\frac{d\theta}{dt} = 2\theta - 4$, if $\theta = 5$ when $t = 2$.

ANSWER:

Solve for θ: Since $\frac{d\theta}{\theta - 2} = 2dt$, we have $\ln|\theta - 2| = 2t + c$, or $\theta - 2 = Ce^{2t}$, i.e.

$$\theta(t) = Ce^{2t} + 2,$$

where C is any constant.

If $\theta = 5$ when $t = 2$, then $5 = Ce^4 + 2$ or $C = 3/e^4$. And the solution is

$$\theta(t) = \frac{3}{e^4}e^{2t} + 2.$$

7. The logistic model for a population's growth is given by $\frac{dP}{dt} = 3P - 3P^2$, where P is the size of the population in millions at any time t, measured in years.

(a) Show with a sketch how $\frac{dP}{dt}$ changes with respect to P.

(b) What is the size of the population when the rate of increase starts to decrease?

(c) Give a neat sketch showing how the size of the population P changes with time. (You can assume that $P = 0$ when $t = 0$.)

ANSWER:

(a) Since $P' = 3P(1 - P)$, we have the sketch given in Figure 10.5.227.

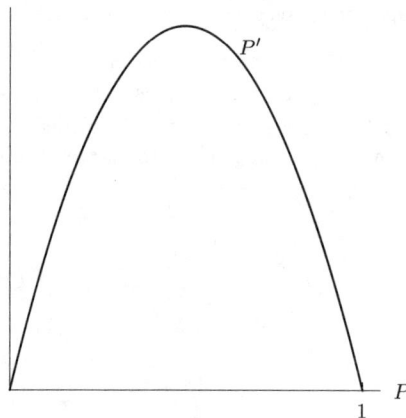

Figure 10.5.227

The growth rate is increasing as P increases from 0 to 0.5 million. P is increasing most rapidly at $P = 0.5$, and then the growth rate decreases, approaching zero as P approaches 1 million.

(b) Half million

(c)

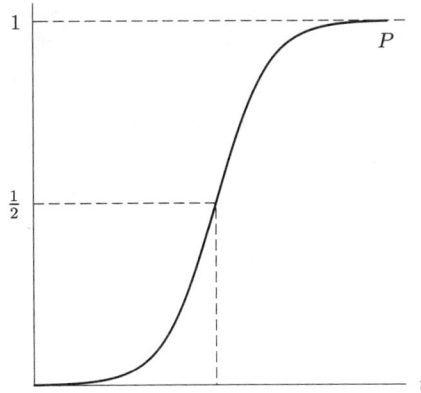

Figure 10.5.228

Problems and Solutions   for Section 10.6

1. Two species of insects coexist with each other. Both would do fine on their own. Species x would do even better in the presence of Species y. Species y does not do well in the presence of Species x. Create a system of differential equations to model this scenario.

ANSWER:

Answers will vary. One possible system:

$$\frac{dx}{dt} = x + xy \qquad \frac{dy}{dt} = y - xy.$$

2. Trout are introduced into a stream. Trout is a predator species and therefore has an influence on the population size of other fish. Figure 10.6.229 shows how the trout and other fish populations vary over time. The progress of time is shown by the direction of the arrow.

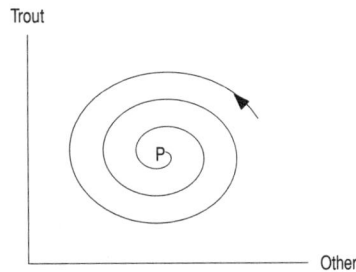

Figure 10.6.229

(a) What happens to the size of the population at point P?

(b) On the same set of axes, graph how the two different populations vary with time.

ANSWER:

(a) The size of the trout population will be stable at point P.

(b)

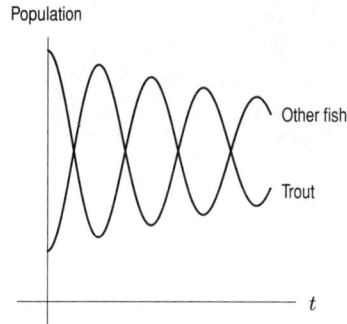

Figure 10.6.230

3. Figure 10.6.231 shows the size of two population groups x and y. Initially the size of group x is zero and the size of group y is b. Describe in words how the sizes of the population groups change with respect to each other on the different intervals.

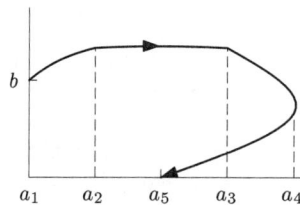

Figure 10.6.231

ANSWER:

Both populations increase until the population of x reaches a_2. Then the population of y remains constant while the population of x increases from a_2 to a_3. Then the population of y starts to decrease. When the population of x is a_4, the population of y is reduced to such a level that the population of x decreases. Both populations decrease until the population of x reaches a_5 and the population of y reaches 0.

4. Romeo and Juliet are star-crossed lovers. In our story, Romeo is attracted to Juliet when she likes him, but loses interest when Juliet dislikes him. Juliet, on the other hand, is contrary, repelled by Romeo when he likes her and attracted to him when he dislikes her. If R represents Romeo's liking of Juliet and J represents Juliet's liking of Romeo, their rocky relationship can be modeled by the differential equations $\dfrac{dR}{dt} = AJ$ and $\dfrac{dJ}{dt} = -BR$ for positive constants A and B. [Take negative liking to be dislike.]

(a) Tell how attraction and repulsion are related to liking, and hence explain how the signs in the equations correspond with the story.

(b) Eliminate t and get a differential equation relating R and J; sketch enough of the slope field in the phase plane to predict what will happen in the long run.

(c) When we first look in on their ill-fated relationship, $R = 1$ and $J = 0$. One week later, $R = 0$ and $J = -1$. Sketch the trajectory of their relationship.

ANSWER:

(a) In general we assume that attraction corresponds to an increasing of liking and repulsion to a decreasing of liking. The first equation says that the rate of change of Romeo's liking is POSITIVELY proportional to Juliet's liking, which corresponds to "Romeo is attracted to Juliet when she likes him, but loses interest when Juliet dislikes him".

The second equation means that the rate of change of Juliet's liking is NEGATIVELY proportional to Romeo's liking, which corresponds to "Juliet is contrary, repelled by Romeo when he likes her and attracted to him when he dislikes her".

(b) Since $\dfrac{dR}{dt}$ is the change of R over a unit of time and $\dfrac{dJ}{dt}$ is the change of J over a unit of time, we have the quotient $\dfrac{dR}{dt} \Big/ \dfrac{dJ}{dt}$ is the change of R over a unit of J; therefore,

$$\frac{dR}{dJ} = \frac{dR/dt}{dJ/dt} = -\frac{AJ}{BR}.$$

Let us assume that $A = B = 1$. We have

$$\frac{dR}{dJ} = -\frac{J}{R}.$$

Since a line of slope $-\dfrac{J}{R}$ is perpendicular to the line of slope $\dfrac{R}{J}$, we have the slope field $\dfrac{dR}{dJ} = -\dfrac{J}{R}$, as shown in Figure 10.6.232.

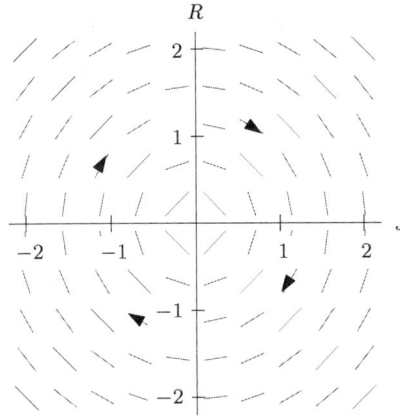

Figure 10.6.232

In the long run their relationship follows a tragic love-hate-love circle.

(c) See (b).

5. Four pairs of species are given below, with descriptions of how they interact, along with three systems of differential equations. Match three of the pairs with the three systems of differential equations given, indicating which species would be x and which would be y, and write a possible system of differential equations for the fourth.

(a) Bees/flowers: each is helped by the other; each needs the other to survive.

(b) Owls/trees: owls need trees to survive, trees don't care one way or the other about owls.

(c) Elk/buffalo: in competition with each other; each would do fine without the other.

(d) Fox/hare: fox eats the hare so it needs it to survive; hare would do fine without the fox.

(I) $\dfrac{dx}{dt} = -0.2x + 0.05xy$ \qquad (II) $\dfrac{dx}{dt} = 0.23x$ \qquad (III) $\dfrac{dx}{dt} = -0.4x + 0.23xy$

$\dfrac{dy}{dt} = 0.5y - 0.05xy$ \qquad\qquad $\dfrac{dy}{dt} = -0.7y + 0.3xy$ \qquad\qquad $\dfrac{dy}{dt} = -0.1y + 0.15xy$

ANSWER:

(a) This situation is best described by system (III), with either species being x and the other y.

(b) This situation is best described by system (II), with the trees as x and the owls as y.

(c) None of the given systems describes this situation accurately. One possible system would be $\dfrac{dx}{dt} = x - xy$ and $\dfrac{dy}{dt} = y - xy$, with either species as x and the other y.

(d) This situation is best described by system (I), with the fox as x and the hare as y.

Problems and Solutions for Section 10.7

1. At time $t = 0$, there are 500 students at a school, 3 of whom have the flu. No one else has been exposed to it yet.

(a) What is I_0? What is S_0?

(b) Use the differential equation $\dfrac{dI}{dt} = 0.0026SI - 0.5I$ and the values given in part (a) to determine whether the flu will spread. Clearly justify your answer.

(c) Assume that 400 of the students have been vaccinated against the flu. Repeat parts (a) and (b) using this additional information.

ANSWER:

(a) I_0 is 3 and S_0 is 497, since initially 3 students have the flu and $500 - 3 = 497$ have not been exposed to it.

(b) We substitute in the values from (a): $\dfrac{dI}{dt} = 0.0026(497)(3) - 0.5(3) = 2.3766$. Since $\dfrac{dI}{dt}$ is positive, the number of infected students will *increase*, and so the disease will spread.

(c) In this case we have $I_0 = 3$ and $S_0 = 500 - 400 - 3 = 97$, so $\dfrac{dI}{dt} = 0.0026(97)(3) - 0.5(3) = -0.7434$. Since $\dfrac{dI}{dt}$ is negative, the number of infected students will *decrease*, and so the disease will *not* spread.

2. For a new strain of flu, the constants are slightly different from those in Problem 1. Assume that the system of differential equations for this flu strain are

$$\frac{dS}{dt} = -0.0038SI \qquad \text{and} \qquad \frac{dI}{dt} = 0.0038SI - 0.8I.$$

(a) Use these differential equations to find $\dfrac{dI}{dS}$.

(b) Find the threshold value for this strain of flu by setting $\dfrac{dI}{dS}$ equal to zero and solving for S.

ANSWER:

(a) $\dfrac{dI}{dS} = \dfrac{dI/dt}{dS/dt} = \dfrac{0.0038SI - 0.8I}{-0.0038SI} = -1 + \dfrac{0.8I}{0.0038SI} = \dfrac{210.5}{S} - 1.$

(b) We have

$$\frac{dI}{dS} = 0$$

$$\frac{210.5}{S} - 1 = 0$$

$$S = 210.5.$$

Thus the threshold value for this strain of flu is about 210; i.e., if S_0 is below 210, there is no epidemic.

3. Consider three strains of the flu modeled by the following sets of differential equations:

(I) $\dfrac{dS}{dt} = -0.01SI \quad \dfrac{dI}{dt} = 0.01SI - 0.3I$

(II) $\dfrac{dS}{dt} = -0.02SI \quad \dfrac{dI}{dt} = 0.02SI - 0.5I$

(III) $\dfrac{dS}{dt} = -0.03SI \quad \dfrac{dI}{dt} = 0.03SI - 0.4I$

(a) Which is the most infectious? The least infectious? Explain.

(b) Which has the infecteds being removed the fastest? The slowest? Explain.

ANSWER:

(a) (III) is the most infectious because it has the highest constant a, which measures how infectious a disease is (in this case $a = 0.03$). Similarly, (I) with $a = 0.01$ is the least infectious.

(b) (II) has the infecteds being removed the fastest because it has the highest constant b, which represents the rate the infected people are removed (in this case, $b = 0.5$). Similarly, (I) with $b = 0.3$ has infecteds being removed the slowest.

4. For each set of equations in Exercise 3, what is the threshold value of S?

ANSWER:

The threshold value is found by computing dI/dS and setting it equal to zero. For (I),

$$\frac{dI}{dS} = \frac{dI/dt}{dS/dt} = \frac{0.01SI - 0.3I}{-0.01SI} \approx -1 + \frac{30}{S}$$

$$-1 + \frac{30}{S} = 0$$

$$S = 30 \text{ people.}$$

For (II),

$$\frac{dI}{dS} = \frac{dI/dt}{dS/dt} = \frac{0.02SI - 0.5I}{-0.02SI} \approx -1 + \frac{25}{S}$$

$$-1 + \frac{25}{S} = 0$$
$$S = 25 \text{ people.}$$

For (III),

$$\frac{dI}{dS} = \frac{dI/dt}{dS/dt} = \frac{0.03SI - 0.4I}{-0.03SI} \approx -1 + \frac{13.33}{S}$$
$$-1 + \frac{13.33}{S} = 0$$
$$S \approx 13 \text{ people.}$$

5. For which sets of equations in Exercise 3 does the disease spread if initially $S_0 = 28$?
    ANSWER:
    We check the sign of dI/dt.
    For (I),

$$\frac{dI}{dt} = 0.01(28)I - 0.3I = -0.02I < 0$$

The disease will not spread.
For (II),

$$\frac{dI}{dt} = 0.02(28)I - 0.5I = 0.06I > 0$$

The disease will spread.
For (II),

$$\frac{dI}{dt} = 0.03(28)I - 0.4I = 0.44I > 0$$

The disease will spread.

Problems and Solutions to Review Problems for Chapter 10

1. Pollutants are being dumped into a lake at a rate of 10 m^3 per day. About 15% of the lake's water leaves the lake each day and fresh, unpolluted, water flows in to replace it. Write a differential equation for the amount of pollutant, Q, in the lake as a function of time, t, in days. Assume the pollutant instantaneously disperses throughout the lake as it is dumped.
    ANSWER:
    The amount of pollutant, Q, is increasing at a constant rate of 10 m^3 per day and decreasing at a rate of $0.15Q$ per day. Thus,

$$\frac{dQ}{dt} = 10 - 0.15Q.$$

2. Find the value(s) of k for which $y = x^2 + kx$ is a solution to the differential equation $xy' - 2y = 3x$.
    ANSWER:
    Differentiating, $y' = 2x + k$. Thus,

$$xy' - 2y = x(2x + k) - 2(x^2 + kx) = -kx$$

Therefore $-kx = 3x$, so $k = -3$.

3. Fill in the missing values in the table given that $dy/dt = -0.4y$. Assume the rate of growth, given by dy/dt, is approximately constant over each unit time interval.

Table 10.7.53

t	0	1	2	3	4
t	125				

    ANSWER:
    For $t = 1$, the change in y is given by $-0.4y = -0.4(125) = -50$. Thus, the new y-value is $125 - 50 = 75$. Similarly, fill in the rest of the table.

Table 10.7.54

t	0	1	2	3	4
t	125	75	45	27	16.2

4. Find the solution to the differential equation $\dfrac{dP}{dt} = -\dfrac{P}{2}$, subject to the initial condition $P(0) = 8$.

ANSWER:

The general solution is of the form $P = Ce^{-\frac{1}{2}t}$, and the initial condition gives $C = 8$, Thus, $P = 8e^{-\frac{1}{2}t}$.

5. Money in a bank account earns interest at a continuous rate of 6% per year, and payments are made continuously out of the account at the rate of $9,000 per year. The account initially contains $100,000.

(a) Write a differential equation for the balance, B, in the account in t years.

(b) When will the account run out of money?

ANSWER:

(a) We know that money goes into the account at a rate of $0.06B$ and out at a constant rate of 9000 per year. Thus,

$$\frac{dB}{dt} = 0.06B - 9000 = 0.06(B - 150,000)$$

(b) The differential equation in part (a) gives us an equation of the form

$$B = 150,000 + Ce^{0.06t}.$$

To find C, use the initial condition that $B(0) = 100,000$.

$$100,000 = 150,000 + Ce^{0.06(0)} = 150,000 + C$$

$$C = -50,000$$

Thus the equation becomes

$$B = 150,000 - 50,000e^{0.06t}.$$

Now solve for t when $B = 0$.

$$0 = 150,000 - 50,000e^{0.06t}$$

$$50,000e^{0.06t} = 150,000$$

$$e^{0.06t} = 3$$

$$0.06t = \ln 3 \approx 1.099$$

$$t \approx 18.3 \text{ years.}$$

6. Suppose x and y are populations of two different species that interact as shown in the following figure. Describe how each population changes with time.

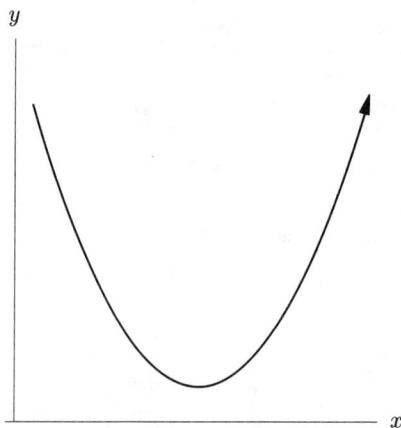

Figure 10.7.233

ANSWER:

At first the population of y is high and the population of x is low. The population of x increases at a fairly constant rate the entire time. The population of y first decreases quite rapidly, then levels off, then increases quite rapidly until it is back at the level it started at.

7. At time $t = 0$, there are 400 students in a school, 5 of whom have the flu. No one else has been exposed yet.

(a) What are I_0 and S_0?

(b) Use the differential equation $\dfrac{dI}{dt} = 0.0014SI - 0.6I$ and your answer to part (a) to determine whether the flu will spread.

(c) Repeat parts (a) and (b) if an additional 200 unexposed students are added to the school.

ANSWER:

(a) $I_0 = 5$ since initially 5 students have the flu, and $S_0 = 400 - 5 = 395$ students who have not been exposed yet.

(b) Substituting the initial values into the differential equation, $\dfrac{dI}{dt} = 0.0014(395)(5) - 0.6(5) = -0.325$. Since dI/dt is negative, the number of infected students will decrease and the flu will not spread.

(c) We still have $I_0 = 5$, but now we have $S_0 = 395 + 200 = 595$ students who have not been exposed. Substituting these into the differential equation, $\dfrac{dI}{dt} = 0.0014(695)(5) - 0.6(5) = 1.865$. Since dI/dt is positive, the number of infected students will increase and the flu will spread.

Chapter 11 Exam Questions

Problems and Solutions   for Section 11.1

1. Find the sum, if it exists, of

$$4 + 4^2 + 4^3 + \cdots + 4^{10}$$

ANSWER:

This is a finite geometric series with $a = 4$, $r = 4$, $n = 11$. Thus

$$S = \frac{4(1 - 4^{11})}{1 - 4} = 5,592,404$$

2. Find the sum, if it exists, of

$$100 + 100(1.03) + 100(1.03)^2 + 100(1.03)^3 + \cdots$$

ANSWER:

This is an infinite geometric series with $a = 100$, $r = 1.03$. Since $r > 1$, the series diverges.

3. Find the sum, if it exists, of

$$500 + 250 + 125 + 62.5 + \cdots$$

ANSWER:

This is an infinite geometric series with $a = 500$, $r = \frac{1}{2}$. Since $-1 < r < 1$, we get

$$S = \frac{a}{1 - r} = \frac{500}{1 - \frac{1}{2}} = 1000$$

4. Find the sum, if it exists, of

$$1 - \frac{1}{2} + \frac{1}{4} - \frac{1}{8} + \cdots$$

ANSWER:

This is an infinite geometric series with $a = 1$, $r = -\frac{1}{2}$. Since $-1 < r < 1$, we get

$$S = \frac{a}{1 - r} = \frac{1}{1 - \left(-\frac{1}{2}\right)} = \frac{2}{3}$$

5. Each quarter, $1000 is deposited into an account earning 1.2% interest a quarter, compounded quarterly.

   (a) How much is in the account right before the 8th deposit?

   (b) How much is in the account right after the 12th deposit?

   ANSWER:

   (a) The quantity in the account after n deposits is represented by the sum

$$B_n = 1000 + 1000(1.012) + 1000(1.012)^2 + \cdots + 1000(1.012)^{n-1}.$$

This is a finite geometric series with $a = 1000$, $r = 1.012$. Right after the 8th deposit,

$$S_7 = \frac{a(1 - r^n)}{1 - r} = \frac{1000(1 - (1.012)^8)}{1 - 1.012} \approx \$8344.12.$$

Therefore, the amount right before the 8th deposit is

$$8344.12 - 1000 = \$7344.12.$$

   (b) Right after the 12th deposit,

$$S_{12} = \frac{a(1 - r^n)}{1 - r} = \frac{1000(1 - (1.012)^{12})}{1 - 1.012} \approx \$12,824.55.$$

6. Estimate the sum

$$10 + 10(0.3) + 10(0.3)^2 + 10(0.3)^3 + \cdots$$

by calculating the partial sums S_n for $n = 1, 5, 10$, and 50. Verify your estimate by calculating the infinite sum directly.

ANSWER:

For a finite geometric series with $a = 10$, $r = 0.3$,

$$S_1 = 10$$

$$S_5 = \frac{10(1 - (0.3)^5)}{1 - 0.3} = 14.251$$

$$S_{10} = \frac{10(1 - (0.3)^{10})}{1 - 0.3} = 14.2856$$

$$S_{50} = \frac{10(1 - (0.3)^{50})}{1 - 0.3} = 14.2857$$

For an infinite series with $a = 10$, $r = 0.3$,

$$S = \frac{10}{1 - 0.3} = 14.2857$$

7. Each week, a patient is given a 40 mg dose of an experimental vaccine, and 25% of the vaccine remains in the body after one week. Find the quantity in the body

(a) Right after the 4th dose.

(b) After 1 year (52 doses).

ANSWER:

(a) The quantity in the body after n doses is represented by the sum

$$Q_n = 40 + 40(0.25) + 40(0.25)^2 + \cdots + 40(0.25)^{n-1}$$

This is a finite geometric series with $a = 40$, $r = 0.25$. Thus

$$Q_4 = \frac{40(1 - (0.25)^4)}{1 - 0.25} = 53.125 \text{mg}$$

(b)

$$Q_{52} = \frac{40(1 - (0.25)^{52})}{1 - 0.25} \approx 53.333 \text{mg}$$

8. Does the infinite series

$$4 + \frac{4}{\sqrt{3}} + \frac{4}{3} + \frac{4}{3^{3/2}} + \frac{4}{3^2} + \cdots$$

converge or diverge?

ANSWER:

$$4 + \frac{4}{\sqrt{3}} + \frac{4}{3} + \frac{4}{3^{3/2}} + \frac{4}{3^2} + \cdots$$

is

$$a + ax + ax^2 + ax^3 + \cdots$$

with $a = 4$ and $x = \dfrac{1}{\sqrt{3}}$.

Since $\left| \dfrac{1}{\sqrt{3}} \right| < 1$, it converges.

9. Find the sum of the first 5 and the sum of the first 10 terms of the series in Exercise 8.

ANSWER:

$$S_5 = \frac{4\left(1 - \left(\frac{1}{\sqrt{3}}\right)^5\right)}{1 - \frac{1}{\sqrt{3}}} \approx 8.86$$

$$S_{10} = \frac{4\left(1 - \left(\frac{1}{\sqrt{3}}\right)^{10}\right)}{1 - \frac{1}{\sqrt{3}}} \approx 9.425$$

10. Find the sum of the series

$$\sum_{n=5}^{15} \left(\frac{4}{3}\right)^n$$

ANSWER:

$$\sum_{n=5}^{15} \left(\frac{4}{3}\right)^n = \left(\frac{4}{3}\right)^5 + \left(\frac{4}{3}\right)^6 + \cdots$$

$$= \left(\frac{4}{3}\right)^5 \left(1 + \frac{4}{3} + \left(\frac{4}{3}\right)^2 + \cdots + \left(\frac{4}{3}\right)^{10}\right)$$

$$= \frac{\left(\frac{4}{3}\right)^5 \left(1 - \left(\frac{4}{3}\right)^{11}\right)}{1 - \frac{4}{3}} \approx 286.68$$

11. If the sum of the series $S = a + \dfrac{a}{3} + \dfrac{a}{9} + \dfrac{a}{27} + \dfrac{a}{81}$ is $\dfrac{1}{2}\left(3^2 - \dfrac{1}{3^3}\right)$, what is a?

ANSWER:

$$S = a\left(1 + \frac{1}{3} + \frac{1}{3^2} + \frac{1}{3^3} + \frac{1}{3^4}\right)$$

$$\text{Sum} = \frac{a\left(1 - \left(\frac{1}{3}\right)^5\right)}{1 - \frac{1}{3}} = \frac{a\left(1 - \frac{1}{3^5}\right)}{\frac{2}{3}} = \frac{1}{2}\left(3a\left(1 - \frac{1}{3^5}\right)\right)$$

$$= \frac{1}{2}\left(3a - \frac{a}{3^4}\right). \text{ So } a = 3.$$

12. A ball is dropped from a height of 14 feet and bounces. Each bounce is $\dfrac{2}{3}$ of the height of the bounce before.

    (a) Find an expression for the height to which the ball rises after it hits the floor for the hth time.

    (b) Find the total vertical distance the ball has traveled when it hits the floor for the 4th time.

        ANSWER:

    (a) Let h_n be the height of the nth bounce after the ball hits the floor for the nth time. Then from Figure 11.1.234,

$$h_0 = \text{height before first bounce} = 14 \text{ feet}$$

$$h_1 = \text{height after first bounce} = 14\left(\frac{2}{3}\right) \text{ feet}$$

$$h_0 = \text{height after second bounce} = 14\left(\frac{2}{3}\right)^2 \text{ feet}$$

Generalizing gives $h_n = 14\left(\frac{2}{3}\right)^n$.

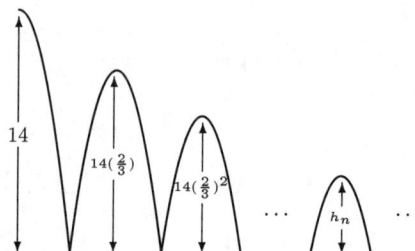

Figure 11.1.234

(b) Total vertical distance when the ball hits the floor for the 4th time is

$$14 + 2\left(14\left(\frac{2}{3}\right)\right) + 2\left(14\left(\frac{2}{3}\right)^2\right) + 2\left(14\left(\frac{2}{3}\right)^3\right) \approx 53.41 \text{ feet}$$

13. A tennis ball is dropped from a height of 40 feet and bounces. Each bounce is $\frac{1}{2}$ the height of the bounce before. A superball has a bounce $\frac{3}{4}$ the height of the bounce before, and is dropped from a height of 30 feet. Which ball bounces a greater total vertical distance?

ANSWER:

Tennis ball:

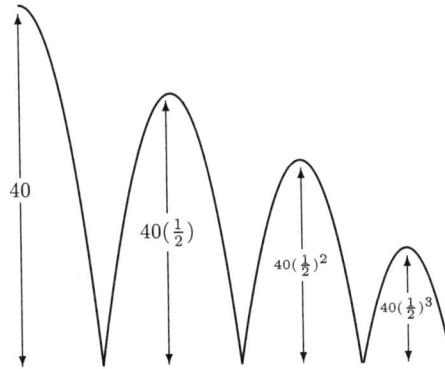

Figure 11.1.235

After the first drop of 40 feet, the total vertical distance is a geometric series with $a = 40$ and $r = \frac{1}{2}$:

$$40 + 2 \cdot 40(\frac{1}{2}) + 2 \cdot 40(\frac{1}{2})^2 + 2 \cdot 40(\frac{1}{2})^3 + \cdots$$

which converges to $40 + 40/(1 - \frac{1}{2}) = 120$ feet.

Superball:

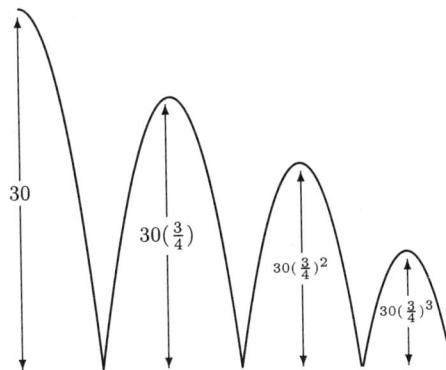

Figure 11.1.236

After the first drop of 30 feet, the total vertical distance is a geometric series with $a = 45$ and $r = \frac{3}{4}$:

$$30 + 2 \cdot 30(\frac{3}{4}) + 2 \cdot 30(\frac{3}{4})^2 + 2 \cdot 30(\frac{3}{4})^3 + \cdots$$

which converges to $30 + 45/(1 - \frac{3}{4}) = 210$ feet. So the superball travels more vertical distance than the tennis ball.

14. Decide which of the following are geometric series. For those which are, give the first term and the ratio between successive terms. For those which are not, explain why not.

    (a) $2 + 2a + 2a^2 + 2a^3 + \cdots$

    (b) $2 + 4a + 6a^2 + 8a^3 + \cdots$

    (c) $2 + 2ak + 2a^2k^2 + 2a^3k^3 + \cdots$

    ANSWER:

    (a) Yes. First term = 2, ratio $= \dfrac{2a}{2} = a$

    (b) No. Ratio between successive terms is not constant: $\dfrac{4a}{2} = 2a$ while $\dfrac{6a^2}{4a} = \dfrac{3}{2}a$

    (c) Yes. First term = 2, ratio $= \dfrac{2ak}{2} = ak$.

15. Find the sum $\displaystyle\sum_{n=2}^{\infty} \left(\frac{4}{5}\right)^n$

    ANSWER:

$$\sum_{n=2}^{\infty} \left(\frac{4}{5}\right)^n = \left(\frac{4}{5}\right)^2 + \left(\frac{4}{5}\right)^3 + \cdots$$

$$= \left(\frac{4}{5}\right)^2 \left(1 + \left(\frac{4}{5}\right) + \left(\frac{4}{5}\right)^2 + \cdots\right)$$

$$\text{Sum} = \frac{\left(\frac{4}{5}\right)^2}{1 - \frac{4}{5}} = \frac{\left(\frac{4}{5}\right)^2}{\frac{1}{5}} = \frac{16}{5}$$

Problems and Solutions   for Section 11.2

1. A yearly deposit of $10, 000$ is made into an account that pays 4.7% interest per year, compounded annually. What is the balance right before and right after the 8th deposit?

    ANSWER:

    The balance right after the 8th deposit is given by

$$B_8 = \frac{10,000(1 - (1.047)^8)}{1 - 1.047} = \$94,472.51$$

The balance right before the 8th deposit is

$$94,472.51 - 10,000 = \$84,472.51.$$

2. A couple wants to establish an annuity for retirement that will make annual payments of $45, 000$ from an account that pays 6% interest per year, compounded annually. If the payments are to start right now, how much should be deposited if they plan to live 25 years? Indefinitely?

    ANSWER:

    For 25 years, the present value is given by

$$P = 45,000 + 45,000(1.06)^{-1} + 45,000((1.06)^{-1})^2 + \cdots + 45,000((1.06)^{-1})^{24}$$

This is a finite geometric series with $a = 45,000$, $r = (1.06)^{-1}$, $n = 25$. Thus

$$P = \frac{45,000(1 - ((1.06)^{-1})^{25})}{1 - (1.06)^{-1}} = \$609,766.$$

The infinite geometric series with $a = 45,000$, $r = (1.06)^{-1}$ gives

$$P = \frac{45,000}{1 - (1.06)^{-1}} = \$795,000.$$

3. The government gives a tax rebate totaling 4 billion dollars. Find the total additional spending resulting from this tax rebate if everyone who receives the money spends

(a) 75% of it.
(b) 85% of it.
(c) 95% of it.

ANSWER:

(a) The recipients of the 4 billion dollars spend 75% of what they receive, for a total of 3 billion dollars. The recipients of this 3 billion dollars spend 75% of that, and so on. Thus, the total additional spending is given by

$$3 + 3(0.75) + 3(0.75)^2 + \cdots \text{ billion dollars}$$

This is an infinite geometric series with $a = 3$ billion, $r = 0.75$. Since $-1 < r < 1$, the total additional spending is given by

$$\frac{a}{1-r} = \frac{3}{1-0.75} = 12 \text{ billion dollars.}$$

(b) 85% of 4 billion dollars = 3.4 billion dollars. Thus, the total additional spending is given by

$$\frac{a}{1-r} = \frac{3.4}{1-0.85} = 22.67 \text{ billion dollars.}$$

(c) 95% of 4 billion dollars = 3.8 billion dollars. Thus, the total additional spending is given by

$$\frac{a}{1-r} = \frac{3.8}{1-0.95} = 76 \text{ billion dollars.}$$

4. Find the market stabilization point if 5,000 new units are manufactured each year and 10% of the total number of units in use fail each year. What if 20% fail?

ANSWER:

If N is the number of units in circulation, then

$$N = 5,000 + 5,000(0.9) + 5,000(0.9)^2 + \cdots$$

This is an infinite geometric series with $a = 5,000$, $r = 0.9$. Since $-1 < r < 1$,

$$N = \frac{5,000}{1-0.9} = 50,000 \text{ units.}$$

If 20% fail, $r = 0.8$, and so

$$N = \frac{5,000}{1-0.8} = 25,000 \text{ units.}$$

5. An employee is offered two options: a fixed annual salary of $50,000, or $1 the first month, $2 the next month, $4 the next month, and so on (doubling each month). If the employee plans to work for two years, which option should the employee choose?

ANSWER:

The first option yields a salary of $2(50,000) = \$100,000$.
The second option yields a salary of

$$S = 1 + 2 + 4 + \cdots + 2^{23} \text{ dollars}$$

This is a finite geometric series with $a = 1$, $r = 2$, $n = 24$. Thus,

$$S = \frac{1(1 - 2^{24})}{1-2} = \$16,777,215.$$

The employee should choose the second option.

6. Your rich uncle leaves you a bequest: a continuous, constant payment stream of $1000 per year for the next ten years. The terms of the bequest require that it be paid into a certain account that will not be available to you until the ten years are up. The account earns 5.65% interest, compounded annually.

You discover that you can get 6.35% interest compounded continuously for a long-term certificate of deposit (CD) at the same bank. You persuade the executor of the estate to buy a CD whose value after ten years will be the same as the amount that would have been available to you under the terms of the bequest, and to pay you now the difference between the present value of the bequest and the cost of the CD. How much do you get now? How much do you get in ten years?

ANSWER:

We assume that the first payment of $1000 would be put into the account earning 5.65% interest inmediately. Then by the end of 10 years we would have a balance of

$$B = 1000(1.0565 + 1.0565^2 + \cdots + 1.0565^{10})$$
$$= \frac{1000(1 - 1.0565^{11})}{1 - 1.0565} - 1000$$
$$= \$13,698.73$$

The cost of a CD for this amount is

$$P = \frac{13,698.73}{e^{0.635}} = \$7259.44.$$

Finally, we calculate the present value of the bequest:

$$\text{Present value} = 1000 + 1000 \left(\frac{1}{1.0565}\right) + \cdots + 1000 \left(\frac{1}{1.0565}\right)^9$$
$$= \frac{1000 \left(1 - \left(\frac{1}{1.0565}\right)^{10}\right)}{1 - \frac{1}{1.0565}}$$
$$= \$7906.52.$$

Therefore the difference between the present value of the bequest and the cost of the CD is $7906.52 - 7259.44 = \$647.08$, and 10 years later you will get $13,698.73.

Problems and Solutions for Section 11.3

1. Twice a day, a patient takes a 25 mg tablet of a drug. At the end of a 12 hour period, 35% of the drug remains in the body. What quantity of the drug remains in the body

   (a) Right after taking the 5th tablet?
   (b) At the steady state level, right after and right before taking a tablet?

   ANSWER:

   (a) Let Q_n be the quantity of the drug, in mg, in the body right after taking the nth tablet. Then

   $$Q_n = 25 + 25(0.35) + 25(0.35)^2 + \cdots + 25(0.35)^{n-1}.$$

   When $n = 5$,

   $$Q_n = \frac{a(1 - r^n)}{1 - r} = \frac{25(1 - 0.35^5)}{1 - 0.35} = 38.26 \text{ mg.}$$

   (b) The steady state level right after a tablet is taken is given by the sum of the infinite geometric series

   $$Q = 25 + 25(0.35) + 25(0.35)^2 + \cdots.$$

   Since $r = 0.35$, and $-1 < r < 1$, the series converges to

   $$Q = \frac{a}{1 - r} = \frac{25}{1 - 0.35} = 38.46 \text{ mg.}$$

   The steady state before a dose is equal to the steady state after a dose minus the size of one dose, which is $38.46 - 25 = 13.46$ mg.

2. A new drug to control high blood pressure is found to have a half life of 2 days. A patient takes two 50 mg tablets of the drug at the same time each day.

   (a) How much of the drug is in the body after the 10th dose?
   (b) What is the steady state level of the drug right after taking the tablets?

ANSWER:

(a) After a single dose of $2 \times 50 = 100$ mg, the quantity, Q, of the drug in the body is given by $Q = 100b^t$, where t is time in hours. Since the half-life is 48 hours, we solve for b in the equation

$$0.5(100) = 100b^{48}$$
$$0.5 = b^{48}$$
$$b = (0.5)^{1/48}$$

Since doses are given every 24 hours, we want to know what fraction of the drug remains after 24 hours. Using $t = 24$,

$$Fraction \ remaining \ after \ 24 \ hours = b^{24} = ((0.5)^{1/48})^{24} = (0.5)^{24/48} = (0.5)^{0.5}.$$

Thus the quantity, Q_{10}, of the drug remaining in the body after the 10th dose is given by

$$Q_{10} = 100 + 100(0.5)^{0.5} + 100((0.5)^{0.5})^2 + \cdots + 100((0.5)^{0.5})^9$$

This is a finite geometric series with $a = 100$, $r = (0.5)^{0.5}$ $n = 10$. Thus,

$$Q_{10} = \frac{100(1 - ((0.5)^{0.5})^{10})}{1 - (0.5)^{0.5}} = 330.75 \text{ mg}.$$

(b) The steady state level right after taking a tablet is given by

$$Q = 100 + 100(0.5)^{0.5} + 100((0.5)^{0.5})^2 + \cdots$$

This is an infinite geometric series with $a = 100$, $r = (0.5)^{0.5}$. Thus,

$$Q = \frac{100}{1 - (0.5)^{0.5}} = 341.42 \text{ mg}.$$

3. In Exercise 2, what quantity of the drug would need to be given daily to achieve a steady state of 1000 mg?

ANSWER:

Let D be the dose taken to achieve a steady state of 1000 mg. Then, as in Exercise 2, we solve for b in the equation

$$0.5D = Db^{48}$$
$$0.5 = b^{48}$$
$$b = (0.5)^{1/48}$$

Similarly, we get

$$Fraction \ remaining \ after \ 24 \ hours = (05)^{0.5}.$$

Thus, the steady state level right after taking a tablet is given by

$$Q = D + D(0.5)^{0.5} + D((0.5)^{0.5})^2 + \cdots$$

This is an infinite geometric series with $a = D$, $r = (0.5)^{0.5}$. Since $-1 < r < 1$, we have

$$Q = \frac{D}{1 - (0.5)^{0.5}}$$

Solving this for D when $Q = 1000$, we have

$$1000 = \frac{D}{1 - (0.5)^{0.5}} = \frac{D}{0.293}, \text{ so}$$
$$D = 293 \text{ mg}.$$

4. Each morning at breakfast, a person consumes 6 micrograms of a toxin found in a pesticide, which leaves the body at a continuous rate of 4% per day. In the long run, how much toxin has accumulated in the body right after breakfast each day?

ANSWER:

Since the toxin leaves the body at a continuous rate of 4% per day, the 6 mg consumed the day earlier has decayed to $6e^{-0.04}$. In the long run, after breakfast each day, we have

$$Total \ accumulation \ of \ toxin = 6 + 6(e^{-0.04}) + 6(e^{-0.04})^2 + \cdots.$$

This is an infinite geometric series with $a = 6$, $r = e^{-0.04} = 0.9608$. Since $-1 < r < 1$, we have

$$Total \ accumulation \ of \ toxin = \frac{6}{1 - e^{-0.04}} \approx 153 \text{ micrograms}.$$

5. At the end of the year 2000, the total reserves of a natural resource was approximately 500,000 m^3. During 2001, 6,000 m^3 of the resource was consumed, and consumption is predicted to increase 9% per year after that. Under these assumptions, how long will the reserves last? How long would the reserves last if consumption remained steady at 6000 m^3 per year?

ANSWER:

Under these assumptions, the amount of resource used in 2002 is predicted to be $6000(1.09)$ m^3, and in 2003 we predict the amount used to be $6000(1.09)^2$ m^3. Thus, if Q_n is the quantity used after n years, we have

$$Q_n = 6000 + 6000(1.09) + 6000(1.09)^2 + \cdots + 6000(1.09)^{n-1}$$

This is a finite geometric series with $a = 6000$, $r = 1.09$. The sum is given by

$$Q_n = \frac{6000(1 - (1.09)^n)}{1 - 1.09}.$$

Since the total reserve is approximately 500,000 m^3, we solve

$$\frac{6000(1 - (1.09)^n)}{1 - 1.09} = 500,000$$

$$6000(1 - (1.09)^n) = -45,000$$

$$1 - (1.09)^n = -7.5$$

$$(1.09)^n = 8.5$$

Taking logarithms,

$$\ln(1.09)^n = \ln 8.5$$

$$n \ln 1.09 = \ln 8.5$$

$$n = \frac{\ln 8.5}{\ln 1.09} = 24.8 \text{ years.}$$

If consumption remained steady, the reserves would last $\frac{500,000}{6,000} = 83.8$ years.

6. A conservation organization sets a goal to decrease usage of the resource in Exercise 5 by 2% per year. If they can accomplish this, how much of the resource will be left after 100 years?

ANSWER:

Using the equation from Exercise 5, but with $r = 0.98$, we have

$$Q_n = \frac{6000(1 - (0.98)^n)}{1 - 0.98}.$$

Thus, when $n = 100$,

$$Q_{100} = \frac{6000(1 - (0.98)^{100})}{1 - 0.98} = 260,214.$$

Therefore, there will be 500,000-260,214=239,786 m^3 left.

Problems and Solutions to Review Problems for Chapter 11 ━━━━━━━━

1. Find the sum, if it exists, of

$$15 + 15(1.25) + 15(1.25)^2 + \cdots + 15(1.25)^{10}$$

ANSWER:

This is a finite geometric series with $a = 15$, $r = 1.25$, $n = 11$. Thus

$$S_{11} = \frac{a(1 - r^n)}{1 - r} = \frac{15(1 - (1.25)^{11})}{1 - 1.25} = 638.49$$

2. Find the sum, if it exists, of

$$25,000 + 25,000 \left(\frac{3}{4}\right) + 25,000 \left(\frac{3}{4}\right)^2 + \cdots$$

ANSWER:

This is an infinite geometric series with $a = 25,000$, $r = \frac{3}{4}$. Since $-1 < r < 1$, we get

$$S = \frac{a}{1-r} = \frac{25,000}{1 - \frac{3}{4}} = 100,000$$

3. Find the sum, if it exists, of

$$0.2 - 0.4 + 0.8 - 1.6 + \cdots$$

ANSWER:

This is an infinite geometric series with $a = 0.2$, $r = -2$. Since $r < -1$, the series diverges.

4. Consider the decimal $0.314314314\ldots$.

(a) Use the fact that

$$0.314314314\ldots = 0.314 + 0.000314 + 0.000000314 + \cdots$$

to write $0.314314314\ldots$ as a geometric series.

(b) Use the formula for the sum of a geometric series to find a fraction equal to $0.314314314\ldots$.

ANSWER:

(a)

$$0.314314314\ldots = 0.314 + .0000314 + 0.000000314 + \cdots$$
$$= 0.314 + 0.314(10^{-3}) + 0.314(10^{-3})^2 + \cdots$$

(b) We have an infinite geometric series with $a = 0.314$, $r = 10^{-3}$. Since $-1 < r < 1$, we get

$$S = \frac{a}{1-r} = \frac{0.314}{1 - 10^{-3}} = \frac{0.314}{\frac{999}{1000}} = \frac{314}{999}.$$

5. A sweepstakes offered a grand prize of $500,000$ per year for 10 years. Suppose all payments are made into a savings account earning 4.5% interest a year, compounded annually. How much money will be in the account after the 10 payments?

ANSWER:

The amount after 10 payments is given by the sum

$$S_{10} = 500,000 + 500,000(1.045) + 500,000(1.045)^2 + \cdots + 500,000(1.045)^9.$$

This is a finite geometric series with $a = 500,000$, $r = 1.045$, $n = 10$. Thus,

$$S_{10} = \frac{a(1-r)^n}{1-r} = \frac{500,000(1 - (1.045)^{10})}{1 - 1.045} = \$6,144,104.$$

6. The government gives a tax rebate totaling 6 billion dollars. Find the total additional spending resulting from this tax rebate if everyone who receives the money spends

(a) 80% of it.

(b) 90% of it.

ANSWER:

(a) The recipients of the 6 billion dollars spend 80% of what they receive, for a total of 4.8 billion dollars. The recipients of this 4.8 billion dollars spend 80% of that, and so on. Thus, the total additional spending =

$$4.8 + 4.8(0.8) + 4.8(0.8)^2 + \cdots \text{ billion dollars}$$

This is an infinite geometric series with $a = 4.8$ billion, $r = 0.8$. Since $-1 < r < 1$, the total additional spending is given by

$$\frac{a}{1-r} = \frac{4.8}{1 - 0.8} = 24 \text{ billion dollars}.$$

(b) 90% of 6 billion dollars = 5.4 billion dollars. Proceeding as in part (a), the total additional spending is given by

$$\frac{a}{1-r} = \frac{5.4}{1 - 0.9} = 54 \text{ billion dollars}.$$

7. An account earns 5% interest per year, compounded annually. Suppose ten payments of $5,000 are to be made once a year, starting now, from the account. How much must be deposited now to cover these payments? What if twenty annual payments are to be made? What if the annual payments are to continue indefinitely?

ANSWER:

For the ten years, the present value, P, is given by

$$P = 5,000 + 5,000(1.05)^{-1} + 5,000((1.05)^{-1})^2 + \cdots + 5,000((1.05)^{-1})^9.$$

This is a finite geometric series with $a = 5,000$, $r = (1.05)^{-1}$, $n = 10$. Thus

$$P = \frac{a(1 - r^n)}{1 - r} = \frac{5,000(1 - ((1.05)^{-1})^{10})}{1 - (1.05)^{-1}} = \$40,539.$$

If the payments are to last twenty years, $n = 20$, and we get

$$P = \frac{a(1 - r^n)}{1 - r} = \frac{5,000(1 - ((1.05)^{-1})^{20})}{1 - (1.05)^{-1}} = \$65,427.$$

If the payments are to continue indefinitely, we have

$$P = 5,000 + 5,000(1.05)^{-1} + 5,000((1.05)^{-1})^2 + \cdots.$$

This is an infinite geometric series with $a = 5,000$, $r = (1.05)^{-1}$. Since $-1 < r < 1$, we have

$$P = \frac{a}{1 - r} = \frac{5,000}{1 - (1.05)^{-1}} = \$105,000.$$

8. Find the market stabilization point if 300 new items are manufactured each year and 30% of the total number of items in use fail each year. What if 50% fail?

ANSWER:

If N is the number of items in circulation, then

$$N = 300 + 300(0.7) + 300(0.7)^2 + \cdots$$

This is an infinite geometric series with $a = 300$, $r = 0.7$. Since $-1 < r < 1$,

$$N = \frac{300}{1 - 0.7} = 1000 \text{ items.}$$

If 50% fail, $r = 0.5$, and so

$$N = \frac{300}{1 - 0.5} = 600 \text{ items.}$$

9. A person takes 250 mg of a pain killer every 4 hours. If the pain killer has a half life of 3 hours, how much of the drug is in the body after 24 hours (right after the 6th dose)? What is the amount of the drug in the body in the long run, right before and after each dose?

ANSWER:

After a single dose of 250 mg, the quantity, Q, of the drug in the body is given by $Q = 250b^t$, where t is time in hours. Since the half-life is 3 hours, we solve for b in the equation

$$0.5(250) = 250b^3$$
$$0.5 = b^3$$
$$b = (0.5)^{1/3}$$

Since doses are given every 4 hours, we want to know what fraction of the drug remains after 4 hours. Using $t = 4$,

$$\text{Fraction remaining after 4 hours} = b^4 = ((0.5)^{1/3})^4 = (0.5)^{4/3}.$$

Thus the quantity, Q_6, of the drug remaining in the body after the 6th doses is given by

$$Q_6 = 250 + 250(0.5)^{4/3} + 250((0.5)^{4/3})^2 + \cdots + 250((0.5)^{4/3})^5$$

This is a finite geometric series with $a = 250$, $r = (0.5)^{4/3}$, $n = 6$. Thus,

$$Q_6 = \frac{250(1 - ((0.5)^{4/3})^6)}{1 - (0.5)^{4/3}} = 406.05 \text{ mg}.$$

The steady state level right after taking a dose is given by

$$Q = 250 + 250(0.5)^{4/3} + 250((0.5)^{4/3})^2 + \cdots$$

This is an infinite geometric series with $a = 250$, $r = (0.5)^{4/3}$. Since $-1 < r < 1$, we get

$$Q = \frac{250}{1 - (0.5)^{4/3}} = 414.49 \text{ mg}.$$

Right before taking a dose, the steady state it given by 414.49 minus the amount of one dose, or 414.49-250=164.49 mg.